STUDENT SOLUTIONS MANUAL

Nancy J. Gardner

California State University, Long Branch

INTRODUCTION TO
CHEMICAL PRINCIPLES

TENTH EDITION

H. STEPHEN STOKER

Prentice Hall

Boston   Columbus   Indianapolis   New York   San Francisco   Upper Saddle River
Amsterdam   Cape Town   Dubai   London   Madrid   Milan   Munich   Paris   Montréal   Toronto
Delhi   Mexico City   São Paulo   Sydney   Hong Kong   Seoul   Singapore   Taipei   Tokyo

Acquisitions Editor: Terry Haugen
Editor in Chief, Chemistry and Geosciences: Nicole Folchetti
Marketing Manager: Erin Gardner
Assistant Editor: Carol G. DuPont
Managing Editor, Chemistry and Geosciences: Gina M. Cheselka
Project Manager: Wendy A. Perez
Operations Specialist: Maura Zaldivar
Supplement Cover Designer: Paul Gourhan
Cover Photo Credit: Digital Art/Corbis

© 2011, 2008, 2005
Pearson Education, Inc.
Pearson Prentice Hall
Upper Saddle River, NJ 07458

All rights reserved. No part of this book may be reproduced, in any form or by any means, without permission in writing from the publisher.

Pearson Prentice Hall™ is a trademark of Pearson Education, Inc.

The author and publisher of this book have used their best efforts in preparing this book. These efforts include the development, research, and testing of the theories and programs to determine their effectiveness. The author and publisher make no warranty of any kind, expressed or implied, with regard to these programs or the documentation contained in this book. The author and publisher shall not be liable in any event for incidental or consequential damages in connection with, or arising out of, the furnishing, performance, or use of these programs.

> This work is protected by United States copyright laws and is provided solely for teaching courses and assessing student learning. Dissemination or sale of any part of this work (including on the World Wide Web) will destroy the integrity of the work and is not permitted. The work and materials from it should never be made available except by instructors using the accompanying text in their classes. All recipients of this work are expected to abide by these restrictions and to honor the intended pedagogical purposes and the needs of other instructors who rely on these materials.

Printed in the United States of America

10 9 8 7 6 5 4 3 2 1

ISBN-13: 978-0-321-67619-1
ISBN-10: 0-321-67619-X

Prentice Hall
is an imprint of

www.pearsonhighered.com

Contents

Chapter 1	The Science of Chemistry	1
Chapter 2	Numbers from Measurements	3
Chapter 3	Unit Systems and Dimensional Analysis	9
Chapter 4	Basic Concepts About Matter	21
Chapter 5	Subatomic Particles, Isotopes, and Nuclear Chemistry	25
Chapter 6	Electronic Structure and Chemical Periodicity	35
Chapter 7	Chemical Bonds	41
Chapter 8	Chemical Nomenclature	53
Chapter 9	Chemical Calculations: The Mole Concept and Chemical Formulas	59
Chapter 10	Chemical Calculations Involving Chemical Equations	85
Chapter 11	States of Matter	101
Chapter 12	Gas Laws	111
Chapter 13	Solutions	137
Chapter 14	Acids, Bases, and Salts	155
Chapter 15	Chemical Equations: Net Ionic and Oxidation–Reduction	165
Chapter 16	Reaction Rates and Chemical Equilibrium	183

CHAPTER ONE
The Science of Chemistry

Practice Problems

Scientific Disciplines (Sec. 1.1)

1.1   a) true   b) false   c) false   d) true

1.3   a) false   b) false   c) true   d) false

Scientific Research and Technology (Sec. 1.2)

1.5   a) basic research   b) applied research   c) basic research   d) applied research

The Scientific Method (Sec. 1.4)

1.7   c, b, e, a, and d

1.9   a) scientific hypothesis   b) nonscientific hypothesis
      c) scientific hypothesis   d) nonscientific hypothesis

1.11 a) scientific hypothesis   b) scientific law
      c) scientific fact   d) scientific hypothesis

1.13 a) false   b) false   c) true   d) false

1.15 While a scientific theory may not be an absolute answer, it is the best answer available. It may be supplanted only if repeated experimental evidence conclusively disproves it and a new theory is developed.

1.17 a) 4 is eliminated   b) 1 is eliminated
      c) 1 and 4 are eliminated d) 1 and 2 are eliminated

1.19 a) qualitative   b) quantitative   c) quantitative   d) qualitative

1.21 The product of the pressure times the volume is a constant; or the pressure of the gas is inversely proportional to its volume; $P_1V_1 = P_2V_2$.

1.23 Scientific laws are discovered by research. Researchers have no control over what the laws turn out to be. Societal laws are arbitrary conventions that can be and are changed by society when necessary.

1.25 Publishing scientific data provides access to that data, enabling scientists to develop new scientific theories based on a wider range of knowledge relating to a particular field.

1.27 Conditions under which an experiment is conducted often affect the results of the experiment. If the conditions are uncontrolled, the data resulting from that experiment is not validated.

1.29 A qualitative observation involves general nonnumerical information about a system under study. A quantitative observation involves numerical measurements.

Answers to Multiple-Choice Practice Test

1.31	e	**1.32**	a	**1.33**	c	**1.34**	d	**1.35**	e
1.36	d	**1.37**	b	**1.38**	b	**1.39**	c	**1.40**	c

CHAPTER TWO
Numbers from Measurements

Practice Problems

Exact and Inexact Numbers (Sec. 2.2)

2.1 a) exact (counting integer)   b) exact (definition)
    c) inexact (measurement)   d) exact (counting integer)

2.3 a) exact (counting integer)   b) inexact (measurement)
    c) exact (counting integer)   d) exact (coins are counted)

2.5 60 seconds in a minute represents an exact number; 60 feet long represents a measured value.

Accuracy and Precision (Sec. 2.3)

2.7 Student A: low precision, low accuracy   Student B: high precision, high accuracy
    Student C: high precision, low accuracy

Uncertainty in Measurements (Sec. 2.4)

2.9 a) Ruler 1 = ±0.1 cm   b) Ruler 2 = ±0.01 cm

2.11 a) Ruler 3 = 27 cm   b) Ruler 4 = 27.0 cm

2.13 a) Ruler 4   b) Ruler 1   c) Ruler 2   d) Ruler 3

2.15 The degree of uncertainty is represented by ±1 in the last digit.
    a) ±0.001   b) ±1   c) ±0.0001   d) ±0.1

2.17 The uncertainty in 3.3 seconds is ±0.1 second and the uncertainty in 3.30 seconds is ±0.01 second.

2.19 a) 40,000–60,000 people   b) 49,000–51,000 people
    c) 49,900–50,100 people   d) 49,990–50,010 people

Significant Figures (Sec. 2.5)

2.21 a) 5   b) 3   c) 6   d) 1

2.23 a) 2   b) 6   c) 5   d) 6

2.25 a) 6   b) 6   c) 3   d) 2

2.27 a) confined zeros = 2, leading zeros = 0, significant trailing zeros = 1, trailing zeros not significant = 0
    b) confined zeros = 2, leading zeros = 3, significant trailing zeros = 2, trailing zeros not significant = 0
    c) confined zeros = 1, leading zeros = 0, significant trailing zeros = 0, trailing zeros not significant = 2
    d) confined zeros = 0, leading zeros = 0, significant trailing zeros = 0, trailing zeros not significant = 6

2.29 a) 3010.2**0**, **0** is the estimated digit b) 0.0030030**0**, **0** is the estimated digit
c) **4**0,400, **4** is the estimated digit d) 3**3**,000,000, **3** is the estimated digit

2.31 a) ±0.01 b) ±0.00000001 c) ±100 d) ±1,000,000

2.33 a) same number of significant figures b) different number of significant figures
c) same number of significant figures d) same number of significant figures

2.35 a) same uncertainty, ±0.01 b) different uncertainties, ±1, ±10
c) same uncertainty, ±0.00001 d) different uncertainties, ±0.000001, ±10,000

2.37 a) 23,000 b) 23,0̄ 0̄0 c) 23,000.0 d) 23,000.000

2.39 a) 3 b) 4 c) 2 d) 5

2.41 A measurement of 1 gram, with an uncertainty of ±0.0001 has two possible answers: 0.9999, which has 4 significant figures, and 1.0001, which has 5 significant figures.

Rounding Off (Sec. 2.6)

2.43 a) 431.2 b) 31.21 c) 8.207 d) 1.021

2.45 a) 42.6 b) 42.6 c) 42.8 d) 42.8

2.47 a) 42,300 b) 42,400 c) 42,500 d) 42,600

2.49 a) 0.00033 b) 0.012 c) 0.20 d) 0.36

2.51 a) 42.3 b) 42.0 c) 42.6 d) 42.3

2.53 a) 0.351 b) 653.9 c) 22.556 d) 0.2777

2.55 a) 30,427.3 b) 30,427 c) 30,430 d) 3 0̄,000

2.57 a) 0.035 b) 2.50 c) 1,500,000 d) 1 0̄0

2.59 a) 0.12 b) 120,000 c) 12 d) 0.00012

Significant Figures in Multiplication and Division (Sec. 2.6)

2.61 a) 2 b) 1 c) 3 d) 3

2.63 a) 3 b) 2 c) 4 or more d) 4 or more

2.65 a) 0.029889922 (calc), 0.0299 (corr) b) 136,900 (calc), 140,000 (corr)
c) 1277.2522 (calc), 1280 (corr) d) 0.98816568 (calc), 0.988 (corr)

2.67 a) 3.9265927 (calc), 3.9 (corr) b) 4.8410309 (calc), 4.84 (corr)
c) 63.13492 (calc), 63 (corr) d) 1.1851852 (calc), 1 (corr)

Numbers from Measurements

Significant Figures in Addition and Subtraction (Sec. 2.6)

2.69 a) tenths, ±0.1   b) tenths, ±0.1   c) ones, ±1   d) hundreds, ±100

2.71 a) 162 (calc and corr)   b) 9.321 (calc), 9.3 (corr)
c) 1260.72 (calc), 1261 (corr)   d) 19.95 (calc), 20.0 (corr)

2.73 a) 957 (calc), 957.0 (corr)   b) 342.63 (calc), 343 (corr)
c) 1250 (calc), 1200 (corr)   d) 131.9927 (calc), 132 (corr)

Calculations Involving Both Significant Figure Rules (Sec. 2.6)

2.75 a) 2.732 (calc), 2.7 (corr)   b) 2.8521 (calc), 2.9 (corr)
c) 0.219 (calc), 0.2 (corr)   d) 46.393464 (calc), 46.4 (corr)

Significant Figures and Exact Numbers (Sec. 2.6)

2.77 a) 267.3 (calc and corr)   b) 257,140 (calc), 260,000 (corr)
c) 201.3 (calc and corr)   d) 3.8038461 (calc), 3.8 (corr)

2.79 a) 168 (calc), 170 (corr)   b) 123.3 (calc), 123 (corr)
c) 131.19 (calc), 131.2 (corr)   d) 0.185 (calc), 0.18 (corr)

2.81 a) 24 + 24 + 24 + 24 + 24 + 24 + 24 = 168 (calc and corr)
b) 13.7 + 13.7 + 13.7 + 13.7 + 13.7 + 13.7 + 13.7 + 13.7 + 13.7 = 123.3 (calc and correct)
c) 43.73 + 43.73 + 43.73 = 131.19 (calc and corr)
d) 0.037 + 0.037 + 0.037 + 0.037 + 0.037 = 0.185 (calc and corr)

Scientific Notation (Sec. 2.7)

2.83 a) negative   b) positive   c) zero   d) positive

2.85 a) Two decimal places to the right   b) Four decimal places to the right
c) Three decimal places to the left   d) Four decimal places to the left

2.87 a) 4 significant figures   b) 5 significant figures
c) 4 significant figures   d) 6 significant figures

2.89 a) 2 digits   b) 5 digits   c) 2 digits   d) 7 digits

2.91 a) 4.732×10^2 b) 1.234×10^{-3} c) 2.3100×10^2 d) 2.31×10^8

2.93 a) 3.00300×10^{-3} (calc), 3.00×10^{-3} (corr)   b) 9.3602×10^5 (calc), 9.36×10^5 (corr)
c) 2.55003×10^1 (calc), 2.51×10^1 (corr)   d) 4.500003×10^8 (calc), 4.50×10^8 (corr)

2.95 a) 7×10^4 b) 6.70×10^4 c) 6.7000×10^4 d) 6.700000×10^4

2.97 a) 0.00230   b) 4350   c) 0.066500   d) 111,000,000

2.99 a) 3.42×10^6 b) 2.36×10^{-3} c) 3.2×10^2 d) 1.2×10^{-4}

2.101 a) smaller b) smaller c) larger d) larger

Uncertainty and Scientific Notation (Sec. 2.7)

2.103 a) $10^{-3} \times 10^4 = \pm 10$ b) $10^{-3} \times 10^6 = \pm 10^3$
c) $10^{-2} \times 10^5 = \pm 10^3$ d) $10^{-1} \times 10^{-2} = \pm 10^{-3}$

2.105 a) 3.65×10^5 b) 3.6500×10^5 c) 3.650000×10^5 d) 3.65000000×10^5

2.107 a) less than b) same c) same d) greater than

Multiplication and Division in Scientific Notation (Sec. 2.8)

2.109 a) 10^8 b) 10^{-8} c) 10^2 d) 10^{-2}

2.111 a) 2.991905×10^8 (calc), 2.992×10^8 (corr) b) 9.129×10^2 (calc), 9.1×10^2 (corr)
c) 2.7×10^{-9} (calc and corr) d) 2.7×10^{11} (calc and corr)

2.113 a) 10^2 b) 10^8 c) 10^{-8} d) 10^{-2}

2.115 a) 2.8649608×10^1 (calc), 2.86×10^1 (corr) b) 8.9991001×10^{17} (calc), 8.999×10^{17} (corr)
c) 3.490449×10^{-2} (calc), 3.49×10^{-2} (corr) d) $1.1112222 \times 10^{-18}$ (calc), 1.111×10^{-18} (corr)

2.117 a) 10^1 b) 10^{-1} c) 10^{20} d) 10^2

2.119 a) 1.5×10^0 (calc and corr) b) 6.6666666×10^{-1} (calc), 6.7×10^{-1} (corr)
c) $8.5073917 \times 10^{-19}$ (calc), 8.51×10^{-19} (corr) d) 7.7775×10^{15} (calc), 8×10^{15} (corr)

Addition and Subtraction in Scientific Notation (Sec. 2.8)

2.121 a) 4.415×10^3 (calc), 4.42×10^3 (corr) b) 9.3×10^{-2} (calc), 9.30×10^{-2} (corr)
c) 9.683×10^5 (calc and corr) d) 1.9189×10^4 (calc), 1.919×10^4 (corr)

2.123 a) 7.713×10^7 (calc and corr) b) 8.253×10^7 (calc and corr)
c) 8.307×10^7 (calc and corr) d) 8.31294×10^7 (calc), 8.313×10^7 (corr)

Additional Problems

2.125 a) yes b) no c) yes d) no

2.127 a) 3 b) 4 c) 4 d) 4

2.129 a) no b) yes c) yes d) yes

2.131 a) $60\overline{\overline{0}}$ pounds b) 600.0 pounds c) $6\overline{0}0$ pounds d) 600.000 pounds

2.133 a) yes b) no c) no d) no

Numbers from Measurements 7

2.135 a) 6.326×10^5 b) 3.13×10^{-1} c) 6.300×10^7 d) 5.000×10^{-1}

2.137 a) 2   b) 4 or more   c) 3   d) 3

2.139 An exact number is a whole number; it cannot possess decimal digits.

2.141 a) $4.2 + 5.30 = 9.5$ (calc and corr)
$28 + 11 = 39$ (exact)
$39 \times 9.5 = 370.5$ (calc), 3.7×10^2 (corr)
b) $28 - 11 = 17$ (exact)
$4.2 \times 5.30 \times 17 = 378.42$ (calc), 3.8×10^2 (corr)
c) 28 (exact) $- 4.2 = 23.8$ (calc and corr)
5.30×11 (exact) $= 58.3$ (calc and corr)
$\dfrac{23.8}{58.3} = 0.40823328$ (calc), 0.408 (corr)
d) 28 (exact) $- 4.2 = 23.8$ (calc and corr)
11 (exact) $- 5.30 = 5.7$ (calc), 5.70 (corr)
$\dfrac{23.8}{5.70} = 4.1754386$ (calc), 4.18 (corr)

2.143 a) 2.07×10^2, 243, 1.03×10^3 b) 2.11×10^{-3}, 0.0023, 3.04×10^{-2}
c) 23,000, 9.67×10^4, 2.30×10^5 d) 0.000014, 0.00013, 1.5×10^{-4}

Answers to Multiple-Choice Practice Test

2.145	d	**2.146**	b	**2.147**	d	**2.148**	d	**2.149**	b	**2.150**	c
2.151	e	**2.152**	a	**2.153**	c	**2.154**	d	**2.155**	c	**2.156**	b
2.157	c	**2.158**	c	**2.159**	e	**2.160**	e	**2.161**	e	**2.162**	b
2.163	d	**2.164**	e								

CHAPTER THREE
Unit Systems and Dimensional Analysis

Practice Problems

Metric System Units (Secs. 3.1–3.4)

3.1   a) no   b) yes   c) no   d) yes

3.3   a) nano-   b) mega-   c) milli-   d) 10^3 e) 10^{-2} f) 10^{-6}

3.5   a) µg   b) km   c) cL   d) decimeter   e) milliliter   f) picogram

3.7

	Metric Prefix	Abbreviation for Prefix	Mathematical Meaning of Prefix
	milli-	m	10^{-3}
a)	tera-	T	10^{12}
b)	nano-	n	10^{-9}
c)	giga-	G	10^{9}
d)	micro-	µ	10^{-6}

3.9

	Metric Unit	Property Being Measured	Abbreviation for Metric Unit
	microliter	volume	µL
a)	kilogram	mass	kg
b)	megameter	length	Mm
c)	nanogram	mass	ng
d)	milliliter	volume	mL

3.11   a) 6.8 dm   b) 3.2 pL   c) 7.23 cL   d) 6.5 Mg

3.13   a) smaller by 100 (10^2) times   b) smaller by 1000 (10^3) times
       c) larger by 10,000 (10^4) times   d) smaller by 1,000,000,000 (10^9) times

3.15   a) length   b) area   c) volume   d) volume

3.17   a) 1 inch   b) 1 meter   c) 1 pound   d) 1 gallon

Area and Volume Measurements (Secs. 3.4 and 3.5)

3.19 a) $A = (4.52 \text{ cm})^2 = 20.4304$ (calc) $= 20.4 \text{ cm}^2$ (corr)
b) $A = (3.5 \text{ m}) \times (9.2 \text{ m}) = 32.2$ (calc) $= 32 \text{ m}^2$ (corr)
c) $A = 3.142 \times (4.579 \text{ mm})^2 = 65.879071$ (calc) $= 65.88 \text{ mm}^2$ (corr)
d) $A = 1/2 \times (5.5 \text{ mm}) \times (3.0 \text{ mm}) = 8.25$ (calc) $= 8.2 \text{ mm}^2$ (corr)

3.21 a) $V = (5.4 \text{ cm}) \times (0.52 \text{ cm}) \times (3.4 \text{ cm}) = 9.5472$ (calc) $= 9.5 \text{ cm}^3$ (corr)
b) $V = 3.142 \times (2.4 \text{ cm})^2 \times (7.5 \text{ cm}) = 135.7344$ (calc) $= 1.4 \times 10^2 \text{ cm}^3$ (corr)
c) $V = 4/3 \times 3.142 \times (87 \text{ mm})^3 = 2.7586886 \times 10^6$ (calc) $= 2.8 \times 10^6 \text{ mm}^3$ (corr)
d) $V = (7.2 \text{ cm})^3 = 373.248$ (calc) $= 3.7 \times 10^2 \text{ cm}^3$

3.23 a) equal b) not equal c) not equal d) equal

Conversion Factors (Sec. 3.6)

3.25 a) 24 hr = 1 day $\dfrac{1 \text{ day}}{24 \text{ hr}}$ $\dfrac{24 \text{ hr}}{1 \text{ day}}$ b) 60 sec = 1 min $\dfrac{1 \text{ min}}{60 \text{ sec}}$ $\dfrac{60 \text{ sec}}{1 \text{ min}}$

c) 10 decades = 1 century $\dfrac{1 \text{ century}}{10 \text{ decades}}$ $\dfrac{10 \text{ decades}}{1 \text{ century}}$

d) 365 days = 1 yr $\dfrac{1 \text{ yr}}{365 \text{ days}}$ $\dfrac{365 \text{ days}}{1 \text{ yr}}$

3.27 a) 1 quart = 2 pints, 1 pint = 2 cups   b) 1 yard = 3 feet
c) 1 ton = 2000 pounds   d) 1 yard = 36 inches

3.29 a) $\dfrac{1 \text{ kL}}{10^3 \text{ L}}$ $\dfrac{10^3 \text{ L}}{1 \text{ kL}}$ b) $\dfrac{1 \text{ mg}}{10^{-3} \text{ g}}$ $\dfrac{10^{-3} \text{ g}}{1 \text{ mg}}$ c) $\dfrac{1 \text{ cm}}{10^{-2} \text{ m}}$ $\dfrac{10^{-2} \text{ m}}{1 \text{ cm}}$ d) $\dfrac{1 \text{ }\mu\text{sec}}{10^{-6} \text{ sec}}$ $\dfrac{10^{-6} \text{ sec}}{1 \text{ }\mu\text{sec}}$

3.31 a) $A = 10^3 \text{ g}$ b) $A = 10^{-6} \text{ m}$ c) $A = 453.6 \text{ g}$ d) $A = 2.540 \text{ cm}$

3.33 a) 4 significant figures   b) exact   c) 4 significant figures   d) exact

3.35 a) exact   b) inexact   c) exact   d) exact

3.37 a) valid   b) not valid   c) valid   d) not valid

Dimensional Analysis—Metric–Metric Unit Conversions (Sec. 3.7)

3.39 a) $37 \text{ L} \times \dfrac{1 \text{ dL}}{10^{-1} \text{ L}} = 3.7 \times 10^2 \text{ dL}$ (calc and corr)

b) $37.0 \text{ mm} \times \dfrac{10^{-3} \text{ m}}{1 \text{ mm}} = 3.70 \times 10^{-2} \text{ m}$ (calc and corr)

c) $0.37 \text{ pg} \times \dfrac{10^{-12} \text{ g}}{1 \text{ pg}} = 3.7 \times 10^{-13} \text{ g}$ (calc and corr)

d) $370 \text{ kL} \times \dfrac{10^3 \text{ L}}{1 \text{ kL}} = 3.7 \times 10^5 \text{ L}$ (calc and corr)

Unit Systems and Dimensional Analysis

3.41 a) $47 \text{ Mg} \times \dfrac{10^6 \text{ g}}{1 \text{ Mg}} \times \dfrac{1 \text{ mg}}{10^{-3} \text{ g}} = 4.7 \times 10^{10}$ mg (calc and corr)

b) $5.00 \text{ nL} \times \dfrac{10^{-9} \text{ L}}{1 \text{ nL}} \times \dfrac{1 \text{ cL}}{10^{-2} \text{ L}} = 5.00 \times 10^{-7}$ cL (calc and corr)

c) $6 \times 10^{-2} \text{ } \mu\text{m} \times \dfrac{10^{-6} \text{ m}}{1 \text{ } \mu\text{m}} \times \dfrac{1 \text{ dm}}{10^{-1} \text{ m}} = 6 \times 10^{-7}$ dm (calc and corr)

d) $37 \text{ pm} \times \dfrac{10^{-12} \text{ m}}{1 \text{ pm}} \times \dfrac{1 \text{ km}}{10^3 \text{ m}} = 3.7 \times 10^{-14}$ km (calc and corr)

3.43 a) $365 \text{ m}^2 \times \left(\dfrac{1 \text{ km}}{10^3 \text{ m}}\right)^2 = 3.65 \times 10^{-4}$ km^2 (calc and corr)

b) $365 \text{ m}^2 \times \left(\dfrac{1 \text{ cm}}{10^{-2} \text{ m}}\right)^2 = 3.65 \times 10^{6}$ cm^2 (calc and corr)

c) $365 \text{ m}^2 \times \left(\dfrac{1 \text{ dm}}{10^{-1} \text{ m}}\right)^2 = 3.65 \times 10^{4}$ dm^2 (calc and corr)

d) $365 \text{ m}^2 \times \left(\dfrac{1 \text{ Mm}}{10^{6} \text{ m}}\right)^2 = 3.65 \times 10^{-10}$ Mm^2 (calc and corr)

3.45 a) $35 \text{ m}^3 \times \left(\dfrac{1 \text{ mm}}{10^{-3} \text{ m}}\right)^3 = 3.5 \times 10^{10}$ mm^3 (calc and corr)

b) $35 \text{ m}^3 \times \left(\dfrac{1 \text{ pm}}{10^{-12} \text{ m}}\right)^3 = 3.5 \times 10^{37}$ pm^3 (calc and corr)

c) $35 \text{ m}^3 \times \left(\dfrac{1 \text{ Gm}}{10^{9} \text{ m}}\right)^3 = 3.5 \times 10^{-26}$ Gm^3 (calc and corr)

d) $35 \text{ m}^3 \times \left(\dfrac{1 \text{ } \mu\text{m}}{10^{-6} \text{ m}}\right)^3 = 3.5 \times 10^{19}$ μm^3 (calc and corr)

3.47 a) $6.0 \text{ cm}^2 \times \left(\dfrac{10^{-2} \text{ m}}{1 \text{ cm}}\right)^2 = 6.0 \times 10^{-4}$ m^2 (calc and corr)

b) $7.2 \text{ mm}^3 \times \left(\dfrac{10^{-3} \text{ m}}{1 \text{ mm}}\right)^3 = 7.2 \times 10^{-9}$ m^3 (calc and corr)

c) $25 \text{ } \mu\text{m}^2 \times \left(\dfrac{10^{-6} \text{ m}}{1 \text{ } \mu\text{m}}\right)^2 \times \left(\dfrac{1 \text{ dm}}{10^{-1} \text{ m}}\right)^2 = 2.5 \times 10^{-9}$ dm^2 (calc and corr)

d) $0.023 \text{ km}^3 \times \left(\dfrac{10^{3} \text{ m}}{1 \text{ km}}\right)^3 \times \left(\dfrac{1 \text{ nm}}{10^{-9} \text{ m}}\right)^3 = 2.3 \times 10^{34}$ nm^3 (calc and corr)

Dimensional Analysis—Metric–English Unit Conversions (Sec. 3.7)

3.49 a) $100.0 \text{ yd} \times \dfrac{3 \text{ ft}}{1 \text{ yd}} \times \dfrac{12 \text{ in.}}{1 \text{ ft}} \times \dfrac{2.540 \text{ cm}}{1 \text{ in.}} \times \dfrac{10^{-2} \text{ m}}{1 \text{ cm}} = 91.44$ m (calc and corr)

b) $100.0 \text{ yd} \times \dfrac{3 \text{ ft}}{1 \text{ yd}} \times \dfrac{12 \text{ in.}}{1 \text{ ft}} \times \dfrac{2.540 \text{ cm}}{1 \text{ in.}} = 9144$ cm (calc and corr)

c) $100.0 \text{ yd} \times \dfrac{3 \text{ ft}}{1 \text{ yd}} \times \dfrac{12 \text{ in.}}{1 \text{ ft}} \times \dfrac{2.540 \text{ cm}}{1 \text{ in.}} \times \dfrac{10^{-2} \text{ m}}{1 \text{ cm}} \times \dfrac{1 \text{ km}}{10^3 \text{ m}} = 0.09144$ km (calc and corr)

d) $100.0 \text{ yd} \times \dfrac{3 \text{ ft}}{1 \text{ yd}} \times \dfrac{12 \text{ in.}}{1 \text{ ft}} = 3600$ (calc) $= 3.600 \times 10^3$ in. (corr)

3.51 a) $75 \text{ mL} \times \dfrac{10^{-3} \text{ L}}{1 \text{ mL}} \times \dfrac{1 \text{ qt}}{0.9463 \text{ L}} = 0.079256$ (calc) $= 0.079$ qt (corr)

b) $75 \text{ mL} \times \dfrac{10^{-3} \text{ L}}{1 \text{ mL}} \times \dfrac{1 \text{ qt}}{0.9463 \text{ L}} \times \dfrac{1 \text{ gal}}{4 \text{ qt}} = 0.019814$ (calc) $= 0.020$ gal (corr)

c) $75 \text{ mL} \times \dfrac{10^{-3} \text{ L}}{1 \text{ mL}} \times \dfrac{1 \text{ qt}}{0.9463 \text{ L}} \times \dfrac{32 \text{ fl oz}}{1 \text{ qt}} = 2.5361936$ (calc) $= 2.5$ fl oz (corr)

d) $75 \text{ mL} \times \dfrac{1 \text{ cm}^3}{1 \text{ mL}} = 75 \text{ cm}^3$ (calc and corr)

3.53 a) $6.6 \times 10^{21} \text{ tons} \times \dfrac{2000 \text{ lb}}{1 \text{ ton}} \times \dfrac{453.6 \text{ g}}{1 \text{ lb}} = 5.98752 \times 10^{27}$ (calc) $= 6.0 \times 10^{27}$ g (corr)

b) $6.6 \times 10^{21} \text{ tons} \times \dfrac{2000 \text{ lb}}{1 \text{ ton}} \times \dfrac{453.6 \text{ g}}{1 \text{ lb}} \times \dfrac{1 \text{ kg}}{10^3 \text{ g}} = 5.98752 \times 10^{24}$ (calc) $= 6.0 \times 10^{24}$ kg (corr)

c) $6.6 \times 10^{21} \text{ tons} \times \dfrac{2000 \text{ lb}}{1 \text{ ton}} \times \dfrac{453.6 \text{ g}}{1 \text{ lb}} \times \dfrac{1 \text{ ng}}{10^{-9} \text{ g}} = 5.98752 \times 10^{36}$ (calc) $= 6.0 \times 10^{36}$ ng (corr)

d) $6.6 \times 10^{21} \text{ tons} \times \dfrac{2000 \text{ lb}}{1 \text{ ton}} \times \dfrac{16 \text{ oz}}{1 \text{ lb}} = 2.112000 \times 10^{26}$ (calc) $= 2.1 \times 10^{26}$ oz (corr)

3.55 a) $61 \text{ cm}^3 \times \left(\dfrac{1 \text{ in.}}{2.540 \text{ cm}}\right)^3 \times \left(\dfrac{1 \text{ ft}}{12 \text{ in.}}\right)^3 = 2.1541947 \times 10^{-3}$ (calc) $= 2.2 \times 10^{-3}$ ft³ (corr)

b) $61 \text{ cm}^3 \times \left(\dfrac{1 \text{ in.}}{2.540 \text{ cm}}\right)^3 \times \left(\dfrac{1 \text{ ft}}{12 \text{ in.}}\right)^3 \times \left(\dfrac{1 \text{ yd}}{3 \text{ ft}}\right)^3 = 7.978507 \times 10^{-5}$ (calc)
$= 8.0 \times 10^{-5}$ yd³ (corr)

c) $61 \text{ cm}^3 \times \left(\dfrac{1 \text{ in.}}{2.540 \text{ cm}}\right)^3 = 3.7224484$ (calc) $= 3.7$ in.³ (corr)

d) $61 \text{ cm}^3 \times \left(\dfrac{1 \text{ in.}}{2.540 \text{ cm}}\right)^3 \times \left(\dfrac{1 \text{ ft}}{12 \text{ in.}}\right)^3 \times \left(\dfrac{1 \text{ mi}}{5280 \text{ ft}}\right)^3 = 1.463468 \times 10^{-14}$ (calc)
$= 1.5 \times 10^{-14}$ mi³ (corr)

3.57 a) $A = 2.1 \text{ cm} \times 2.5 \text{ cm} = 5.25$ (calc) $= 5.2$ cm² (corr)

b) $A = 2.1 \text{ cm} \times 2.5 \text{ cm} \times \left(\dfrac{1 \text{ in.}}{2.540 \text{ cm}}\right)^2 = 0.81375163$ (calc) $= 0.81$ in.² (corr)

3.59 $V = (95 \times 105 \times 145) \text{ cm}^3 \times \left(\dfrac{1 \text{ in.}}{2.540 \text{ cm}}\right)^3 \times \left(\dfrac{1 \text{ ft}}{12 \text{ in.}}\right)^3 = 51.078251$ (calc) $= 51$ ft³ (corr)

Dimensional Analysis—Units Involving Two Types of Measurements (Sec. 3.7)

3.61 a) $\dfrac{55\text{ L}}{1\text{ sec}} \times \dfrac{60\text{ sec}}{1\text{ min}} \times \dfrac{60\text{ min}}{1\text{ hr}} = 198{,}000 \text{ (calc)} = 2.0 \times 10^5 \text{ L/hr (corr)}$

b) $\dfrac{55\text{ L}}{1\text{ sec}} \times \dfrac{1\text{ kL}}{10^3\text{ L}} = 5.5 \times 10^{-2} \text{ kL/sec (calc and corr)}$

c) $\dfrac{55\text{ L}}{1\text{ sec}} \times \dfrac{1\text{ dL}}{10^{-1}\text{ L}} \times \dfrac{60\text{ sec}}{\text{min}} = 33{,}000 \text{ (calc)} = 3.3 \times 10^4 \text{ dL/min (corr)}$

d) $\dfrac{55\text{ L}}{1\text{ sec}} \times \dfrac{1\text{ mL}}{10^{-3}\text{ L}} \times \dfrac{60\text{ sec}}{1\text{ min}} \times \dfrac{60\text{ min}}{1\text{ hr}} \times \dfrac{24\text{ hr}}{1\text{ day}} = 4.752 \times 10^9 \text{ (calc)} = 4.8 \times 10^9 \text{ mL/day (corr)}$

3.63 a) $\dfrac{0.0057\text{ μg}}{1\text{ mL}} \times \dfrac{1\text{ mL}}{10^{-3}\text{ L}} = 5.7 \text{ μg/L (calc and corr)}$

b) $\dfrac{0.0057\text{ μg}}{1\text{ mL}} \times \dfrac{10^{-6}\text{ g}}{1\text{ μg}} = 5.7 \times 10^{-9} \text{ g/mL (calc and corr)}$

c) $\dfrac{0.0057\text{ μg}}{1\text{ mL}} \times \dfrac{1\text{ mL}}{10^{-3}\text{ L}} \times \dfrac{10^{-6}\text{ L}}{1\text{ μL}} \times \dfrac{10^{-6}\text{ g}}{1\text{ μg}} \times \dfrac{1\text{ mg}}{10^{-3}\text{ g}} = 5.7 \times 10^{-9} \text{ mg/μL (calc and corr)}$

d) $\dfrac{0.0057\text{ μg}}{1\text{ mL}} \times \dfrac{1\text{ mL}}{10^{-3}\text{ L}} \times \dfrac{10^3\text{ L}}{1\text{ kL}} \times \dfrac{10^{-6}\text{ g}}{1\text{ μg}} \times \dfrac{1\text{ kg}}{10^3\text{ g}} = 5.7 \times 10^{-6} \text{ kg/kL (calc and corr)}$

Density (Sec. 3.8)

3.65 density = mass (in grams) divided by volume (in mL)

a) $\dfrac{25.0\text{ g}}{31.7\text{ mL}} = 0.78864353 \text{ (calc)} = 0.789 \text{ g/mL (corr)}$

b) $\dfrac{25.0\text{ g}}{3.48\text{ cm}^3} \times \dfrac{1\text{ cm}^3}{1\text{ mL}} = 7.1839080 \text{ (calc)} = 7.18 \text{ g/mL (corr)}$

c) $\dfrac{22.9\text{ g}}{25.0\text{ mL}} = 0.916\ \dfrac{\text{g}}{\text{mL}}$ (calc and corr)

d) $\dfrac{37{,}200\text{ g}}{25.0\text{ L}} \times \dfrac{10^{-3}\text{ L}}{1\text{ mL}} = 1.488 \text{ (calc)} = 1.49 \text{ g/mL (corr)}$

3.67 Use density as a conversion factor.

a) $47.6 \text{ mL} \times \dfrac{0.791\text{ g}}{1\text{ mL}} = 37.6516 \text{ (calc)} = 37.7 \text{ g (corr)}$

b) $47.6 \text{ cm}^3 \times \dfrac{10.40\text{ g}}{1\text{ cm}^3} = 495.04 \text{ (calc)} = 495 \text{ g (corr)}$

c) $47.6 \text{ L} \times \dfrac{1.25\text{ g}}{1\text{ L}} = 59.5 \text{ g (calc and corr)}$

d) $47.6 \text{ cm}^3 \times \dfrac{2.18\text{ g}}{1\text{ cm}^3} = 103.768 \text{ (calc)} = 104 \text{ g (corr)}$

3.69 Use density as a conversion factor.

a) $17.6 \text{ g} \times \dfrac{1 \text{ mL}}{1.027 \text{ g}} = 17.13729309$ (calc) $= 17.1$ mL (corr)

b) $17.6 \text{ g} \times \dfrac{1 \text{ cm}^3}{19.3 \text{ g}} \times \dfrac{1 \text{ mL}}{1 \text{ cm}^3} = 0.9119170984$ (calc) $= 0.912$ mL (corr)

c) $17.6 \text{ g} \times \dfrac{1 \text{ L}}{1.29 \text{ g}} \times \dfrac{1 \text{ mL}}{10^{-3} \text{ L}} = 13643.41085$ (calc) $= 1.36 \times 10^4$ mL (corr)

d) $17.6 \text{ g} \times \dfrac{1 \text{ cm}^3}{1.008 \text{ g}} \times \dfrac{1 \text{ mL}}{1 \text{ cm}^3} = 17.460317$ (calc) $= 17.5$ mL (corr)

3.71 $(5.261 - 3.006) \text{ g} = 2.255$ g red liquid (calc and corr) $d = \dfrac{2.255 \text{ g}}{2.171 \text{ mL}} = 1.0386918$ (calc)
$= 1.039$ g/mL (corr)

3.73 pathway: gal → qt → L → mL → g → lb

$13.0 \text{ gal} \times \dfrac{4 \text{ qt}}{1 \text{ gal}} \times \dfrac{0.9463 \text{ L}}{1 \text{ qt}} \times \dfrac{1 \text{ mL}}{10^{-3} \text{ L}} \times \dfrac{0.56 \text{ g}}{1 \text{ mL}} \times \dfrac{1 \text{ lb}}{453.6 \text{ g}} = 60.750123$ (calc) $= 61$ lb (corr)

3.75 Al = aluminum, Cr = chromium

$100.0 \text{ g Al} \times \dfrac{1 \text{ cm}^3 \text{ Al}}{2.70 \text{ g Al}} \times \dfrac{1 \text{ cm}^3 \text{ Cr}}{1 \text{ cm}^3 \text{ Al}} \times \dfrac{7.18 \text{ g Cr}}{1 \text{ cm}^3 \text{ Cr}} = 265.92593$ (calc) $= 266$ g (corr)

3.77 a) float, less dense than water   b) sink, more dense than water

3.79 $16.0 \text{ gal} \times \dfrac{4 \text{ qt}}{1 \text{ gal}} \times \dfrac{0.9463 \text{ L}}{1 \text{ qt}} \times \dfrac{1 \text{ mL}}{10^{-3} \text{ L}} \times \dfrac{1 \text{ cm}^3}{1 \text{ mL}} \times \left(\dfrac{1 \text{ in.}}{2.540 \text{ cm}}\right)^3 \times \left(\dfrac{1 \text{ yd}}{36 \text{ in.}}\right)^3$

$= 0.0792136 \text{ yd}^3 \times \dfrac{1011 \text{ lb}}{1 \text{ yd}^3} \times \dfrac{453.6 \text{ g}}{1 \text{ lb}} = 3.6326568 \times 10^4$ (calc) $= 3.63 \times 10^4$ g (corr)

Equivalence Conversion Factors (Sec. 3.9)

3.81 a) equivalence   b) equality   c) equality   d) equivalence

3.83 $15.9 \text{ kg} \times \dfrac{32 \text{ mg antibiotic}}{1 \text{ kg}} = 508.8$ (calc) $= 5.1 \times 10^2$ mg antibiotic (corr)

3.85 $375 \text{ mg} \times \dfrac{1 \text{ kg}}{6.00 \text{ mg}} \times \dfrac{10^3 \text{ g}}{1 \text{ kg}} \times \dfrac{1 \text{ lb}}{453.6 \text{ g}} = 137.7866$ (calc) $= 138$ lb (corr)

3.87 a) $2.30 \text{ L} \times \dfrac{2.30 \text{ μg}}{1 \text{ L}} \times \dfrac{10^{-6} \text{ g}}{1 \text{ μg}} = 5.29 \times 10^{-6}$ g (calc and corr)

b) $2.30 \text{ mL} \times \dfrac{10^{-3} \text{ L}}{1 \text{ mL}} \times \dfrac{2.30 \text{ μg}}{1 \text{ L}} \times \dfrac{10^{-6} \text{ g}}{1 \text{ g}} = 5.29 \times 10^{-9}$ g (calc and corr)

Unit Systems and Dimensional Analysis

c) $2.30 \text{ qt} \times \dfrac{0.9463 \text{ L}}{1 \text{ qt}} \times \dfrac{2.30 \text{ μg}}{1 \text{ L}} \times \dfrac{10^{-6} \text{ g}}{1 \text{ μg}} = 5.005927 \times 10^{-6}$ g (calc) $= 5.01 \times 10^{-6}$ g (corr)

d) $2.30 \text{ fl oz} \times \dfrac{1 \text{ qt}}{32 \text{ fl oz}} \times \dfrac{0.9463 \text{ L}}{1 \text{ qt}} \times \dfrac{2.30 \text{ μg}}{1 \text{ L}} \times \dfrac{10^{-6} \text{ g}}{1 \text{ μg}} = 0.156435218 \times 10^{-6}$ (calc)

$= 1.56 \times 10^{-7}$ g (corr)

3.89 a) $16.0 \text{ gal} \times \dfrac{1 \text{ sec}}{0.20 \text{ gal}} \times \dfrac{1 \text{ min}}{60 \text{ sec}} = $ (calc) $= 1.33333333$ min (calc) $= 1.3$ min (corr)

b) $13.0 \text{ qt} \times \dfrac{1 \text{ gal}}{4 \text{ qt}} \times \dfrac{1 \text{ sec}}{0.20 \text{ gal}} \times \dfrac{1 \text{ min}}{60 \text{ sec}} = 0.270833333$ min (calc) $= 0.27$ min (corr)

c) $7.00 \text{ L} \times \dfrac{1 \text{ qt}}{0.9463 \text{ L}} \times \dfrac{1 \text{ gal}}{4 \text{ qt}} \times \dfrac{1 \text{ sec}}{0.20 \text{ gal}} \times \dfrac{1 \text{ min}}{60 \text{ sec}} = 0.154108986$ min (calc) $= 0.15$ min (corr)

d) $5355 \text{ mL} \times \dfrac{10^{-3} \text{ L}}{1 \text{ mL}} \times \dfrac{1 \text{ qt}}{0.9463 \text{ L}} \times \dfrac{1 \text{ gal}}{4 \text{ qt}} \times \dfrac{1 \text{ sec}}{0.20 \text{ gal}} \times \dfrac{1 \text{ min}}{60 \text{ sec}} = 0.117893374$ (calc)

$= 0.12$ min (corr)

3.91 $4.2 \text{ miles} \times \dfrac{1.609 \text{ km}}{1 \text{ mi}} \times \dfrac{1 \text{ hr}}{45 \text{ km}} \times \dfrac{60 \text{ min}}{1 \text{ hr}} = 9.0104$ (calc) $= 9.0$ min (corr)

3.93 a) $\$1000.00 \times \dfrac{1 \text{ bushel}}{\$9.20} = 108.6956522$ (calc) $= 109$ bushels (corr)

b) $45 \text{ bushels} \times \dfrac{\$9.20}{1 \text{ bushel}} = \$414.00$ (calc and corr)

3.95 a) $3.75 \text{ g} \times \dfrac{75 \text{ cents}}{1 \text{ g}} \times \dfrac{\$1.00}{100 \text{ cents}} = 2.8125$ (calc) $= \$2.81$ (corr)

b) $3.75 \text{ cg} \times \dfrac{10^{-2} \text{ g}}{1 \text{ cg}} \times \dfrac{75 \text{ cents}}{1 \text{ g}} \times \dfrac{\$1.00}{100 \text{ cents}} = 0.028125$ (calc) $= \$0.03$ (corr)

c) $6.20 \text{ lb} \times \dfrac{453.6 \text{ g}}{1 \text{ lb}} \times \dfrac{75 \text{ cents}}{1 \text{ g}} \times \dfrac{\$1.00}{100 \text{ cents}} = 2109.24$ (calc) $= \$2109.24$ (corr)

d) $6.20 \text{ oz} \times \dfrac{1 \text{ lb}}{16 \text{ oz}} \times \dfrac{453.6 \text{ g}}{1 \text{ lb}} \times \dfrac{75 \text{ cents}}{1 \text{ g}} \times \dfrac{\$1.00}{100 \text{ cents}} = 131.8275$ (calc) $= \$131.83$ (corr)

Percentage and Percent Error (Sec. 3.10)

3.97 a) % nickels $= \dfrac{6 \text{ nickels}}{34 \text{ coins (total)}} \times 100 = 17.6470588$ (calc) $= 17.6\%$ (corr)

b) % quarters $= \dfrac{10 \text{ quarters}}{34 \text{ coins (total)}} \times 100 = 29.4117647$ (calc) $= 29.4\%$ (corr)

c) % 10¢ or less $= \dfrac{15 \text{ pennies} + 6 \text{ nickels} + 3 \text{ dimes}}{34 \text{ coins total}} \times 100 = 70.5882352$ (calc) $= 70.6\%$ (corr)

d) % smaller than nickel $= \dfrac{15 \text{ pennies} + 3 \text{ dimes}}{34 \text{ coins (total)}} \times 100 = 52.9411764$ (calc) $= 52.9\%$ (corr)

3.99 a) % copper = $\dfrac{2.902 \text{ g copper}}{3.053 \text{ g penny}} \times 100$ = 95.054045 (calc) = 95.05% copper (corr)

b) mass zinc = 3.053 g penny − 2.902 g copper = 0.151 g zinc

% zinc = $\dfrac{0.151 \text{ g zinc}}{3.053 \text{ g penny}} \times 100$ = 4.945948 (calc) = 4.95% zinc (corr)

Alternate method: % zinc = 100 − % copper = 100 − 95.05 = 4.95% zinc (calc and corr)

3.101 65.3 g mixture × $\dfrac{34.2 \text{ g water}}{100 \text{ g mixture}}$ = 22.3326 (calc) = 22.3 g water (corr)

3.103 437 g solution × $\dfrac{15.3 \text{ g salt}}{100 \text{ g solution}}$ = 66.861 (calc) = 66.9 g salt (corr)

3.105 1.00 gal solution × $\dfrac{4 \text{ qt}}{1 \text{ gal}} \times \dfrac{0.9463 \text{ L}}{1 \text{ qt}} \times \dfrac{1 \text{ mL}}{10^{-3} \text{ L}} \times \dfrac{1.013 \text{ g}}{1 \text{ mL}}$ = 3834.4076 g of solution

3834.4076 g solution × $\dfrac{9.00 \text{ g acetic acid}}{100 \text{ g solution}}$ = 345.09668 (calc) = 345 g acetic acid (corr)

3.107 661 Gummi × $\dfrac{30.9 \text{ bears}}{100 \text{ Gummi}} \times \dfrac{23.0 \text{ orange}}{100 \text{ bears}} \times \dfrac{6.4 \text{ one-ear}}{100 \text{ orange}}$ = 3.0065453 (calc) = 3 one-ear (corr)

3.109 student 1: $\dfrac{(78.0 - 78.5)°C}{78.5°C} \times 100$ = −0.63694267% (calc) = −0.6% (corr)

student 2: $\dfrac{(77.9 - 78.5)°C}{78.5°C} \times 100$ = −0.76433121% (calc) = −0.8% (corr)

student 3: $\dfrac{(79.7 - 78.5)°C}{78.5°C} \times 100$ = 1.5286624% (calc) = 1.5% (corr)

3.111 72.6 yr − 53.6 yr = $\dfrac{19.0 \text{ yr}}{53.6 \text{ yr}} \times 100$ = 35.4477611% (calc) = 35.4% (corr)

Temperature Scales (Sec. 3.11)

3.113 a) 32°F   b) 0°C   c) 273 K

3.115 20°C × $\dfrac{9°F}{5°C}$ = 36°F (calc and corr)

3.117 a) 1352°C = 1352°C above FP × $\dfrac{9°F}{5°C}$ = 2433.6 (calc) = 2434 (corr) °F above FP (32°F)

= 2434 + 32 = 2466°F (corr)

b) 37.6°C = 37.6°C above FP × $\dfrac{9°F}{5°C}$ = 67.68 (calc) = 67.7 (corr) °F above FP (32.0°F)

= 67.7 + 32.0 = 99.7°F (corr)

Unit Systems and Dimensional Analysis

c) $-7°C = 7°C$ below FP $\times \dfrac{9°F}{5°C} = 12.6$ (calc) $= 13$ (corr) °F below FP (32°F)

$\qquad = 32 - 13 = 19°F$ (calc) $= 19°F$ (corr)

d) $295°C = 295°C$ above FP $\times \dfrac{9°F}{5°C} = 531$ (calc and corr) °F above FP (32°F)

$\qquad = 32 + 531 = 563°F$ (corr)

3.119 a) $1352°F = 1352 - 32 = 1320°F$ above FP $\times \dfrac{5°C}{9°F} = 733.3333$ (calc)

$\qquad = 733$ (corr) °C above FP (0°C) $= 733°C$

b) $37.6°F = 37.6 - 32 = 5.6°F$ above FP $\times \dfrac{5°C}{9°F} = 3.11111$ (calc)

$\qquad = 3.1$ (corr) °C above FP (0°C) $= 3.1°C$

c) $-7°F = -7 - 32 = 39°F$ below FP $\times \dfrac{5°C}{9°F} = 21.666667$ (calc)

$\qquad = 22$ (corr) °C below FP (0°C) $= -22°C$

d) $295°F = 295 - 32 = 263°F$ above FP $\times \dfrac{5°C}{9°F} = 146.111111$ (calc)

$\qquad = 146$ (corr) °C above FP (0°C) $= 146°C$

3.121 a) $K = 275°C + 273 = 548$ K (calc and corr)
b) $K = 275.2°C + 273.2 = 548.4$ K (calc and corr)
c) $K = 275.73°C + 273.15 = 548.88$ K (calc and corr)
d) $275°F = 275 - 32 = 243°F$ above FP $\times \dfrac{5°C}{9°F} = 135$ (calc and corr) °C above FP

FP (0°C) $= 135°C$, K $= 135°C + 273 = 408$ K (calc and corr)

3.123 a) $101°F = 101 - 32 = 69°F$ above FP $\times \dfrac{5°C}{9°F} = 38.33333$ (calc)

$\qquad = 38$ (corr) °C above FP (0°C) $= 38°C$

b) $-218.4°C = 218.4°C$ below FP $\times \dfrac{9°F}{5°C} = 393.12$ (calc) $= 393.1$ (corr) °F below FP

$\qquad (32°F) = (32 - 393.1) = -361.1°F$

c) $804°C + 273 = 1077$ K (calc and corr)

d) 77 K $- 273 = -196°C$ (calc and corr) $-196°C = 196°C$ below FP $\times \dfrac{9°F}{5°C}$

$\qquad = 352.8$ (calc) $= 353$ (corr) °F below FP (32°F) $= (32 - 353) = -321°F$

3.125 $-10°C = 10°C$ below the ice point

$10°C \times \dfrac{9°F}{5°C} = 18°F$ below the ice point (calc and corr)

$32°F - 18°F = 14°F$ ∴ $-10°C$ is a higher temperature than $10°F$

3.127 223 K $- 273 = -50°C$ (calc and corr) $-50°C = 50°C$ below the ice point

$50°C \times \dfrac{9°F}{5°C} = 90°F$ below the ice point (calc and corr) $32°F - 90°F = -58°F$

∴ $-60°F$ is a lower temperature than 223 K

Additional Problems

3.129 a) 2   b) 3   c) 5   d) 4

3.131 a) 6.301 km   b) 1.442 sec   c) 1.327 mg   d) 2.1 cL

3.133 a) 10^3 L   b) 10^{-12} L   c) 10^{-9} L   d) 10^{-1} L

3.135 Before adding, all values must have the same units; convert them all to centimeters.

(1) $20.9 \text{ dm} \times \dfrac{10^{-1} \text{ m}}{1 \text{ dm}} \times \dfrac{1 \text{ cm}}{10^{-2} \text{ m}} = 209 \text{ cm}$ (2) $2030 \text{ mm} \times \dfrac{10^{-3} \text{ m}}{1 \text{ mm}} \times \dfrac{1 \text{ cm}}{10^{-2} \text{ m}} = 203 \text{ cm}$

(3) $1.90 \text{ m} \times \dfrac{1 \text{ cm}}{10^{-2} \text{ m}} = 190 \text{ cm}$ (4) $0.00183 \text{ km} \times \dfrac{10^3 \text{ m}}{1 \text{ km}} \times \dfrac{1 \text{ cm}}{10^{-2} \text{ m}} = 183 \text{ cm}$

(5) $203 \text{ cm} = 203 \text{ cm}$ $209 \text{ cm} + 203 \text{ cm} + 190 \text{ cm} + 183 \text{ cm} + 203 \text{ cm} = 988 \text{ cm}$

Average $= \dfrac{988 \text{ cm}}{5} = 197.6$ (calc) $= 198$ cm (corr)

3.137 a) $3.72 \text{ gal} \times \dfrac{4 \text{ qt}}{1 \text{ gal}} \times \dfrac{0.9463 \text{ L}}{1 \text{ qt}} = 14.080944$ (calc) $= 14.1$ L (corr)

b) $3.720 \text{ gal} \times \dfrac{4 \text{ qt}}{1 \text{ gal}} \times \dfrac{0.9463 \text{ L}}{1 \text{ qt}} = 14.080944$ (calc) $= 14.08$ L (corr)

c) $3.7200 \text{ gal} \times \dfrac{4 \text{ qt}}{1 \text{ gal}} \times \dfrac{0.9463 \text{ L}}{1 \text{ qt}} = 14.080944$ (calc) $= 14.081$ L (corr)

d) $3.72000 \text{ gal} \times \dfrac{4 \text{ qt}}{1 \text{ gal}} \times \dfrac{0.9463 \text{ L}}{1 \text{ qt}} = 14.080944$ (calc) $= 14.0809$ L (corr)

3.139 $8 \text{ yr} \times \dfrac{365 \text{ days}}{1 \text{ yr}} \times \dfrac{24 \text{ hr}}{1 \text{ day}} \times \dfrac{60 \text{ min}}{1 \text{ day}} \times \dfrac{69 \text{ beats}}{1 \text{ min}} = 2.90131200 \times 10^8$ beats (calc)

$\qquad = 2.9 \times 10^8$ beats (corr)

3.141 pathway: pt → qt → L → mL → g

$1.00 \text{ pt} \times \dfrac{1 \text{ qt}}{2 \text{ pt}} \times \dfrac{0.9463 \text{ L}}{1 \text{ qt}} \times \dfrac{1 \text{ mL}}{10^{-3} \text{ L}} \times \dfrac{1.05 \text{ g}}{1 \text{ mL}} = 496.8075$ (calc) $= 497$ g (corr)

3.143 $325 \text{ g acid} \times \dfrac{100 \text{ g solution}}{38.1 \text{ g acid}} \times \dfrac{1 \text{ cm}^3 \text{ solution}}{1.29 \text{ g solution}} \times \dfrac{1 \text{ mL solution}}{1 \text{ cm}^3 \text{ solution}} = 661.25455$ (calc) $= 661$ mL (corr)

3.145

a) pathway: μg → g → mg ↗   ↘ mL → L → dL

$5000 \dfrac{1 \text{ μg}}{1 \text{ mL}} \times \dfrac{10^{-6} \text{ g}}{1 \text{ μg}} \times \dfrac{1 \text{ mg}}{10^{-3} \text{ g}} \times \dfrac{1 \text{ mL}}{10^{-3} \text{ L}} \times \dfrac{10^{-1} \text{ L}}{1 \text{ dL}} = 500 \dfrac{\text{mg}}{\text{dL}}$ (calc and corr)

∴ Yes, a life-threatening situation

Unit Systems and Dimensional Analysis

b) pathway $\begin{matrix} \nearrow g \to mg \\ \searrow L \to dL \end{matrix}$

$$0.5 \frac{1\ g}{1\ L} \times \frac{1\ mg}{10^{-3}\ g} \times \frac{10^{-1}\ L}{1\ dL} = 50 \frac{mg}{dL}\ (\text{calc and corr}) \quad \therefore \text{No, not a life-threatening situation}$$

3.147 pathway: dimensions → volume → mass $\quad 3.5\ m \times 3.0\ m \times 3.2\ m = 33.6\ (\text{calc}) = 34\ m^3\ (\text{corr})$

$$34\ m^3 \times \left(\frac{1\ cm}{10^{-2}\ m}\right)^3 \times 5.7 \times 10^{-3} \frac{1\ \mu g}{1\ cm^3} \times \frac{10^{-6}\ g}{1\ \mu g} = 0.1938\ (\text{calc}) = 0.19\ g\ (\text{corr})$$

3.149 $\text{Area (cm}^2) = 4.0\ in. \times 4.0\ in. \times \left(\frac{2.540\ cm}{1\ in.}\right)^2 = 103.2256\ cm^2\ (\text{calc}) = 1.0 \times 10^2\ cm^2\ (\text{corr})$

$\text{Volume (cm}^3) = 0.466\ g \times \left(\frac{1\ cg}{10^{-2}\ g} \times \frac{1\ cm^3}{269\ cg}\right) = 0.1732342\ (\text{calc}) = 0.173\ cm^3\ (\text{corr})$

$\text{Thickness} = \frac{0.173\ cm^3}{1.0 \times 10^2\ cm^2} = 0.00173\ cm = 0.0173\ mm\ (\text{calc}) = 1.7 \times 10^{-2}\ mm\ (\text{corr})$

3.151 Volume of soil $= 1.0\ m \times (2.00 \times 10^3\ m) \times (0.50 \times 10^3\ m) = 1.0 \times 10^6\ m^3\ (\text{calc and corr})$

$1.0 \times 10^6\ m^3 \times \frac{22\ worms}{1\ m^3} = 22 \times 10^6\ (\text{calc}) = 2.2 \times 10^7\ worms\ (\text{calc})$

3.153 Conversion factor: $(212 - 32)°F = [200 - (-200)]°H; \quad 180°F = 400°H \quad 20°H = 9°F$

$50°F = 50 - 32 = 18°F\ \text{above FP} \times \frac{20°H}{9°F} = 40°H\ \text{above FP}\ (-200°H) = 40 + (-200)$
$\hspace{12cm} = -160°H$

3.155 Conversion factors: (1) $\frac{30.3\ orange}{100\ bears}$ (2) $\frac{8.10\ one\text{-}ear\ orange}{100\ orange}$ (3) $\frac{3\ one\text{-}ear\ orange}{1\ bag}$

$\frac{3\ one\text{-}ear\ orange}{1\ bag} \times \frac{100\ orange}{8.10\ one\text{-}ear\ orange} \times \frac{100\ bears}{30.3\ orange} = 122.23\ (\text{calc}) = 122 \frac{bears}{bag}\ (\text{corr})$

Cumulative Problems

3.157 a) $d = \frac{2.000\ g}{4.000\ mL} = 0.5\ (\text{calc}) = 5.000 \times 10^{-1} \frac{g}{mL}\ (\text{corr})$

b) $d = \frac{2.00\ g}{4.0\ mL} = 0.5\ (\text{calc}) = 5.0 \times 10^{-1} \frac{g}{mL}\ (\text{corr})$

c) $d = \frac{2.0000\ g}{4.0000\ mL} = 0.5\ (\text{calc}) = 5.0000 \times 10^{-1} \frac{g}{mL}\ (\text{corr})$

d) $d = \frac{2.000\ g}{4.0000\ mL} = 0.5\ (\text{calc}) = 5.000 \times 10^{-1} \frac{g}{mL}\ (\text{corr})$

3.159 a) $V = 1\overline{0}$ cm $\times$ $2\overline{00}$ cm $\times$ 4 cm = 8000 (calc) = 8×10^3 cm^3 (corr)
b) $V = 10.0$ cm $\times$ 200.0 cm $\times$ 4.0 cm = 8000 (calc) = 8.0×10^3 cm^3 (corr)
c) $V = 10.00$ cm $\times$ 200.00 cm $\times$ 4.00 cm = 8000 (calc) = 8.00×10^3 cm^3 (corr)
d) $V = 1\overline{0}$ cm $\times$ $2\overline{0}$ cm $\times$ 4.0 cm = 8000 (calc) = 8.0×10^3 cm^3 (corr)

3.161 Measurements as a decimal number with $\pm$ uncertainty, using the same units for the set.

a) 3.256×10^3 g = 3256 $\pm$ 1 g 3.256×10^4 g = 32,560 $\pm$ 10 g
3.256×10^5 g = 325,600 $\pm$ 100 g $\therefore$ 3.256×10^3 g has the greatest precision (least uncertainty).

b) 3.34 g = 3.34 $\pm$ 0.01 g   3.34 kg $\times \dfrac{10^3 \text{ g}}{1 \text{ kg}}$ = 3.34×10^3 g = 3,340 $\pm$ 10 g

3.34 mg $\times \dfrac{10^{-3} \text{ g}}{1 \text{ mg}}$ = 3.34×10^{-3} g = 0.00334 $\pm$ 0.00001 g

$\therefore$ 3.34 mg has the greatest precision (least uncertainty).

c) 4.31 $\pm$ 0.01 g 4.31×10^{-3} kg $\times \dfrac{10^3 \text{ g}}{1 \text{ kg}}$ = 4.31 $\pm$ 0.01 g

4.31×10^3 mg $\times \dfrac{10^{-3} \text{ g}}{1 \text{ mg}}$ = 4.31 $\pm$ 0.01 g $\therefore$ Each has the same precision.

d) 325.0 cg $\times \dfrac{10^{-2} \text{ g}}{1 \text{ cg}}$ = 3.250 $\pm$ 0.001 g, 3.2500 g = 3.2500 $\pm$ 0.0001 g

0.00325 kg $\times \dfrac{10^3 \text{ g}}{1 \text{ kg}}$ = 3.25 g = 3.25 $\pm$ 0.01 g

$\therefore$ 3.2500 g has the greatest precision (least uncertainty).

3.163 1.2120 $\pm$ 0.0001 g

a) 0.00121 kg $\times \dfrac{10^3 \text{ g}}{1 \text{ kg}}$ = 1.21 $\pm$ 0.01 g $\therefore$ Therefore is not equivalent

b) 121.20 cg $\times \dfrac{10^{-2} \text{ g}}{1 \text{ cg}}$ = 1.2120 $\pm$ 0.0001 g $\therefore$ Therefore is equivalent

c) 12120 mg $\times \dfrac{10^{-3} \text{ g}}{1 \text{ mg}}$ = 12.120 g $\pm$ 0.001 g $\therefore$ Therefore is not equivalent

d) 1212.0 μg $\times \dfrac{10^{-6} \text{ g}}{1 \text{ } \mu\text{g}}$ = 0.0012120 $\pm$ 0.0000001 g $\therefore$ Therefore is not equivalent

Answers to Multiple-Choice Practice Test

3.165	b	3.166	e	3.167	d	3.168	c	3.169	d	3.170	a
3.171	c	3.172	a	3.173	c	3.174	c	3.175	d	3.176	b
3.177	b	3.178	b	3.179	e	3.180	d	3.181	c	3.182	e
3.183	e	3.184	a								

CHAPTER FOUR
Basic Concepts About Matter

Practice Problems

Physical States of Matter (Sec. 4.2)

4.1 a) shape (indefinite vs. definite)   b) indefinite shape

4.3 a) Does not take shape of the container, definite volume.
     b) Takes on shape of container, indefinite volume.
     c) Does not take shape of the container, definite volume.
     d) Takes on shape of container, definite volume.

4.5 a) solid   b) liquid   c) liquid   d) gas

Properties of Matter (Sec. 4.3)

4.7 a) chemical   b) chemical   c) physical   d) physical

4.9 a) chemical   b) physical   c) physical   d) physical

4.11 a) extensive   b) intensive   c) intensive   d) intensive

4.13 a) differ in extensive properties (amount)
     b) differ in intensive properties (substance)
     c) differ in intensive properties (temperature)
     d) differ in extensive (amount) and intensive properties (temperature)

Changes in Matter (Sec. 4.4)

4.15 a) physical   b) physical   c) chemical   d) physical

4.17 a) chemical   b) physical   c) chemical   d) physical

4.19 a) physical   b) physical   c) chemical   d) chemical

4.21 a) physical   b) physical   c) chemical   d) physical

4.23 a) freezing   b) condensation   c) sublimation   d) evaporation

Pure Substances and Mixtures (Secs. 4.5 and 4.6)

4.25 a) heterogeneous and homogeneous mixtures   b) pure substance, homogeneous mixture

4.27 a) false   b) true   c) false   d) true

4.29 a) heterogeneous mixture   b) homogeneous mixture
     c) homogeneous mixture   d) heterogeneous mixture

4.31 a) homogeneous mixture, 1 phase b) homogeneous mixture, 1 phase
c) heterogeneous mixture, 2 phases d) heterogeneous mixture, 2 phases

4.33 a) chemically homogeneous, physically homogeneous
b) chemically heterogeneous, physically homogeneous
c) chemically heterogeneous, physically heterogeneous
d) chemically heterogeneous, physically heterogeneous

Elements and Compounds (Sec. 4.7)

4.35 a) compound b) compound c) no classification possible d) no classification possible

4.37 a) true b) false c) false d) false

4.39 a) A (no classification possible), B (no classification possible), C (compound)
b) D (compound), E (no classification possible), F (no classification possible),
G (no classification possible)

4.41 First box, mixture; second box, compound

4.43 a) applies b) applies c) applies d) does not apply

Discovery and Abundance of the Elements (Sec. 4.8)

4.45 a) false b) true c) false d) true

4.47 a) true b) true c) false d) true

4.49 a) Yes, silicon is more abundant. b) No, hydrogen is more abundant.
c) No, oxygen is more abundant. d) Yes, sodium is more abundant.

Names and Chemical Symbols of the Elements (Sec. 4.9)

4.51 a) nitrogen b) nickel c) lead d) tin
e) Al f) Ne g) H h) U

4.53 a) Na and S b) Mg and Mn c) Ca and Cd d) As and Ar

4.55 a) boron b) barium c) beryllium d) bismuth e) berkelium f) bromine

4.57 a) fluorine b) zinc c) potassium d) sulfur

4.59 a) iron, Fe b) tin, Sn c) sodium, Na d) gold, Au

4.61 a) Re-Be-C-Ca b) Ra-Y-Mo-Nd c) Na-N-C-Y
d) Br-U-Ce or B-Ru-Ce e) S-H-Ar-O-N f) Al-I-Ce

Atoms and Molecules (Secs. 4.10 and 4.11)

4.63 a) consistent b) consistent c) not consistent d) consistent

4.65 a) heteroatomic, compound b) heteroatomic, compound
b) homoatomic, triatomic, element d) heteroatomic, compound

Basic Concepts About Matter

4.67 a) False, molecules must contain two or more atoms.
b) true
c) False, some compounds have molecules as their basic structural unit and others have ions as their basic structural unit.
d) False, the diameter is approximately 10^{-10} meters.

4.69 a) one substance, homoatomic—element
b) two substances, molecules are homoatomic—mixture
c) one phase present, two kinds of triatomic heteroatomic molecules—mixture
d) two kinds of molecules—mixture

Chemical Formulas (Sec. 4.13)

4.71 a) AB_2 b) BC_2 c) B_2C_2 d) A_3B_2C

4.73 a) compound b) compound c) element d) element

4.75 a) $C_{20}H_{30}$ b) H_2SO_4 c) HCN d) $KMnO_4$

4.77 a) same, both 3 b) same, both 5 c) fewer, 5 and 7 d) more, 17 and 13

4.79 a) NO = 1 atom nitrogen, 1 atom oxygen
No = 1 atom nobelium
b) Cs_2 = 2 atoms cesium
CS_2 = 1 atom carbon, 2 atoms sulfur
c) $CoBr_2$ = 1 atom cobalt, 2 atoms bromine
$COBr_2$ = 1 atom carbon, 1 atom oxygen, 2 atoms bromine
d) H = 1 atom hydrogen
H_2 = 2 atoms hydrogen

4.81 a) H_3PO_4 b) $SiCl_4$ c) NO_2 d) H_2O_2

4.83 a) HCN b) H_2SO_4

Additional Problems

4.85 a) colorless gas, odorless gas, colorless liquid, boils at 43°C
b) toxic to humans, Ni reacts with CO

4.87 a) compound b) mixture c) compound d) mixture

4.89 a) homogeneous mixture b) element
c) heterogeneous mixture d) homogeneous mixture

4.91 a) heterogeneous mixture b) heterogeneous mixture
c) homogeneous mixture d) heterogeneous, but not a mixture

4.93 (b) and (c)

4.95 Alphabetized: Bh, Bi, Bk, Cf, Co, Cs, Cu, Hf, Ho, Hs, In, Nb, Ni, No, Np, Os, Pb, Po, Pu, Sb, Sc, Si, Sn, Yb

4.97 a) $2 + x + 3 = 7; x = 2$
b) $1 + 2(1 + x) = 9; 3 + 2x = 9; x = 3$
c) $x + x + 10 = 16; 2x = 6; x = 3$
d) $x + 2x + x = 24; 4x = 24; x = 6$

4.99 a) 4 (N_2O, H_2O, CCl_4, CH_2Br_2)
b) 2 (CCl_4 and CH_2Br_2)
c) 4 (N_2O, H_2O, CCl_4, CH_2Br_2)
d) 6 (O, N, H, C, Cl, and Br)
e) $(3 \times 2) + (3 \times 3) + (3 \times 3) + (3 \times 5) + (3 \times 5) = 54$

Cumulative Problems

4.101 a) solid b) state determination not possible (boiling point not known)
c) gas d) state determination not possible (two states present)

4.103 a) $\% = \dfrac{88}{117 \text{ total}} \times 100 = 75.21367521$ (calc) $= 75.2\%$ (corr)

b) $\% = \dfrac{14}{111 \text{ total}} \times 100 = 12.61261$ (calc) $= 12.6\%$ (corr)

c) $\% = \dfrac{32}{117 \text{ total}} \times 100 = 27.35042735$ (calc) $= 27.4\%$ (corr)

d) $\% = (20.1 + 2.4 + 2.2)\% = 24.7\%$ (calc and corr)

4.105 a) density(I) $= \dfrac{4.32 \text{ g}}{3.78 \text{ mL}} = 1.1428571$ (calc) $= 1.14$ gm/L (corr)

density(II) $= \dfrac{5.73 \text{ g}}{5.02 \text{ mL}} = 1.1414342$ (calc) $= 1.14$ g/mL (corr)

density(III) $= \dfrac{1.52 \text{ g}}{1.33 \text{ mL}} = 1.1428571$ (calc) $= 1.14$ g/mL (corr)

It is likely that all the students were working with the same substance.
b) No, density alone will not distinguish between elements and compounds.

4.107 The substance analyzed is likely a compound. The percent composition to three significant figures is the same in each case. The substance cannot be an element because two substances, Q and X, were produced in the analysis.

4.109 $3.22 \text{ cm}^3 \times \dfrac{19.3 \text{ g}}{1 \text{ cm}^3} \times \dfrac{1 \text{ Au atom}}{3.27 \times 10^{-22} \text{ g}} = 1.900489 \times 10^{23}$ (calc) $= 1.90 \times 10^{23}$ Au atoms (corr)

4.111 pathway: atoms → cm → in. → ft → mi

3.17×10^{22} atoms $\times \dfrac{1.44 \times 10^{-8} \text{ cm}}{1 \text{ atom}} \times \dfrac{1 \text{ in.}}{2.540 \text{ cm}} \times \dfrac{1 \text{ ft}}{12 \text{ in.}} \times \dfrac{1 \text{ mi}}{5280 \text{ ft}}$

$= 2.8364352 \times 10^9$ (calc) $= 2.84 \times 10^9$ miles (corr)

Answers to Multiple-Choice Practice Test

4.113	e	4.114	c	4.115	d	4.116	c	4.117	a	4.118	d
4.119	d	4.120	c	4.121	d	4.122	a	4.123	a	4.124	b
4.125	d	4.126	c	4.127	a	4.128	b	4.129	d	4.130	c
4.131	e	4.132	a								

CHAPTER FIVE
Subatomic Particles, Isotopes, and Nuclear Chemistry

Practice Problems

Subatomic Particles (Sec. 5.1)

5.1 a) electron b) proton c) proton, neutron d) neutron

5.3 a) false b) false c) false d) true

5.5 Relative mass of proton to electron
$$= \frac{\text{mass of proton}}{\text{mass of electron}} = \frac{1.673 \times 10^{-24} \text{ g}}{9.109 \times 10^{-28} \text{ g}} = 1836.6450763 \text{ (calc)} = 1837 \text{ (corr)}$$

Atomic Number and Mass Number (Sec. 5.2)

5.7 a) 50 b) 47 c) nickel d) iodine

5.9 a) 24, 53 b) 44, 103 c) 101, 256 d) 16, 34

5.11 atomic number = number protons a) 5 b) 8 c) 13 d) 18

5.13 mass number = number protons + number neutrons
a) 5 + 6 = 11 b) 8 + 8 = 16 c) 13 + 14 = 27 d) 18 + 22 = 40

5.15 number of nucleons = number protons + number neutrons
a) 5 + 6 = 11 b) 8 + 8 = 16 c) 13 + 14 = 27 d) 18 + 22 = 40

5.17 charge on nucleus = number protons
a) 5 b) 8 c) 13 d) 18

5.19 a) $^{11}_{5}\text{B}$ b) $^{16}_{8}\text{O}$ c) $^{27}_{13}\text{Al}$ d) $^{40}_{18}\text{Ar}$

5.21 a) atomic number b) atomic number and mass number
c) mass number d) atomic number and mass number

5.23 a) Nitrogen, N b) Aluminum, Al c) Barium, Ba d) Gold, Au

5.25

	Symbol	Atomic Number	Mass Number	Number of Protons	Number of Neutrons
	$_{2}^{3}He$	2	3	2	1
a)	$_{28}^{60}Ni$	28	60	28	32
b)	$_{18}^{37}Ar$	18	37	18	19
c)	$_{38}^{90}Sr$	38	90	38	52
d)	$_{92}^{235}U$	92	235	92	143

5.27 a) 27 protons, 27 electrons, 32 neutrons   b) 45 protons, 45 electrons, 58 neutrons
c) 69 protons, 69 electrons, 100 neutrons   d) 9 protons, 9 electrons, 10 neutrons

5.29 a) total subatomic particles = 11 protons + 11 electrons + 12 neutrons = 34
b) total subatomic particles in the nucleus = 11 protons + 12 neutrons = 23
c) total number nucleons = total subatomic particles in the nucleus = 11 protons + 12 neutrons = 23
d) total charge in the nucleus = +11, +1 for each proton, neutrons have no charge

5.31 a) atom has 15 protons; mass number = 15 + 16; $_{15}^{31}P$
b) oxygen has 8 protons; mass number is 8 + 10 = 18; $_{8}^{18}O$
c) chromium has 24 protons; $_{24}^{54}Cr$
d) A gold atom has 79 protons and 79 electrons.
276 subatomic particles − 79 protons − 79 electrons = 118 neutrons.
∴ mass number = 79 + 118 = 197; $_{79}^{197}Au$

5.33 a) 12 protons + 12 electrons = 24 subatomic particles that carry a charge
b) 37 subatomic particles − 24 subatomic particles that carry a charge (protons + electrons)
 = 13 subatomic particles with no charge (neutrons)
c) 12 protons + 13 neutrons = 25 particles in the nucleus
d) 37 subatomic particles − 25 particles in the nucleus
 = 12 particles not found in the nucleus (electrons)

5.35 a) atomic number = number protons = 20, number of neutrons = 20 + 3 = 23
 mass number = 20 protons + 23 neutrons = 43
b) atomic number of 20 = number protons = number electrons,
 number of neutrons = 20 + 3 = 23, mass number = 20 protons + 23 neutrons = 43
c) atomic number of 30 = number protons, number of neutrons = 30,
 mass number = 30 protons + 30 neutrons = 60
d) atomic number of 30 = number protons = number electrons, number of neutrons = 30,
 mass number = 30 protons + 30 neutrons = 60

5.37 a) same total number of subatomic particles, 60
b) same number of neutrons, 16
c) same number of neutrons, 12
d) same number of electrons, 3

5.39 a) scandium   b) chlorine   c) strontium   d) arsenic

5.41 a) no   b) yes   c) no   d) yes

Subatomic Particles, Isotopes, and Nuclear Chemistry

Isotopes (Sec. 5.3)

5.43 $^{54}_{26}Fe$ $^{56}_{26}Fe$ $^{57}_{26}Fe$ $^{58}_{26}Fe$

5.45 $^{96}_{40}Zr$ $^{94}_{40}Zr$ $^{92}_{40}Zr$ $^{91}_{40}Zr$ $^{90}_{40}Zr$

5.47 a) To determine the mass numbers, we round the amu to the nearest whole number.
53.940 amu = mass number of 54, thus our symbol is ^{54}Fe
55.935 amu = mass number of 56, thus our symbol is ^{56}Fe
56.935 amu = mass number of 57, thus our symbol is ^{57}Fe
57.933 amu = mass number of 58, thus our symbol is ^{58}Fe
b) The percent abundance for the third isotope can be calculated from the percent abundance of the other three isotopes.
First isotope = 5.82%
Second isotope = 91.66%
Fourth isotope = 0.33%
Total = 97.81%   By definition the percent abundance must = 100, thus the percent abundance of the third isotope = 100% − 97.81% = 2.19%

5.49 a) not isotopes, different atomic numbers
b) not isotopes, different atomic numbers
c) not isotopes, different atomic numbers

5.51 a) no   b) yes   c) no   d) yes

5.53 a) isotopes   b) isobars   c) neither   d) isotopes

5.55 a) isotopes   b) isobars   c) neither   d) isotopes

5.57 a) $^{31}_{15}P$ $^{33}_{15}P$ $^{35}_{15}P$ b) $^{31}_{15}P$ $^{32}_{15}P$ $^{33}_{15}P$
c) $^{31}_{15}P$ $^{32}_{15}P$ $^{34}_{15}P$ d) $^{30}_{15}P$ $^{32}_{15}P$ $^{34}_{15}P$

5.59 $^{40}_{18}Ar$ $^{40}_{19}K$ $^{40}_{20}Ca$

5.61 a) relative mass of protons and neutrons = 1 amu, therefore mass number = 59
b) mass number = 20 protons + 20 neutrons = 40
c) isobars have the same mass numbers, mass number = 18, atomic number is 9, $^{18}_{9}F$
d) O has 8 protons + 10 neutrons = mass number = 18, $^{18}_{8}O$

Atomic Masses (Sec. 5.4)

5.63 a) 55.85 amu   b) 14.01 amu   c) calcium   d) iodine

5.65 Values on this hypothetical relative mass scale are Q = 8.00 bebs, X = 4.00 bebs, and Z = 2.00 bebs.

5.67 a) Z = ¾(12) = 9 amu; X = 3Z = 27 amu; Q = 9(12) = 108 amu
b) Z is Be, X is Al, Q is Ag

5.69   0.220 × 271 lb   = 59.62 (calc)   = 59.6 lb (corr)
0.190 × 175 lb   = 33.25 (calc)   = 33.2 lb (corr)
0.260 × 263 lb   = 68.38 (calc)   = 68.4 lb (corr)
0.150 × 182 lb   = 27.3 (calc)   = 27.3 lb (corr)
0.190 × 191 lb   = 36.29 (calc)   = 36.3 lb (corr)
                                      = 224.8 lb (calc and corr)

5.71 0.7553 × 34.9689 amu = 26.41201 (calc)   = 26.41 amu (corr)
0.2447 × 36.9659 amu = 9.0455557 (calc)   = 9.046 amu (corr)
                                           = 35.456 amu (calc)
                                           = 35.46 amu (corr)

5.73 0.0793 × 45.95263 amu   = 3.6440435 (calc)   = 3.64 amu (corr)
0.0728 × 46.95176 amu   = 3.418088128 (calc)   = 3.42 amu (corr)
0.7394 × 47.94795 amu   = 35.452714 (calc)   = 35.45 amu (corr)
0.0551 × 48.94787 amu   = 2.6970276 (calc)   = 2.70 amu (corr)
0.0534 × 49.94479 amu   = 2.667051786 (calc)   = 2.67 amu (corr)
                                                = 47.88 amu (calc and corr)

5.75 a) ^{14}N   b) ^{51}V   c) ^{121}Sb   d) ^{193}Ir

5.77 a) possible   b) possible   c) possible   d) not possible

5.79

	Isotope A	Isotope B	Atomic Mass	Abundance % A	Abundance % B
	46 amu	48 amu	47 amu	50%	50%
a)	46 amu	50 amu	47 amu	75%	25%
b)	46 amu	50 amu	49 amu	25%	75%
c)	45 amu	50 amu	46 amu	80%	20%
d)	45 amu	50 amu	49 amu	20%	80%

5.81 a) $\dfrac{35.45 \text{ amu}}{85.47 \text{ amu}} = 0.414765$ (calc) = 0.415 (corr)

b) $\dfrac{2(35.45 \text{ amu})}{2(85.47 \text{ amu})} = 0.414765$ (calc) = 0.415 (corr)

c) $\dfrac{107.87 \text{ amu}}{2(63.55 \text{ amu})} = 0.848702$ (calc) = 0.849 (corr)

d) $\dfrac{10(1.01 \text{ amu})}{9.01 \text{ amu}} = 1.1209767$ (calc) = 1.12 (corr)

5.83 a) iron (Fe)   b) strontium (Sr)   c) beryllium (Be)   d) helium (He)

5.85 The number 12.0000 amu applies only to the ^{12}C isotope. The number 12.01 amu applies to naturally occurring carbon, a mixture of ^{12}C, ^{13}C, and ^{14}C.

Evidence Supporting the Existence and Arrangement of Subatomic Particles (Sec. 5.5)

5.87 a) true   b) false   c) false   d) false

Nuclear Stability and Radioactivity (Sec. 5.6)

5.89 The spontaneous emission of radiation from the nucleus indicates the nucleus is unstable.

5.91 The largest stable nucleus contains 209 nucleons.

Subatomic Particles, Isotopes, and Nuclear Chemistry

5.93 A radioactive atom is an atom with an unstable nucleus from which radiation is spontaneously emitted whereas a radioactive nuclide is an atom with a specific atomic number and mass number.

5.95 a) $^{9}_{5}B$ or boron-9 b) $^{44}_{19}K$ or potassium-44
 c) $^{96}_{45}Rh$ or rhodium-96 d) $^{182}_{73}Ta$ or tantalum-182

5.97 a) $^{14}_{7}N$ b) $^{197}_{79}Au$ c) Rubidium-92 d) Tin-121

Half-life (Sec. 5.7)

5.99 a) $5.2 \text{ days} \times \dfrac{1 \text{ half-life}}{2.6 \text{ days}} = 2.0$ half-lives $\therefore \dfrac{1}{2^2} = \dfrac{1}{4}$ remains unreacted (calc and corr)

 b) $13 \text{ days} \times \dfrac{1 \text{ half-life}}{2.6 \text{ days}} = 5.0$ half-lives $\therefore \dfrac{1}{2^5} = \dfrac{1}{32}$ remains unreacted (calc and corr)

 c) 3.0 half-lives $\therefore \dfrac{1}{2^3} = \dfrac{1}{8}$ remains unreacted (calc and corr)

 d) 6.0 half-lives $\therefore \dfrac{1}{2^6} = \dfrac{1}{64}$ remains unreacted (calc and corr)

5.101 a) $\dfrac{1}{2^n} = \dfrac{1}{16}$ $\therefore n = 4$ half-lives (calc and corr)

 b) $\dfrac{1}{2^n} = \dfrac{1}{64}$ $\therefore n = 6$ half-lives (calc and corr)

 c) $\dfrac{1}{2^n} = \dfrac{1}{32}$ $\therefore n = 5$ half-lives (calc and corr)

 d) $\dfrac{1}{2^n} = \dfrac{1}{512}$ $\therefore n = 9$ half-lives (calc and corr)

5.103 Amount left = initial amount × fraction left

 a) $4.8 \text{ min} \times \dfrac{1 \text{ half-life}}{2.4 \text{ min}} = 2$ half-lives. Amount left $= 8.0 \text{ g} \times \dfrac{1}{2^2} = 2$ (calc) $= 2.0$ g (corr)

 b) $9.6 \text{ min} \times \dfrac{1 \text{ half-life}}{2.4 \text{ min}} = 4$ half-lives. Amount left $= 8.0 \text{ g} \times \dfrac{1}{2^4} = 0.5$ (calc) $= 0.50$ g (corr)

 c) $16.8 \text{ min} \times \dfrac{1 \text{ half-life}}{2.4 \text{ min}} = 7$ half-lives. Amount left $= 8.0 \text{ g} \times \dfrac{1}{2^7}$
 $= 0.0625$ (calc) $= 0.062$ g (corr)

 d) $24.0 \text{ min} \times \dfrac{1 \text{ half-life}}{2.4 \text{ min}} = 10$ half-lives. Amount left $= 8.0 \text{ g} \times \dfrac{1}{2^{10}}$
 $= 0.0078125$ (calc) $= 0.0078$ g (corr)

5.105 Fraction decayed = (1 − fraction left)
 Amount decayed = initial amount × fraction decayed

 a) $6.4 \text{ hr} \times \dfrac{1 \text{ half-life}}{3.2 \text{ hr}} = 2$ half-lives. Fraction left $= \dfrac{1}{2^2} = \dfrac{1}{4}$. Fraction decayed $= \dfrac{3}{4}$.
 Amount decayed $= 10.0 \text{ g} \times \dfrac{3}{4} = 7.5$ g (calc) $= 7.50$ g (corr)

 b) $19.2 \text{ hr} \times \dfrac{1 \text{ half-life}}{3.2 \text{ hr}} = 6$ half-lives. Fraction left $= \dfrac{1}{2^6} = \dfrac{1}{64}$. Fraction decayed $= \dfrac{63}{64}$.
 Amount decayed $= 10.0 \text{ g} \times \dfrac{63}{64} = 9.84375$ (calc) $= 9.84$ g (corr)

c) $28.8 \text{ hr} \times \dfrac{1 \text{ half-life}}{3.2 \text{ hr}} = 9$ half-lives. Fraction left $= \dfrac{1}{2^9} = \dfrac{1}{512}$. Fraction decayed $= \dfrac{511}{512}$.

Amount decayed $= 10.0 \text{ g} \times \dfrac{511}{512} = 9.9804688$ (calc) $= 9.98$ g (corr)

d) $48 \text{ hr} \times \dfrac{1 \text{ half-life}}{3.2 \text{ hr}} = 15$ half-lives. Fraction left $= \dfrac{1}{2^{15}} = \dfrac{1}{32{,}768}$. Fraction decayed $= \dfrac{32{,}767}{32{,}768}$.

Amount decayed $= 10.0 \text{ g} \times \dfrac{32{,}767}{32{,}768} = 9.999694824$ (calc) $= 10.0$ g (corr)

5.107 a) If $\dfrac{3}{4}$ decayed, $\dfrac{1}{4}$ left $= \dfrac{1}{2^2}$. 2 half-lives $\times \dfrac{8.0 \text{ hr}}{\text{half-life}} = 16$ hr (calc and corr)

b) If $\dfrac{15}{16}$ decayed, $\dfrac{1}{16}$ left $= \dfrac{1}{2^4}$. 4 half-lives $\times \dfrac{8.0 \text{ hr}}{\text{half-life}} = 32$ hr (calc and corr)

c) If $\dfrac{31}{32}$ decayed, $\dfrac{1}{32}$ left $= \dfrac{1}{2^5}$. 5 half-lives $\times \dfrac{8.0 \text{ hr}}{\text{half-life}} = 40.$ hr (calc and corr)

d) If $\dfrac{127}{128}$ decayed, $\dfrac{1}{128}$ left $= \dfrac{1}{2^7}$. 7 half-lives $\times \dfrac{8.0 \text{ hr}}{\text{half-life}} = 56$ hr (calc and corr)

The Nature of Natural Radioactive Emissions (Sec. 5.8)

5.109 a) $^{4}_{2}\alpha$ b) $^{0}_{-1}\beta$ c) $^{0}_{0}\gamma$

5.111 2 protons and 2 neutrons

Equations for Radioactive Decay (Sec. 5.9)

5.113 a) $^{200}_{84}\text{Po} \rightarrow {}^{4}_{2}\alpha + {}^{196}_{82}\text{Pb}$

b) $^{244}_{96}\text{Cm} \rightarrow {}^{4}_{2}\alpha + {}^{240}_{94}\text{Pu}$

c) $^{240}_{96}\text{Cm} \rightarrow {}^{4}_{2}\alpha + {}^{236}_{94}\text{Pu}$

d) $^{238}_{92}\text{U} \rightarrow {}^{4}_{2}\alpha + {}^{234}_{90}\text{Th}$

5.115 a) $^{10}_{4}\text{Be} \rightarrow {}^{0}_{-1}\beta + {}^{10}_{5}\text{B}$

b) $^{77}_{32}\text{Ge} \rightarrow {}^{0}_{-1}\beta + {}^{77}_{33}\text{As}$

c) $^{60}_{26}\text{Fe} \rightarrow {}^{0}_{-1}\beta + {}^{60}_{27}\text{Co}$

d) $^{25}_{11}\text{Na} \rightarrow {}^{0}_{-1}\beta + {}^{25}_{12}\text{Mg}$

5.117 a) mass number is decreased by 4
b) mass number remains unchanged
c) mass number remains unchanged

5.119 a) $^{190}_{78}\text{Pt} \rightarrow \underline{} + {}^{186}_{76}\text{Os}$ to balance this equation, we need an alpha particle $^{4}_{2}\alpha$

b) $^{19}_{8}\text{O} \rightarrow \underline{} + {}^{19}_{9}\text{F}$ to balance this equation, we need a beta particle $^{0}_{-1}\beta$

5.121 a) $^{0}_{-1}\beta$ b) $^{125}_{52}\text{Te}$ c) $^{4}_{2}\alpha$ d) $^{229}_{90}\text{Th}$

5.123 a) $^{199}_{79}\text{Au} \rightarrow {}^{0}_{-1}\beta + {}^{199}_{80}\text{Hg}$ b) $^{120}_{48}\text{Cd} \rightarrow {}^{0}_{-1}\beta + {}^{120}_{49}\text{In}$

c) $^{152}_{67}\text{Ho} \rightarrow {}^{4}_{2}\alpha + {}^{148}_{65}\text{Tb}$ d) $^{226}_{88}\text{Ra} \rightarrow {}^{4}_{2}\alpha + {}^{222}_{86}\text{Rn}$

Subatomic Particles, Isotopes, and Nuclear Chemistry

Transmutation and Bombardment Reactions (Sec. 5.10)

5.125   a) $^{4}_{2}\alpha$ b) $^{2}_{1}H$ c) $^{81}_{34}Se$ d) $^{9}_{4}Be$

5.127   a) $^{9}_{4}Be + ^{4}_{2}\alpha \rightarrow ^{1}_{0}n + ^{12}_{6}C$ b) $^{58}_{28}Ni + ^{1}_{1}H \rightarrow ^{4}_{2}\alpha + ^{55}_{27}Co$
c) $^{113}_{48}Cd + ^{1}_{0}n \rightarrow ^{114}_{48}Cd + ^{0}_{0}\gamma$ d) $^{27}_{13}Al + ^{4}_{2}\alpha \rightarrow ^{30}_{15}P + ^{1}_{0}n$

5.129   nine

Positron Emission and Electron Capture (Sec. 5.11)

5.131   The mass number remains the same; the atomic number decreases by 1

5.133   Beta decay, $^{0}_{-1}\beta$

5.135   a) $^{29}_{15}P \rightarrow ^{0}_{1}\beta + ^{29}_{14}Si$ b) $^{112}_{51}Sb \rightarrow ^{0}_{1}\beta + ^{112}_{50}Sn$
c) $^{46}_{23}V \rightarrow ^{0}_{1}\beta + ^{46}_{22}Ti$ d) $^{132}_{58}Ce \rightarrow ^{0}_{1}\beta + ^{132}_{57}La$

5.137   a) $^{76}_{36}Kr + ^{0}_{-1}e \rightarrow ^{76}_{35}Br$ b) $^{122}_{54}Xe + ^{0}_{-1}e \rightarrow ^{122}_{53}I$
c) $^{100}_{46}Pd + ^{0}_{-1}e \rightarrow ^{100}_{45}Rh$ d) $^{175}_{73}Ta + ^{0}_{-1}e \rightarrow ^{175}_{72}Hf$

5.139   a) $^{0}_{1}\beta$ b) $^{0}_{-1}e$ c) $^{103}_{47}Ag$ d) $^{133}_{55}Cs$

Neutron-to-Proton Ratio and Type of Radioactive Decay (Sec. 5.12)

5.141   Beta emission

5.143   a) $^{65}_{28}Ni \rightarrow ^{0}_{-1}\beta + ^{65}_{29}Cu$
^{65}Ni: 28p, 37n   n/p = 37/28 = 1.32 before
^{65}Cu: 29p, 36n   n/p = 36/29 = 1.24 after

b) $^{192}_{78}Pt \rightarrow ^{4}_{2}\alpha + ^{188}_{76}Os$
^{192}Pt: 78p, 114n   n/p = 114/78 = 1.46 before
^{188}Os: 76p, 112n   n/p = 112/76 = 1.47 after

c) $^{165}_{69}Tm + ^{0}_{-1}e \rightarrow ^{165}_{68}Er$
^{165}Tm: 69p, 96n   n/p = 96/69 = 1.39 before
^{165}Er: 68p, 97n   n/p = 97/68 = 1.43 after

d) $^{107}_{49}In \rightarrow ^{0}_{1}\beta + ^{107}_{48}Cd$
^{107}In: 49p, 58n   n/p = 58/49 = 1.18 before
^{107}Cd: 48p, 59n   n/p = 59/48 = 1.23 after

5.145   Beta emission occurs in nuclides with too high neutron-proton ratio. Positron emission occurs in nuclides with too low neutron-proton ratio. Radioactive nuclides with a mass number greater than the atomic mass of that element (the mass of the stable nuclides) will decay by beta emission. Those with a mass number less than the atomic mass will decay by positron emission.

a) $^{74}_{34}Kr$ = positron; $^{87}_{36}Kr$ = beta   b) $^{68}_{33}As$ = positron; $^{84}_{34}Se$ = beta
c) $^{74}_{31}Ga$ = beta; $^{64}_{31}Ga$ = positron   d) $^{99}_{41}Nb$ = beta; $^{99}_{46}Pd$ = positron

Radioactive Decay Series (Sec. 5.13)

5.147 Stable. The series terminates when a stable nuclide is reached.

5.149 $^{238}_{92}U \rightarrow ^{4}_{2}\alpha + ^{234}_{90}Th \rightarrow ^{0}_{-1}\beta + ^{234}_{91}Pa$

5.151 (1) $^{220}_{86}Rn \rightarrow ^{4}_{2}\alpha + ^{216}_{84}Po$ (2) $^{216}_{84}Po \rightarrow ^{4}_{2}\alpha + ^{212}_{82}Pb$
(3) $^{212}_{82}Pb \rightarrow ^{0}_{-1}\beta + ^{212}_{83}Bi$ (4) $^{212}_{83}Bi \rightarrow ^{0}_{-1}\beta + ^{212}_{84}Po$

Additional Problems

5.153 a) false b) false c) true d) true

5.155 a) $^{37}_{18}Ar, ^{39}_{19}K, ^{42}_{20}Ca, ^{44}_{21}Sc, ^{43}_{22}Ti$ b) $^{44}_{21}Sc, ^{42}_{20}Ca, ^{43}_{22}Ti, ^{39}_{19}K, ^{37}_{18}Ar$
c) $^{37}_{18}Ar, ^{39}_{19}K, ^{42}_{20}Ca, ^{44}_{21}Sc, ^{43}_{22}Ti$ d) $^{44}_{21}Sc, ^{43}_{22}Ti, ^{42}_{20}Ca, ^{39}_{19}K, ^{37}_{18}Ar$

5.157 a) $^{8}_{5}B$ b) $^{12}_{5}B$ c) $^{12}_{5}B$ d) $^{16}_{5}B$

5.159 a) 29 and 29 b) 29 and 29 c) 34 and 36

5.161 a) 13 b) 27 c) 14 d) 27.0 amu

5.163 a) 1.679×30.97 amu = 51.99863 (calc) = 52.00 amu = Cr
b) $30e \times 3 = 90e = Th$
c) $4p + 20p = 24p = Cr$
d) $[20e + 20p = 40 \text{ e\&p}] \times 2 = 80 \text{ e\&p} = 40p = Zr$

5.165 Nickel weighs 20.000 amu on the new scale and 58.6934 amu on the old carbon-12 scale. Therefore, 20.000 new amu units is equal to 58.6934 old amu units.

a) 107.87 old amu $\times \dfrac{20.000 \text{ new amu}}{58.6934 \text{ old amu}}$ = 36.757114 (calc) = 36.757 new amu (corr)

b) 196.97 old amu $\times \dfrac{20.000 \text{ new amu}}{58.6934 \text{ old amu}}$ = 67.118279 (calc) = 67.118 new amu (corr)

5.167 a) $^{221}_{88}Ra \rightarrow ^{4}_{2}He + ^{217}_{86}Rn$ thus X = ^{221}Ra
b) $^{47}_{19}K \rightarrow ^{0}_{-1}\beta + ^{47}_{20}Ca$ thus X = ^{47}K
c) $^{78}_{36}Kr \rightarrow ^{0}_{1}\beta + ^{78}_{35}Br$ thus X = ^{78}Kr
d) $^{37}_{18}Ar + ^{0}_{-1}e \rightarrow ^{37}_{17}Cl$ thus X = ^{37}Ar

5.169 a) $^{232}_{90}Th \xrightarrow{\alpha} ^{228}_{88}Ra \xrightarrow{\beta^-} ^{228}_{89}Ac \xrightarrow{\beta^-} ^{228}_{90}Th$

b) $^{228}_{89}Ac \xrightarrow{\beta^-} ^{228}_{90}Th \xrightarrow{\alpha} ^{224}_{88}Ra \xrightarrow{\alpha} ^{220}_{86}Rn$

5.171 ^{28}P has too low a neutron-to-proton ratio; the decay mode should be positron emission or electron capture. ^{34}P has too high a neutron-to-proton ratio; the decay mode should be beta emission.

Subatomic Particles, Isotopes, and Nuclear Chemistry

5.173 For the decay $A \rightarrow B$, the half-life = 3.2 minutes.
This decay will be complete (to 3 significant figures) in less than 10 half-lives, 32 minutes. The amount of A left is 0 mole. For the decay $B \rightarrow C$, the half-life = 25 days.
We can assume that the 1.00 mole B was all made at day 0 (to 3 significant figures), so after 50 days (2 half-lives), 1/4 of B will be left, and 3/4 of B will have decayed.
For the decay $C \rightarrow D$, the half-life is 9 seconds.
Compared to the 50-days duration, this half-life is so small that we can assume all the C produced decays instantly. The amount of $C = 0$ mole and D will be the amount of C made (or B lost) in 50 days.
The amount of $D = 0.750$ mole.
$A = 0$; $B = 0.250$ mole; $C = 0$; $D = 0.750$ mole.

5.175 $\dfrac{4.000 \text{ g}}{0.125 \text{ g}} = 32.0$, $\therefore \dfrac{1}{32}$ of the sample remains after 35 days

a) $\dfrac{1}{32} = \dfrac{1}{2^5}$, $\therefore$ 5 half-lives have elapsed. $\dfrac{35 \text{ days}}{5 \text{ half-lives}} = 7$ (calc) $= \dfrac{7 \text{ days}}{\text{half-life}}$ (corr)

b) $4.000 - 0.125 \text{ g} = 3.875 \text{ g X}$ that has decayed
Assuming that isotopic masses can be approximated by using mass numbers

$$3.875 \text{ g X} \times \dfrac{206 \text{ g Q}}{210 \text{ g X}} = 3.8011904 \text{ (calc)} = 3.80 \text{ g Q (corr)}$$

5.177 a) $^A_Z Q \rightarrow {}^0_1 \beta + {}^A_{Z-1} R$ Q and R are isobars.

b) $^A_Z Q \rightarrow {}^4_2 \alpha + {}^{A-4}_{Z-2} R$ Q and R are neither isobars nor isotopes.

c) $^A_Z Q + {}^1_0 n \rightarrow {}^{A+1}_Z R$ Q and R are isotopes.

d) $^A_Z Q \rightarrow {}^4_2 \alpha + {}^{A-4}_{Z-2} R$
$^{A-4}_{Z-2} R \rightarrow {}^0_{-1} \beta + {}^{A-4}_{Z-1} S$ Q and T are isotopes.
$^{A-4}_{Z-1} S \rightarrow {}^0_{-1} \beta + {}^{A-4}_Z T$

Cumulative Problems

5.179 82

5.181 $C_6H_{12}O_6$ 6 carbon atoms, 6 protons in each carbon atom, $6 \times 6 = 36$ protons
12 hydrogen atoms, 1 proton in each hydrogen atom $12 \times 1 = 12$ protons
6 oxygen atoms, 8 protons in each oxygen atom $6 \times 8 = \underline{48 \text{ protons}}$

Total number protons in 1 molecule $C_6H_{12}O_6$ $= 96$ protons
Therefore 7 molecules of $C_6H_{12}O_6$ contain $7 \times 96 = 672$ protons (calc and corr)

5.183 pathway: in. (nucleus) $\rightarrow$ in. (atom) $\rightarrow$ ft (atom) $\rightarrow$ mi (atom)

$$1.5 \text{ in. (nucleus)} \times \dfrac{10^5 \text{ in. (atom)}}{1 \text{ in. (nucleus)}} \times \dfrac{1 \text{ ft}}{12 \text{ in.}} \times \dfrac{1 \text{ mile}}{5280 \text{ ft}} = 2.36742 \text{ (calc)} = 2.4 \text{ miles (corr)}$$

Answers to Multiple-Choice Practice Test

5.185	e	**5.186**	d	**5.187**	e	**5.188**	c	**5.189**	c	**5.190**	d
5.191	c	**5.192**	c	**5.193**	e	**5.194**	c	**5.195**	e	**5.196**	c
5.197	c	**5.198**	e	**5.199**	b	**5.200**	b	**5.201**	c	**5.202**	b
5.203	a	**5.204**	a								

CHAPTER SIX
Electronic Structure and Chemical Periodicity

Practice Problems

Periodic Law and Periodic Table (Secs. 6.1 and 6.2)

6.1   a) Al   b) Be   c) Sn   d) K

6.3   a) (1) same group   b) (2) same period
        c) (3) neither   d) (2) same period

6.5   a) $_{19}$K and $_{37}$Rb   b) $_{15}$P and $_{33}$As
        c) $_9$F and $_{53}$I   d) $_{11}$Na and $_{55}$Cs

6.7   a) group   b) periodic law   c) periodic law   d) period

6.9   a) bromine   b) lithium   c) argon   d) strontium

6.11   a) 3   b) 4   c) 4   d) 4

6.13   a) 3   b) 1   c) 2   d) 2

Terminology Associated with Electron Arrangements (Secs. 6.3–6.6)

6.15   a) 2   b) 10   c) 6   d) 14

6.17   Any orbital has a maximum capacity of 2 electrons.
        a) 2   b) 2   c) 2   d) 2

6.19   a) 1   b) 1   c) 5   d) 3

6.21   a) 3d subshell can hold 10; second shell can hold 8; ∴ 3d subshell can hold more.
        b) $n = 1$ shell can hold 2; 2p subshell can hold 6; ∴ 2p subshell can hold more.
        c) 3p orbital can hold 2; 3p subshell can hold 6; ∴ 3p subshell can hold more.
        d) 4f subshell can hold 14; third shell can hold 18; ∴ third shell can hold more.

6.23   a) true   b) false   c) true   d) true   e) false

6.25   a) spherical   b) dumbbell   c) cloverleaf   d) spherical

6.27   a) allowed   b) allowed
        c) not allowed, 2p is lowest allowed   d) not allowed, 4f is lowest allowed

Electron Configurations (Sec. 6.7)

6.29   a) 3p   b) 5s   c) 4d   d) 4p

6.31   a) 3s   b) 3p   c) 4f   d) 3d

6.33 a) 1s, 2s, 3s, 4s   b) 4s, 4p, 4d, 4f
      c) 2p, 6s, 4f, 5d   d) 3p, 4s, 3d, 5p

6.35 a) $1s^2 2s^2 2p^6 3s^2 3p^1$ b) $1s^2 2s^2 2p^3$
      c) $1s^2 2s^2 2p^6$ d) $1s^2 2s^2 2p^6 3s^2 3p^3$

6.37 a) $1s^2 2s^2 2p^6 3s^2 3p^6 4s^2 3d^6$
      b) $1s^2 2s^2 2p^6 3s^2 3p^6 4s^2 3d^{10} 4p^6 5s^1$
      c) $1s^2 2s^2 2p^6 3s^2 3p^6 4s^2 3d^{10} 4p^6 5s^2 4d^{10} 5p^5$
      d) $1s^2 2s^2 2p^6 3s^2 3p^6 4s^2 3d^{10} 4p^6 5s^2 4d^{10} 5p^6 6s^2 4f^{14} 5d^{10} 6p^6$

6.39 a) $_{10}$Ne   b) $_{19}$K   c) $_{22}$Ti   d) $_{30}$Zn

6.41 a) [Ne] $3s^2 3p^1$ b) [Ar] $4s^1$
      c) [Ar] $4s^2 3d^{10}$ d) [Kr] $5s^2 4d^{10} 5p^2$

6.43 a) [He] $2s^2 2p^4$ b) [Ne] $3s^2$
      c) [Ar] $4s^2$ d) [Ar] $4s^2 3d^{10} 4p^5$

6.45 a) magnesium   b) chlorine   c) tellurium   d) barium

6.47 a) 10 core electrons   b) 10 core electrons
      c) 36 core electrons   d) 54 core electrons

6.49 a) 2 outer electrons   b) 7 outer electrons
      c) 16 outer electrons d) 2 outer electrons

Orbital Diagrams (Sec. 6.8)

6.51 a) ↑↓   ↑
          1s   2s

      b) ↑↓   ↑↓   ↑↓ ↑↓ ↑
          1s   2s   2p

      c) ↑↓   ↑↓   ↑↓ ↑↓ ↑↓   ↑↓   ↑ ↑ ↑
          1s   2s   2p   3s   3p

      d) ↑↓   ↑↓   ↑↓ ↑↓ ↑↓   ↑↓   ↑↓ ↑↓ ↑↓   ↑↓   ↑↓ ↑↓ ↑↓ ↑ ↑
          1s   2s   2p   3s   3p   4s   3d

6.53 a) ↑↓   ↑↓   ↑ ↑
          1s   2s   2p

      b) ↑↓   ↑↓   ↑↓ ↑↓ ↑↓
          1s   2s   2p

      c) ↑↓   ↑↓   ↑↓ ↑↓ ↑↓   ↑
          1s   2s   2p   3s

      d) ↑↓   ↑↓   ↑↓ ↑↓ ↑↓   ↑↓   ↑ ↑ ↑
          1s   2s   2p   3s   3p

Electronic Structure and Chemical Periodicity

6.55 a) [Ne] ↑↓   ↑↓ ↑↓ ↑
              3s   3p

b) [Ar] ↑↓   ↑ ↑ ↑ ↑ ↑
          4s   3d

c) [Kr] ↑↓
          5s

d) [Kr] ↑↓   ↑↓ ↑↓ ↑↓ ↑↓ ↑↓   ↑ ↑ ↑
          5s   4d   5p

6.57 a) Mg, $3s^2$ ↑↓
                      3s

b) Cl, $3p^5$ ↑↓ ↑↓ ↑
                3p

c) Ca, $4s^2$ ↑↓
                4s

d) Co, $3d^7$ ↑↓ ↑↓   ↑ ↑ ↑
                   3d

6.59 a) carbon has 2 unpaired electrons   b) sodium has 1 unpaired electron
      c) argon has no unpaired electrons   d) titanium has 2 unpaired electrons

6.61 a) paramagnetic (unpaired)   b) paramagnetic (unpaired)
      c) diamagnetic (paired)   d) paramagnetic (unpaired)

Electron Configurations and the Periodic Law (Sec. 6.9)

6.63 a) no   b) yes   c) no   d) yes

Electron Configurations and the Periodic Table (Sec. 6.10)

6.65 a) *p* area   b) *d* area   c) *d* area   d) *s* area

6.67 a) *p* subshell   b) *d* subshell   c) *d* subshell   d) *s* subshell

6.69 a) 3*p* subshell   b) 3*d* subshell   c) 4*d* subshell   d) 6*s* subshell

6.71 a) 3, Cr, Ru, and Pt   b) 3, S, Xe, and element 114
      c) 2, Ca and element 114   d) 2, Ca and Ru

6.73 a) Al   b) Li   c) La   d) Sc

6.75 a) Kr   b) Li   c) K   d) Lu

6.77 a) $1s^22s^22p^63s^23p^64s^23d^{10}4p^3$
      b) $1s^22s^22p^63s^23p^64s^23d^{10}4p^65s^24d^{10}5p^3$
      c) $1s^22s^22p^63s^23p^64s^23d^{10}4p^65s^1$
      d) $1s^22s^22p^63s^23p^64s^23d^{10}4p^65s^24d^{10}$

6.79 a) $1s^22s^22p^63s^2$ b) $1s^22s^22p^63s^23p^64s^2$
      c) $1s^22s^22p^63s^23p^64s^23d^{10}$ d) $1s^22s^22p^63s^23p^64s^23d^{10}4p^65s^24d^{10}5p^2$

6.81 a) Ti, $1s^2 2s^2 2p^6 3s^2 3p^6 4s^2 3d^2$ ∴ 2 electrons in $3d$
b) Ni, $1s^2 2s^2 2p^6 3s^2 3p^6 4s^2 3d^8$ ∴ 8 electrons in $3d$
c) Se, $1s^2 2s^2 2p^6 3s^2 3p^6 4s^2 3d^{10} 4p^4$ ∴ 10 electrons in $3d$
d) Pd, $1s^2 2s^2 2p^6 3s^2 3p^6 4s^2 3d^{10} 4p^6 5s^2 4d^8$ ∴ 10 electrons in $3d$

6.83 a) 4   b) 9   c) 1   d) 0

Classification Systems for the Elements (Sec. 6.11)

6.85 a) metallic   b) nonmetallic   c) metallic   d) nonmetallic

6.87 Only c and d contain pairs in which both elements are metals.

6.89 a) (1) Group IIIA contains more metals than nonmetals.
b) (1) Group IIIB contains only metals.
c) (2) Group VIA contains more nonmetals than metals.
d) (2) Group VIIIA contains only nonmetals.

6.91 a) Li   b) K   c) Fe   d) Hg

6.93 a) 1 is a metal (Rb).   b) 3 is a nonmetal (S).
c) 2 is a good conductor of electricity (Al).   d) 4 is a good conductor of heat (Sn).

6.95 a) transition element   b) representative element
c) noble gas element   d) transition element

6.97 a) $_1$H   b) $_2$He   c) $_3$Li   d) $_{58}$Ce

6.99 a) transition elements—$3d, 4d, 5d, 6d,$ = 4 periods of transition elements in the current periodic table
b) representative elements, s and p subshells, $s^1, s^2, p^1, p^2, p^3, p^4, p^5$ = 7 groups
c) periods that contain metals, Period 2, 3, 4, 5, 6, 7 = 6 periods
d) no periods consist of only metals, all periods contain at least one noble gas

6.101 a) 1 is a noble gas element (Xe).
b) 1 is an inner transition element (U).
c) 3 are nonmetal elements (H, S, and Xe).
d) 2 are metallic representative elements (element 114 and Ca).

Chemical Periodicity (Sec. 6.12)

6.103 a) false   b) true   c) false

6.105 a) Ge   b) B   c) Po   d) Te

6.107 a) $_{37}$Rb   b) $_{22}$Ti   c) $_{34}$Se   d) $_{56}$Ba

6.109 a) $_9$F   b) $_{15}$P   c) $_{30}$Zn   d) $_{17}$Cl

6.111 a) Bi is the most metallic in Group VA.
b) Beryllium is the most nonmetallic in Group IIA.
c) Potassium is the most metallic in Period 4.
d) Neon is the most nonmetallic in Period 2.

Electronic Structure and Chemical Periodicity

6.113 a) B   b) Ga   c) K   d) Rb

6.115 a) Nitrogen has the smallest atomic radius in Group VA.
b) Radium has the largest atomic radius in Group IIA.
c) Krypton has the smallest atomic radius in Period 4.
d) Lithium has the largest atomic radius in Period 2.

Additional Problems

6.117 a) The 2s subshell is more than filled. It has a maximum occupancy of 2 electrons.
b) The 2p subshell must be filled before the 3s. It can accommodate 6 electrons.
c) The 2p subshell, which fills after the 2s, is omitted.
d) The 3p and 4s subshells fill after the 3s and before the 3d subshell.

6.119 a) Period 3, Group IA   b) Period 3, Group IIIA
c) Period 4, Group IIIB   d) Period 4, Group VIIA

6.121 a) 1 unpaired electron   b) 1 unpaired electron
c) 1 unpaired electron   d) 1 unpaired electron

6.123 a) paramagnetic   b) paramagnetic
c) paramagnetic   d) paramagnetic

6.125 a) $x = 6, y = 2$ b) $x = 2, y = 1$ c) $x = 2, y = 10$ d) $x = 2, y = 6$

6.127 a) above: [Ne] $3s^2$, below: [Kr] $5s^2$
b) above: [Ar] $4s^2 3d^3$, below: [Xe] $6s^2 4f^{14} 5d^3$
c) above: [He] $2s^2 2p^5$, below: [Ar] $4s^2 3d^{10} 4p^5$
d) above: [Ar] $4s^2 3d^{10} 4p^1$, below: [Xe] $6s^2 4f^{14} 5d^{10} 6p^1$

6.129 a) before: [Ar] $4s^1$, after: [Ar] $4s^2 3d^1$
b) before: [Kr] $5s^2 4d^2$, after: [Kr] $5s^2 4d^4$
c) before: [Ne] $3s^2 3p^4$, after: [Ne] $3s^2 3p^6$
d) before: [Kr] $5s^2 4d^{10}$, after: [Kr] $5s^2 4d^{10} 5p^2$

6.131 a) [Noble Gas] ns^2 = Be, Mg, Ca, Sr, Ba, Ra = 6 elements
b) [Noble Gas] $ns^2 np^3$ = N, P = 2 elements
c) [Noble Gas] $ns^2 (n-1)d^3$ = V, Nb = 2 elements
d) [Noble Gas] $ns^2 (n-1)d^{10} np^5$ = Br, I = 2 elements

6.133 a) $1s^2 2s^2 2p^6 3s^2 3p^2$ (Si)   b) $1s^2 2s^2 2p^2$ (C)   c) $1s^2 2s^2 2p^3$ (N)   d) $1s^2 2s^2 2p^3$ (N)

6.135 a) Be   b) Be   c) Ne   d) Ar

6.137 a) 2   b) 4   c) 1   d) 2

6.139 a) Groups IVA and VIA   b) Group VB and middle column of Group VIIIB
c) Group IVB and last column of Group VIIIB   d) Group IA

6.141 a) Po   b) Cr   c) Elements 88–116, 118   d) Elements 12–18

6.143 a) F   b) Ag

Cumulative Problems

6.145 a) O, Li, He, B, K
  b) O, He, B (the nonmetals)
  c) O, Li, He, B, Sr, K (all of them)
  d) K

6.147 the same

6.149 Since the isotopic mass is 36.96590 amu, the mass number is 37. Since 17 electrons are present, the atomic number is 17.
protons = electrons = 17; neutrons = 37 − 17 = 20
17 + 17 + 20 = 54

6.151 +38 (Sr)

6.153 a) 22   b) 16/22 × 100 = 72.7%   c) 12/22 × 100 = 54.5%   d) 50% (#e + #p)

6.155 P, S, Cl, Se, Br

Answers to Multiple-Choice Practice Test

6.157	b	6.158	a	6.159	a	6.160	d	6.161	e	6.162	d
6.163	c	6.164	c	6.165	d	6.166	e	6.167	b	6.168	c
6.169	a	6.170	d	6.171	d	6.172	e	6.173	c	6.174	b
6.175	c	6.176	c								

CHAPTER SEVEN
Chemical Bonds

Practice Problems

Valence Electrons (Sec. 7.2)

7.1 Valence electrons are all of the *s* and *p* electrons in the highest shell.
    a) 2 + 3 = 5   b) 2   c) 2   d) 2 + 4 = 6

7.3 a) 2 + 2 = 4   b) 2 + 1 = 3   c) 2 + 4 = 6   d) 2

7.5 a) (1) Be has more valence electrons than Li.
    b) (3) N has the same number of valence electrons as P.
    c) (1) F has more valence electrons than O.
    d) (2) Al has fewer valence electrons than Cl.

7.7 a) 1 element has 5 valence electrons (Bi).
    b) 1 element has 7 valence electrons (I).
    c) 2 elements have 2 valence electrons (Mg and Ba).
    d) 1 element has 8 valence electrons (Ne).

7.9 a) Group IA, 1 valence electron   b) Group VIIIA, 8 valence electrons
    c) Group IIA, 2 valence electrons   d) Group VIIA, 7 valence electrons

7.11 a) C   b) F   c) Mg   d) P

Lewis Symbols for Atoms (Sec. 7.2)

7.13 a) ·C·   b) ·Si·   c) ·Cl: d) ·Ba·

7.15 a) B   b) C   c) F   d) Li

7.17 a) Incorrect, O has 6 valence electrons.   b) Correct, Al has 3 valence electrons.
    c) Incorrect, Cl has 7 valence electrons.   d) Correct, Se has 6 valence electrons.

7.19 a) Li · Be :, ∴ (2) Li has fewer dots   b) Mg : Ca :, ∴ (3) same number dots

    c) ·Al· K·, ∴ (1) Al has more dots   d) : Ar : : Ne :, ∴ (3) same number dots

Notation for Ions (Sec. 7.4)

7.21 a) Li^+ b) P^{3-} c) Br^- d) Ba^{2+}

7.23

	Chemical Symbol	Ion Formed	Number of Electrons in Ion	Number of Protons in Ion
	Ca	Ca^{2+}	18	20
a)	Be	Be^{2+}	2	4
b)	I	I^-	54	53
c)	Al	Al^{3+}	10	13
d)	S	S^{2-}	18	16

7.25 a) (1) A neutral species, number of protons = number of electrons
b) (3) A positively charged species, contains less electrons than protons
c) (3) A positively charged species, contains less electrons than protons
d) (2) A negatively charged species, contains more electrons than protons

7.27 a) Al^{3+} b) O^{2-} c) Mg^{2+} d) Be^{2+}

7.29 a) (2), anion   b) (1), cation   c) (3), not an ion   d) (2), anion

The Sign and Magnitude of Ionic Charge (Sec. 7.5)

7.31 a) Nitrogen is predicted to form an ion with a negative charge.
b) Beryllium is predicted to form an ion with a positive charge.
c) Magnesium is predicted to form an ion with a positive charge.
d) Chlorine is predicted to form an ion with a negative charge.

7.33 a) Nitrogen is predicted to form an anion.
b) Beryllium is predicted to form a cation.
c) Magnesium is predicted to form a cation.
d) Chlorine is predicted to form an anion.

7.35 a) Nitrogen is predicted to gain 3 electrons.
b) Beryllium is predicted to lose 2 electrons.
c) Magnesium is predicted to lose 2 electrons.
d) Chlorine is predicted to gain 1 electron.

7.37 a) loss   b) loss   c) loss   d) gain

7.39 a) $1s^2 2s^2 2p^6$ b) $1s^2 2s^2 2p^6$ c) $1s^2 2s^2 2p^6 3s^2 3p^6$ d) $1s^2 2s^2 2p^6 3s^2 3p^6$

7.41 a) [Ne]   b) [Ne]   c) [Ar]   d) [Ar]

7.43 a) 4 elements form positively charged ions, (Li, Al, Ca, and Rb).
b) 3 elements form anions, (N, F, and Br).
c) 1 element forms an ion with a charge magnitude of 2 (Ca).
d) 2 elements form ions that involve the loss of 2 or more electrons (Ca and Al).

7.45 a) Group IA   b) Group VIA   c) Group VA   d) Group IVA

7.47 a) $_2$He   b) $_{54}$Xe   c) $_{36}$Kr   d) $_{54}$Xe

Chemical Bonds

7.49 Neon has 10 electrons. a) Na^+ b) F^- c) O^{2-} d) Mg^{2+}

7.51 a) yes, isoelectronic   b) no, not isoelectronic
c) yes, isoelectronic   d) no, not isoelectronic

Lewis Structures for Ionic Compounds (Sec. 7.6)

7.53 a) $\cdot\ddot{B}r\!:$ $\quad$ $:\!\ddot{B}r\!:^-$
$Ca\!\cdot\quad\quad\quad\rightarrow\quad Ca^{2+}$
$\cdot\ddot{B}r\!:\quad\quad\quad\quad:\!\ddot{B}r\!:^-$

b) $Mg\cdot\cdot\ddot{S}\!:\quad\rightarrow\quad Mg^{2+}\;:\!\ddot{\ddot{S}}\!:^{2-}$

c) $Be\cdot$
$\quad\cdot\ddot{N}\!:\quad\quad Be^{2+}\quad:\!\ddot{\ddot{N}}\!:^{3-}$
$Be\cdot\quad\quad\rightarrow\quad Be^{2+}$
$\quad\cdot\ddot{N}\!:\quad\quad Be^{2+}\quad:\!\ddot{\ddot{N}}\!:^{3-}$
$Be\cdot$

d) $Na\cdot$
$\quad Na\cdot\cdot\ddot{P}\!:\quad\rightarrow\quad Na^+\quad:\!\ddot{\ddot{P}}\!:^{3-}$
$Na\cdot\quad\quad\quad\quad\quad\quad Na^+$

7.55 a) Lithium has 1 valence electron (lose 1); nitrogen has five valence electrons (gain 3).

$Li\cdot\quad\quad\quad\quad Li^+$
$Li\cdot\cdot\ddot{N}\!:\quad\rightarrow\quad Li^+\;:\!\ddot{\ddot{N}}\!:^{3-}$
$Li\cdot\quad\quad\quad\quad Li^+$

b) Mg has 2 valence electrons (lose 2); O has 6 valence electrons (gain 2).

$Mg\cdot\cdot\ddot{O}\!:\quad\rightarrow\quad Mg^{2+}\;:\!\ddot{\ddot{O}}\!:^{2-}$

c) Cl has 7 valence electrons (gain 1); barium has 2 valence electrons (lose 2). Write Ba first.

$\quad\quad\cdot\ddot{C}l\!:\quad\quad\quad:\!\ddot{\ddot{C}}l\!:^-$
$Ba\cdot\quad\quad\rightarrow\quad Ba^{2+}$
$\quad\quad\cdot\ddot{C}l\!:\quad\quad\quad:\!\ddot{\ddot{C}}l\!:^-$

d) F has 7 valence electrons (gain 1); K has 1 valence electron (lose 1). Write K first.

$K\cdot\cdot\ddot{F}\!:\quad\rightarrow\quad K^+\;:\!\ddot{\ddot{F}}\!:^-$

Chemical Formulas for Ionic Compounds (Sec. 7.7)

7.57 a) $CaCl_2$ b) BeO c) AlN d) K_2S

7.59

	F⁻	O²⁻	N³⁻	C⁴⁻
	Na⁺	NaF	Na₂O	Na₃N	Na₄C
a)	Ca²⁺	CaF₂	CaO	Ca₃N₂	Ca₂C
b)	Al³⁺	AlF₃	Al₂O₃	AlN	Al₄C₃
c)	Ag⁺	AgF	Ag₂O	Ag₃N	Ag₄C
d)	Zn²⁺	ZnF₂	ZnO	Zn₃N₂	Zn₂C

7.61 a) Mg^{2+}, S^{2-} b) Al^{3+}, N^{3-} c) $2\,Na^+, O^{2-}$ d) $3\,Ca^{2+}, 2\,N^{3-}$

7.63 a) X_2Z b) XZ_3 c) X_3Z d) ZX_2

7.65 a) Be_3N_2 b) $NaBr$ c) SrF_2 d) Al_2S_3

Polyatomic-ion-containing Ionic Compounds (Sec. 7.9)

7.67 a) $Mg(CN)_2$ b) $CaSO_4$ c) $Al(OH)_3$ d) NH_4NO_3

7.69 a) $AlPO_4$ b) $Al_2(CO_3)_3$ c) $Al(ClO_3)_3$ d) $Al(C_2H_3O_2)_3$

7.71

		OH⁻	CN⁻	NO₃⁻	SO₄²⁻
	Na⁺	NaOH	NaCN	NaNO₃	Na₂SO₄
a)	K⁺	KOH	KCN	KNO₃	K₂SO₄
b)	Mg²⁺	Mg(OH)₂	Mg(CN)₂	Mg(NO₃)₂	MgSO₄
c)	Al³⁺	Al(OH)₃	Al(CN)₃	Al(NO₃)₃	Al₂(SO₄)₃
d)	NH₄⁺	NH₄OH	NH₄CN	NH₄NO₃	(NH₄)₂SO₄

Lewis Structures for Covalent Compounds (Secs. 7.10–7.15)

7.73 a) 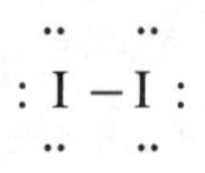 :Ï—Ï: b) :C̈l—F̈: c) H—S̈—H d) 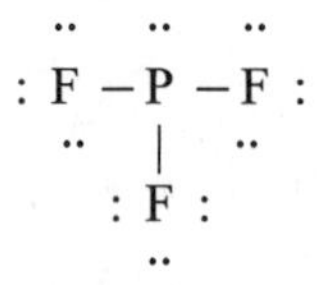 :F̈—P̈—F̈: with :F̈: below

7.75 a) 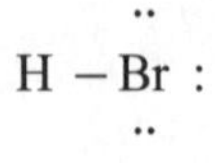 H—B̈r: b) F̈—Ö—F̈ c) 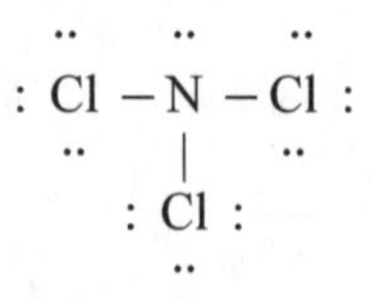 :C̈l—N̈—C̈l: with :C̈l: below d) 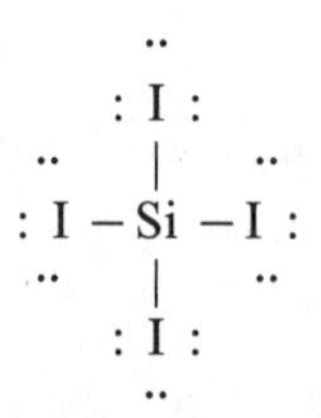 :Ï—Si—Ï: with :Ï: above and :Ï: below

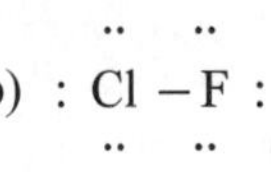

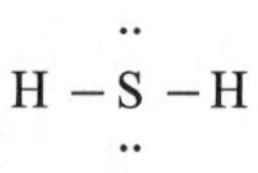

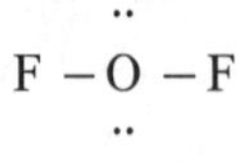

7.77 a) 3 bonding pairs, 2 nonbonding pairs
    b) 6 bonding pairs, 0 nonbonding pairs
    c) 4 bonding pairs, 4 nonbonding pairs
    d) 7 bonding pairs, 2 nonbonding pairs

7.79 a) 1 triple bond b) 4 single, 1 double c) 2 double d) 2 triple, 1 single

Chemical Bonds

7.81 a) yes   b) yes   c) no, C has to form 4 bonds   d) yes

7.83 A coordinate covalent bond is a bond in which both of the shared pair electrons come from one of the two atoms.

7.85 a) the N—O bond   b) none   c) the O—Cl bond   d) the two O—Br bonds

7.87 Resonance structures are two or more Lewis structures for a molecule or ion that have the same arrangement of atoms, contain the same number of electrons, and differ only in the location of the electrons.

7.89

$$\left[\begin{array}{c} :\ddot{O}=N-\ddot{O}: \\ | \\ :\ddot{O}: \end{array}\right]^{-} \leftrightarrow \left[\begin{array}{c} :\ddot{O}-N=\ddot{O}: \\ | \\ :\ddot{O}: \end{array}\right]^{-} \leftrightarrow \left[\begin{array}{c} :\ddot{O}-N-\ddot{O}: \\ \| \\ :\ddot{O}: \end{array}\right]^{-}$$

Systematic Procedures for Drawing Lewis Structures (Sec. 7.16)

7.91 a) $1 + 5 + (3 \times 6) = 24$ b) $5 + (3 \times 7) = 26$
c) $5 + (3 \times 6) + 1 = 24$ d) $6 + (4 \times 6) + 2 = 32$

7.93 a) 24 dots   b) 26 dots   c) 24 dots   d) 32 dots

7.95 a) 1 bonding and 3 nonbonding electron pairs   b) 3 bonding and 2 nonbonding electron pairs
c) 1 bonding and 6 nonbonding electron pairs   d) 1 bonding and 3 nonbonding electron pairs

7.97 a) 1. There are 14 valence electrons
   $(6 \times 1) + (2 \times 4)$.
2. The two C atoms are central atoms.
3. Attach hydrogens (all 14 electrons used).

b) 1. There are 14 valence electrons
   $(4 \times 1) + (2 \times 5)$.
2. The two N atoms are central.
3. Attach H atoms on N's (8 electrons used).
4. Complete octet on each N by putting a nonbonding pair on each N (all 14 electrons used).

c) 1. There are 20 electrons
   $(2 \times 7) + 4 + (2 \times 1)$.
2. The C is central.
3. Attach H and F atoms (8 electrons used).
4. Complete octet on F (all 20 electrons used).

d) 1. There are 32 electrons
   $(3 \times 1) + (2 \times 4) + (3 \times 7)$.
2. The C atoms are central.
3. Attach H and Cl atoms (12 electrons used).
4. Complete octets on chlorine (all 32 electrons used).

7.99 a) 1. There are 8 electrons
$[5 + (4 \times 1) - 1]$.
2. The N atom is central.
3. Attach H atoms (8 electrons used).

b) 1. There are 8 electrons
$[2 + (4 \times 1) + 2]$.
2. The Be atom is central.
3. Attach H atoms (all 8 electrons used).

c) 1. There are 26 electrons
$[7 + (3 \times 6) + 1]$.
2. The Cl atom is central.
3. Attach O atoms (6 electrons used).
4. Complete octet on O atoms (24 electrons used).
5. Complete octet on Cl with a nonbonding pair (all 26 electrons used).

d) 1. There are 32 electrons
$[7 + (4 \times 6) + 1]$.
2. The I atom is central.
3. Attach O atoms (8 electrons used).
4. Complete octet on O atoms (all 32 electrons used).

$$\left[\begin{array}{c} H \\ | \\ H-N-H \\ | \\ H \end{array}\right]^+$$

$$\left[\begin{array}{c} H \\ | \\ H-Be-H \\ | \\ H \end{array}\right]^{2-}$$

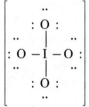

7.101 a) 1. There are 24 electrons
$[(2 \times 4) + (2 \times 1) + (2 \times 7)]$.
2. Join 2 C atoms (2 electrons used).
3. Attach H and Cl atoms (8 electrons).
4. Complete octet on Cl atoms (12 electrons used).
5. Complete octet on C atoms by double bond (all 24 electrons used).

b) 1. There are 16 electrons
$[(2 \times 4) + (3 \times 1) + 5]$.
2. Join 2 C atoms (2 electrons used).
3. Attach 3 H atoms to a C (6 electrons used).
4. Attach the N atom to other C (2 more electrons used).
5. Complete octet on other C atom by triple bond (4 electrons used to make a triple bond).
6. Complete octet on N atom (16 electrons used).

c) 1. There are 16 electrons
$[(3 \times 4) + (4 \times 1)]$.
2. Join 3 C atoms (4 electrons used).
3. Attach the 4 H atoms (8 electrons).
4. Complete octet on C atoms by triple bond (all 16 electrons used).

d) 1. There are 24 electrons
$[(2 \times 5) + (2 \times 7)]$.
2. Join 2 N atoms (2 electrons).
3. Attach 2 F atoms (4 electrons).
4. Complete octet on 2 F atoms (12 electrons).
5. Complete octet on 2 N atoms by making a double bond (2 electrons) and 1 nonbonding pair on each (all 24 electrons used).

Chemical Bonds

7.103 Follow steps used in Problem 7.97.

a) H–C(H)(H)–N=O with :Ö: ↔ H–C(H)(H)–N–Ö: with :O: (double bond to O)

b) :N≡N–Ö: ↔ :N̈=N=Ö: ↔ :N̈–N≡O:

c) [O=C(–Ö:)(–Ö:)]²⁻ ↔ [:Ö–C(–Ö:)(=O)]²⁻ ↔ [:Ö–C(=Ö:)(–Ö:)]²⁻

d) [:S̈–C≡N:]⁻ ↔ [S̈=C=N̈]⁻ ↔ [:S≡C–N̈:]⁻

7.105 Total electrons in structure = 32; total valence electrons = $4(6) + 7 = 31$; $31 - 32 = -1$; $\therefore n = 1$.

Molecular Geometry (VSEPR Theory) (Sec. 7.17)

7.107
a) The electron arrangement about the central atom involves 2 single bonds and 2 nonbonding locations. This arrangement produces an angular geometry.
b) The electron arrangement about the central atom involves 1 single bond and 1 triple bond, and 0 nonbonding locations. This arrangement produces a linear geometry.
c) The electron arrangement about the central atom involves 1 single bond, 1 double bond, and 1 nonbonding location. This arrangement produces angular geometry.
d) The electron arrangement of one double bond between two atoms, and two nonbonding locations, produces a linear arrangement between two atoms.

7.109
a) The electron arrangement about the central atom involves 2 bonding locations and 2 nonbonding locations. This arrangement produces an angular geometry.
b) The electron arrangement about the central atom involves 2 bonding locations and no nonbonding locations. This arrangement produces a linear geometry.
c) The electron arrangement about the central atom involves 2 bonding locations and no nonbonding locations. This arrangement produces a linear geometry.
d) The electron arrangement about the central atom involves 2 bonding locations and 1 nonbonding location. This arrangement produces an angular geometry.

7.111
a) The electron arrangement about the central atom involves 3 bonding locations and 1 nonbonding location. This arrangement produces a trigonal pyramidal geometry.
b) The electron arrangement about the central atom involves 2 bonding locations and no nonbonding locations. This arrangement produces a linear geometry.
c) The electron arrangement about the central atom involves 2 bonding locations and 1 nonbonding location. This arrangement produces an angular geometry.
d) The electron arrangement about the central atom involves 4 bonding locations and no nonbonding locations. This arrangement produces a tetrahedral geometry.

7.113 a) The electron arrangement about the central atom involves 4 bonding locations and no nonbonding locations. This arrangement produces a tetrahedral geometry.
b) The electron arrangement about the central atom involves 2 bonding locations and no nonbonding locations. This arrangement produces a linear geometry.
c) The electron arrangement about the central atom involves 3 bonding locations and 1 nonbonding location. This arrangement produces a trigonal pyramidal geometry.
d) The electron arrangement about the central atom involves 3 bonding locations and 1 nonbonding location. This arrangement produces a trigonal pyramidal geometry.

7.115 a) Each central atom has 3 bonding locations and 1 nonbonding location, giving a trigonal pyramid for each center. The centers are joined by one axis of each trigonal pyramid.
b) The C has 4 bonding locations and no nonbonding locations and is thus tetrahedral. The O has two bonding locations and 2 nonbonding locations and is thus angular.

Electronegativity (Sec. 7.18)

7.117 Electronegativity increases across a period; decreases down a group.

7.119 a) Be and O are in the same period; O is farther to the right and therefore more electronegative.
b) Be and Ca are in the same group; Be is above Ca and is therefore more electronegative.
c) H is in the first period and C is in the second period. Since C is much farther to the right and it is more electronegative.
d) Ca is in the second group in the fourth period, while Cs is in the sixth period and in the first group. The larger gap between periods prevails and therefore Ca is more electronegative.

7.121 Arranged in order of decreasing electronegativity, from left to right:
a) F, O, B, Li   b) F, Cl, S, P   c) N, P, As, Sb   d) F, Si, Mg, Fr

7.123 a) C = 2.5, ∴ elements N (3.0), O (3.5), F (4.0), Cl (3.0), and Br (2.8) are more electronegative.
b) 1.0 ∴ elements that are less electronegative include Na (0.9), K (0.8), Rb (0.8), Cs (0.7), Ba (0.9), Fr (0.7), and Ra (0.9).
c) 4 most electronegative elements include F (4.0), O (3.5), N and Cl (3.0).
d) For period 2, elements increase from left to right by 0.5 in electronegative values.

7.125 a) True, Be has a higher electronegativity value than Mg.
b) False, N has a higher electronegativity value than B.
c) False, Br has a higher electronegativity value than K.
d) False, F has a higher electronegativity value than As.

Bond Polarity (Sec. 7.19)

7.127 a) N = 3, S = 2.5; thus S will have the partial positive charge in the N—S bond.
b) N = 3, Br = 2.8; thus Br will have the partial positive charge in the N—Br bond.
c) N = 3, O = 3.5; thus N will have the partial positive charge in the N—O bond.
d) N = 3, F = 4; thus N will have the partial positive charge in the N—F bond.

7.129 Calculating the difference in electronegativity between the two atoms in each pair gives:
a) H—Cl = 0.9; H—O = 1.4; H—Br = 0.7 ∴ H—O has the greatest polarity
b) O—F = 0.5; O—P = 1.4; O—Al = 2.0 ∴ O—Al has the greatest polarity
c) H—Cl = 0.9; Br—Br = 0; B—N = 1.0 ∴ B—N has the greatest polarity
d) Al—Cl = 1.5; C—N = 0.5; Cl—F = 1.0 ∴ Al—Cl has the greatest polarity

Chemical Bonds

7.131 Calculating the difference in electronegativity between the two atoms in each pair gives:
  a) H—Cl = 0.9; H—O = 1.4; H—Br = 0.7 ∴ H—O has the greatest ionic character
  b) O—F = 0.5; O—P = 1.4; O—Al = 2.0 ∴ O—Al has the greatest ionic character
  c) H—Cl = 0.9; Br—Br = 0; B—N = 1.0 ∴ B—N has the greatest ionic character
  d) Al—Cl = 1.5; C—N = 0.5; Cl—F = 1.0 ∴ Al—Cl has the greatest ionic character

7.133 Calculating the difference in electronegativity between the two atoms in each pair gives:
  a) H—Cl = 0.9; H—O = 1.4; H—Br = 0.7 ∴ H—O has the least covalent character
  b) O—F = 0.5; O—P = 1.4; O—Al = 2.0 ∴ O—Al has the least covalent character
  c) H—Cl = 0.9; Br—Br = 0; B—N = 1.0 ∴ B—N has the least covalent character
  d) Al—Cl = 1.5; C—N = 0.5; Cl—F = 1.0 ∴ Al—Cl has the least covalent character

7.135 Calculating the difference in electronegativity between the two atoms in each bond gives:
  a) H—Cl = 0.9; H—O = 1.4; H—Br = 0.7, H—O has a larger ς+ on the less electronegative atom.
  b) O—F = 0.5; O—P = 1.4; O—Al = 2.0, O—Al has a larger ς+ on the less electronegative atom.
  c) H—Cl = 0.9; Br—Br = 0; B—N = 1.0, B—N has a larger ς+ on the less electronegative atom.
  d) Al—Cl = 1.5; C—N = 0.5; Cl—F = 1.0, Al—Cl has a larger ς+ on the less electronegative atom.

7.137
  a) C—N = 0.5 difference in electronegativity and is thus considered polar covalent.
  b) Be—O = 2.0 difference in electronegativity and is thus considered ionic.
  c) P—Cl = 0.9 difference in electronegativity and is thus considered polar covalent.
  d) Si—Si = 0 difference in electronegativity and is thus considered nonpolar covalent.

7.139
  a) 0.3 difference; bonds with an electronegativity difference less than 0.4 are nonpolar covalent bonds.
  b) 1.2 difference; bonds with an electronegativity difference more than 0.4 but less than 1.5 are considered polar covalent bonds.
  c) 1.9 difference; bonds with an electronegativity difference between 1.5 and 2.0 are considered to be ionic if the bond involves a metal and a nonmetal and polar covalent if the bond involves two nonmetals.
  d) 2.1 difference; bonds with an electronegativity difference of 2.0 or greater are considered ionic bonds.

7.141
  a) 0.3 difference; bonds have more covalent character.
  b) 1.2 difference; bonds have more covalent character.
  c) 1.9 difference; bonds have more ionic character.
  d) 2.1 difference; bonds have more ionic character.

7.143
  a) P—I = 0.4 difference in electronegativity and thus has more covalent character.
  b) H—F = 1.9 difference in electronegativity and thus has more ionic character.
  c) B—O = 1.5 difference in electronegativity and thus has more covalent character.
  d) Na—Cl = 2.1 difference in electronegativity and thus has more ionic character.

Molecular Polarity (Sec. 7.20)

7.145 a) nonpolar b) polar c) polar d) nonpolar

7.147 a) nonpolar b) polar c) polar d) nonpolar

7.149
  a) F_2 is nonpolar, BrF is polar
  b) HOCl and HCN are both polar
  c) CH_4 and CCl_4 are both nonpolar
  d) SO_3 is nonpolar and NF_3 is polar

7.151   a) A polar molecule with tetrahedral geometry requires a minimum of 3 different elements.
b) A nonpolar molecule with trigonal planar geometry requires a minimum of 2 different elements.
c) A nonpolar molecule with linear geometry requires a minimum of 2 different elements.
d) A polar molecule with angular geometry requires a minimum of 2 different elements.

Additional Problems

7.153   a) Mg: $1s^2 2s^2 2p^6 3s^2$ Mg^{2+}: $1s^2 2s^2 2p^6$
b) F: $1s^2 2s^2 2p^5$ F^-: $1s^2 2s^2 2p^6$
c) N: $1s^2 2s^2 2p^3$ N^{3-}: $1s^2 2s^2 2p^6$
d) Ca^{2+}: $1s^2 2s^2 2p^6 3s^2 3p^6$ S^{2-}: $1s^2 2s^2 2p^6 3s^2 3p^6$

7.155   a) nonisoelectronic cations   b) nonisoelectronic anions
c) isoelectronic cations   d) nonisoelectronic anions

7.157   a) ionic, Al is a metal, O is a nonmetal   b) molecular, H and O are both nonmetals
c) ionic, K is a metal, S is a nonmetal   d) molecular, N and H are both nonmetals

7.159   a) and d) are molecules; b) and c) are ionic compounds for which the formula unit is the simplest ratio of ions.

7.161   a) (1), CaF_2 contains only monoatomic ions   b) (3), $NaNO_3$ contains monoatomic and polyatomic ions
c) (2) $(NH_4)_2SO_4$ contains only polyatomic ions   d) (1) BaO contains only monoatomic ions

7.163   a) CaO, Ca has charge of +2   b) MgF_2, Mg has a charge of +2
c) Na_2S, neither ion contains charge of +2   d) NO, neither ion contains charge of +2

7.165   a) O   b) N   c) O   d) F

7.167   a) correct number of electron dots, but improperly placed
b) not enough electron dots
c) correct number of electron dots, but improperly placed
d) correct number of electron dots, but improperly placed

7.169   BA, CA, DB, and DA

7.171   a) same, both single   b) different, single and triple
c) different, triple, and double   d) different, triple and double

7.173   a) There are 26 electrons shown. The 3 oxygen and 1 hydrogen provide 19, so X must provide 7: X = Cl
b) There are 20 electrons shown. The 2 oxygen provide 12, the ionic charge provides 1, so X must provide 7: X = Cl

7.175   a) Ca^{2+} [:Ö–S̈–Ö:]$^{2-}$ (with double-bonded O)   b) [H–N̈H–H]$^+$ (with H above and below)   [:Ö–N=Ö:]$^-$

Note that other resonance structures are possible for the SO_4^{2-} and NO_3^- ions!

Chemical Bonds

7.177 a) tetrahedral electron pair geometry; tetrahedral molecular geometry
b) tetrahedral electron pair geometry; trigonal pyramidal molecular geometry
c) linear electron pair geometry; linear molecular geometry
d) tetrahedral electron pair geometry; tetrahedral molecular geometry

7.179 a) 109.5° because the electron pair arrangement about the oxygen atom is tetrahedral
b) 120° because the electron pair arrangement about the carbon atom is trigonal planar

Cumulative Problems

7.181 a) O contains 2 unpaired electrons. b) O^{2-} contains 0 unpaired electrons.
c) Ca contains 0 unpaired electrons. d) Ca^{2+} contains 0 unpaired electrons.

7.183 a) Sr^{2+} is an Alkaline Earth Metal ion isoelectronic with Kr.
b) N^{3-} is isoelectronic with Ne.
c) Ca^{2+} ion contains 18 electrons.
d) Ga^{3+} ion is a Group IIIA ion with an electron configuration of $[Ar] 3d^{10}$.

7.185 A = Al, D = N, formula = AlN

7.187 D = Mg, A = S, formula = MgS

7.189 A = O and D = F. The Lewis structure of the compound is :F̈—Ö—F̈:

7.191 A = H and D = O. x is 2 and y is 1. The formula is H_2O.

7.193 A = Be and D = F. The Lewis structure is $\left[:\ddot{F}-\underset{:\ddot{F}:}{\overset{:\ddot{F}:}{Be}}-\ddot{F}: \right]^{2-}$

Answers to Multiple-Choice Practice Test

7.195	a	7.196	d	7.197	c	7.198	b	7.199	c	7.200	c
7.201	b	7.202	e	7.203	c	7.204	d	7.205	c	7.206	c
7.207	d	7.208	d	7.209	b	7.210	c	7.211	e	7.212	b
7.213	d	7.214	b								

CHAPTER EIGHT
Chemical Nomenclature

Practice Problems

Nomenclature Classifications for Compounds (Sec. 8.1)

8.1   a) molecular   b) ionic   c) ionic   d) molecular

8.3   a) ionic   b) ionic   c) ionic   d) ionic

8.5   a) yes   b) yes   c) no, both molecular   d) no, both ionic

Types of Binary Ionic Compounds (Sec. 8.2)

8.7   a) binary   b) not binary c) binary   d) binary

8.9   a) not ionic   b) not ionic   c) ionic   d) not ionic

8.11 a) not binary ionic   b) not binary ionic
      c) binary ionic   d) not binary ionic

8.13 a) fixed-charge   b) variable-charge   c) fixed-charge   d) variable-charge

8.15 a) and d)

8.17 a) variable-charge   b) variable-charge   c) fixed-charge   d) fixed-charge

Nomenclature for Binary Ionic Compounds (Sec. 8.3)

8.19 a) magnesium ion   b) potassium ion   c) zinc ion   d) silver ion

8.21 a) copper(II) ion b) copper(I) ion   c) cobalt(III) ion d) cobalt(II) ion

8.23 a) bromide ion   b) nitride ion   c) sulfide ion   d) chloride ion

8.25 a) Zn^{2+} b) Pb^{2+} c) Ca^{2+} d) N^{3-}

8.27 a) magnesium oxide b) lithium sulfide c) silver chloride d) zinc bromide

8.29 a) fixed-charge   b) variable-charge c) fixed-charge   d) variable-charge

8.31 a) no Roman numeral needed   b) yes, Roman numeral needed
      c) no Roman numeral needed   d) yes, Roman numeral needed

8.33 a) +2   b) +2   c) +3   d) +2

Copyright © 2011 Pearson Education, Inc.

8.35 a) iron(II) oxide   b) nickel(II) chloride   c) gold(III) oxide   d) cobalt(III) nitride

8.37 a) iron(II) bromide, iron(III) bromide   b) copper(I) sulfide, copper(II) sulfide
c) tin(II) sulfide, tin(IV) sulfide   d) nickel(II) oxide, nickel(III) oxide

8.39 a) aluminum oxide   b) cobalt(III) fluoride   c) silver nitride   d) barium sulfide

8.41 a) plumbic oxide   b) auric chloride   c) iron(III) iodide   d) tin(II) bromide

8.43 In b) and c) both names denote the same compound

8.45 a) Li_2S b) ZnS c) Al_2S_3 d) Ag_2S

8.47 a) CuS b) Cu_3N_2 c) SnO d) SnO_2

8.49

Formula of Positive Ion	Formula of Negative Ion	Compound Formula	Compound Name
Mg^{2+}	Cl^-	$MgCl_2$	magnesium chloride
Al^{3+}	O^{2-}	Al_2O_3	aluminum oxide
Pb^{2+}	Br^-	$PbBr_2$	lead(II) bromide
Fe^{2+}	S^{2-}	FeS	iron(II) sulfide
Zn^{2+}	Br^-	$ZnBr_2$	zinc bromide

Nomenclature for Ionic Compounds Containing Polyatomic Ions (Sec. 8.4)

8.51 a) OH^- b) NH_4^+ c) NO_3^- d) ClO_4^-

8.53 a) peroxide   b) thiosulfate   c) oxalate   d) chlorate

8.55 a) SO_4^{2-}, SO_3^{2-} b) PO_4^{3-}, HPO_4^{2-} c) OH^-, O_2^{2-} d) CrO_4^{2-}, $Cr_2O_7^{2-}$

8.57 a) Ag has a charge of +1   b) Zn has a charge of +2
c) Pb has a charge of +2   d) Cu has a charge of +1

8.59 a) fixed-charge ionic compound   b) fixed-charge ionic compound
c) variable-charge ionic compound   d) variable-charge ionic compound

8.61 a) silver nitrate   b) zinc sulfate   c) lead(II) cyanide   d) copper(I) phosphate

8.63 a) iron(III) carbonate, iron(II) carbonate   b) gold(I) sulfate, gold(III) sulfate
c) tin(II) hydroxide, tin(IV) hydroxide   d) chromium(III) acetate, chromium(II) acetate

8.65 a) Ag_2CO_3 b) $AuNO_3$ c) $Fe_2(SO_4)_3$ d) $CuCN$

8.67 a) sulfate has a −2 charge   b) cyanide has a −1 charge
c) peroxide has a −2 charge   d) azide has a −1 charge

8.69 a) $(NH_4)_2SO_4$ b) NH_4CN c) Na_2O_2 d) KN_3

Chemical Nomenclature

8.71 a) (3) ionic compound with 4 different elements   b) (2) ternary ionic compound
c) (1) binary ionic compound   d) (1) binary ionic compound

8.73

Formula of Positive Ion	Formula of Negative Ion	Compound Formula	Compound Name
Mg^{2+}	NO_3^-	$Mg(NO_3)_2$	magnesium nitrate
Al^{3+}	SO_4^{2-}	$Al_2(SO_4)_3$	aluminum sulfate
Cu^{2+}	CN^-	$Cu(CN)_2$	copper(II) cyanide
Fe^{2+}	OH^-	$Fe(OH)_2$	iron(II) hydroxide
Zn^{2+}	N_3^-	$Zn(N_3)_2$	zinc azide

Nomenclature for Binary Molecular Compounds (Sec. 8.5)

8.75 a) CO: O is to the right in the order given in the textbook on page 318.
b) Cl_2O: O is to the right in the order given on page 318.
c) N_2O: O is to the right in the order given on page 318.
d) HI: I is to the right in the order given on page 318.

8.77 a) 7   b) 5   c) 3   d) 10

8.79 a) tetraphosphorus decoxide   b) sulfur tetrafluoride
c) carbon tetrabromide   d) chlorine dioxide

8.81 a) SO   b) S_2O c) SO_2 d) S_7O_2

8.83 a) hydrogen sulfide   b) hydrogen fluoride   c) ammonia   d) methane

8.85 a) PH_3 b) HBr   c) C_2H_6 d) H_2Te

8.87 Ionic compounds with variable-charge metal are named by first naming the metal, then the charge on the metal using Roman numerals in parenthesis, then the name of the anion. The correct name for Cu_2SO_3 is therefore copper(I) sulfite.

Nomenclature for Acids (Sec. 8.6)

8.89 a) no   b) yes   c) yes   d) no

8.91 a) NO_3^- b) I^- c) ClO^- d) PO_3^{3-}

8.93 a) HCN   b) H_2SO_4 c) HNO_2 d) H_3BO_3

8.95 a) nitric acid   b) hydroiodic acid
c) hypochlorous acid   d) phosphorous acid

8.97 a) hydrocyanic acid   b) sulfuric acid
c) nitrous acid   d) boric acid

8.99 a) HCl b) H_2CO_3 c) $HClO_3$ d) H_2SO_4

8.101

Positive Ion Present in Solution	Negative Ion Present in Solution	Chemical Formula of Acid	Name of Acid
H^+	SO_3^{2-}	H_2SO_3	sulfurous acid
H^+	NO_2^-	HNO_2	nitrous acid
H^+	PO_4^{3-}	H_3PO_4	phosphoric acid
H^+	CN^-	HCN	hydrocyanic acid
H^+	ClO^-	HClO	hypochlorous acid

8.103 a) arsenous acid b) periodic acid c) hypophosphorous acid d) bromous acid

8.105 a) hydrogen bromide b) hydrocyanic acid c) hydrogen sulfide d) hydroiodic acid

Additional Problems

8.107 $X = Se^{2-}$, $Y = Br^-$ therefore compound is $SeBr_2$

8.109 a) HF b) MgS c) Li_3N d) SBr_2

8.111 a) no b) no c) yes d) no

8.113 a) yes b) no c) no d) yes

8.115 a) sulfate b) perchlorate c) peroxide d) dichromate

8.117 a) Ca_3N_2 b) $Ca(NO_3)_2$ c) $Ca(NO_2)_2$ d) $Ca(CN)_2$

8.119 a) K_3P b) K_3PO_4 c) K_2HPO_4 d) KH_2PO_4

8.121 a) N_2O, CO_2 b) NO_2, SO_2 c) SF_2, SCl_2 d) N_2O_3

8.123 a) $CaCO_3, HNO_2$ b) $NaClO_4, NaClO_3$
c) $HClO_2, HClO$ d) Li_2CO_3, Li_3PO_4

8.125 a) sodium nitrate b) aluminum sulfide, magnesium nitride, beryllium phosphide
c) iron(III) oxide d) gold(I) chlorate

8.127 a) $Ni_2(SO_4)_3$ b) Ni_2O_3 c) $Ni_2(C_2O_4)_3$ d) $Ni(NO_3)_3$

8.129 The superoxide ion is O_2^-. The nitronium ion is NO_2^+. The formula of nitronium superoxide is NO_2O_2.

8.131 a) HCN, (4) ternary molecular b) HBr, (3) binary molecular
c) Na_2O_2, (1) binary ionic d) H_2O_2, (3) binary molecular

Cumulative Problems

8.133 a) X is a metal b) Y is a nonmetal
c) formula = X_3Y_2 d) X = Be, Y = P, beryllium phosphide

Chemical Nomenclature

8.135 a) X = S, sodium sulfate
b) Y = O, sodium oxide
c) Z protons = 40 − 3(11) = 7 = N, sodium nitride
d) Q⁻ electrons = 46 − 10 = 36 = Br⁻, sodium bromide

8.137 a) magnesium chloride, $MgCl_2$ b) oxygen difluoride, OF_2

8.139 a) $x = 4$; silicon tetrachloride   b) $x = 2$; magnesium chloride
c) $x = 3$; potassium nitride   d) $x = 3$; nitrogen trichloride

8.141 beryllium bromate

8.143 aluminum nitride

8.145 carbon dioxide

8.147 beryllium cyanide

Answers to Multiple-Choice Practice Test

8.149	d	**8.150**	a	**8.151**	d	**8.152**	c	**8.153**	b	**8.154**	c
8.155	d	**8.156**	c	**8.157**	a	**8.158**	d	**8.159**	a	**8.160**	e
8.161	b	**8.162**	c	**8.163**	d	**8.164**	e	**8.165**	d	**8.166**	a
8.167	c	**8.168**	d								

CHAPTER NINE
Chemical Calculations: The Mole Concept and Chemical Formulas

Practice Problems

Law of Definite Proportions (Sec. 9.1)

9.1 The same: 42.9% C and 57.1% O. Any size sample of CO will have that composition.

9.3 According to the law of definite proportions, the percentage A in each sample should be the same and the percentage D should also be the same.

Sample I: $\%A = \dfrac{10.03 \text{ g}}{17.35 \text{ g}} \times 100 = 57.809798 \text{ (calc)} = 57.81\% \text{ (corr)}$

$\%D = \dfrac{7.32 \text{ g}}{17.35 \text{ g}} \times 100 = 42.190202 \text{ (calc)} = 42.2\% \text{ (corr)}$

Sample II: $\%A = \dfrac{13.17 \text{ g}}{22.78 \text{ g}} \times 100 = 57.813872 \text{ (calc)} = 57.81\% \text{ (corr)}$

$\%D = \dfrac{9.61 \text{ g}}{22.78 \text{ g}} \times 100 = 42.186128 \text{ (calc)} = 42.2\% \text{ (corr)}$

9.5 The mass ratio between X and Q present in the sample will be the same for samples of the same compound.

Experiment	Q/X mass ratio
1	$\dfrac{8.90 \text{ g}}{3.37 \text{ g}} = 2.6409496 \text{ (calc)} = 2.64 \text{ (corr)}$
2	$\dfrac{1.711 \text{ g}}{0.561 \text{ g}} = 3.0499109 \text{ (calc)} = 3.05 \text{ (corr)}$
3	$\dfrac{71.0 \text{ g}}{26.9 \text{ g}} = 2.6394052 \text{ (calc)} = 2.64 \text{ (corr)}$

∴ Experiments 1 and 3 produced the same compound.

9.7 If all 50.1 g S is used, the mass of SO_2 produced would be

$$50.1 \text{ g S} \times \frac{100 \text{ g } SO_2}{50.1 \text{ g S}} = 100 \text{ (calc)} = 100. \text{ g } SO_2 \text{ (corr)}$$

If all 75.0 g O is used, the mass of SO_2 produced would be

$$75.0 \text{ g O} \times \frac{100 \text{ g } SO_2}{49.1 \text{ g O}} = 152.7494908 \text{ (calc)} = 153 \text{ g } SO_2 \text{ (corr)}$$

But after making 100. g SO_2, all the S is gone. ∴ The maximum possible is 100. g SO_2.

Formula Masses (Secs. 9.2 and 9.3)

9.9
a) Sn 1×118.71 amu =   118.71 amu
   F 2×19.00 amu =   38.00 amu
          SnF_2 =   156.71 amu

b) Fe 1×55.85 amu =   55.85 amu
   S 1×32.07 amu =   32.07 amu
   O 4×16.00 amu =   64.00 amu
          $FeSO_4$ =   151.92 amu

c) C 12×12.01 amu =   144.12 amu
   H 22×1.01 amu =   22.22 amu
   O 11×16.00 amu =   176.00 amu
          $C_{12}H_{22}O_{11}$ =   342.34 amu

d) C 14×12.01 amu =   168.14 amu
   H 9×1.01 amu =   9.09 amu
   Cl 5×35.45 amu =   177.25 amu
          $C_{14}H_9Cl_5$ =   354.48 amu

9.11
a) C 2×12.01 amu =   24.02 amu
   H 6×1.01 amu =   6.06 amu
   O 2×16.00 amu =   32.00 amu
          $C_2H_4(OH)_2$ =   62.08 amu

b) H 4×1.01 amu =   4.04 amu
   N 2×14.01 amu =   28.02 amu
   C 1×12.01 amu =   12.01 amu
   O 1×16.00 amu =   16.00 amu
          $(H_2N)_2CO$ =   60.07 amu

c) Mg 3×24.31 amu =   72.93 amu
   Si 4×28.09 amu =   112.36 amu
   O 12×16.00 amu =   192.00 amu
   H 2×1.01 amu =   2.02 amu
          $Mg_3(Si_2O_5)_2(OH)_2$ =   379.31 amu

d) Al 2×26.98 amu =   53.96 amu
   Si 2×28.09 amu =   56.18 amu
   O 7×16.00 amu =   112.00 amu
   H 2×1.01 amu =   2.02 amu
          $Al_2Si_2O_5(OH)_2$ =   224.16 amu

Chemical Calculations: The Mole Concept and Chemical Formulas 61

9.13 a) ethylene glycol = $C_2H_6O_2$

    C 2×12.01 amu =   24.02 amu
    H 6×1.01 amu =   6.06 amu
    O 2×16.00 amu =   32.00 amu
    $C_2H_6O_2$ =   62.08 amu

b) lactic acid = $C_3H_6O_3$

    C 3×12.01 amu =   36.03 amu
    H 6×1.01 amu =   6.06 amu
    O 3×16.00 amu =   48.00 amu
    $C_3H_6O_3$ =   90.09 amu

9.15 $y(12.01 \text{ amu}) + 8(1.01 \text{ amu}) + 32.07 \text{ amu}$ = 88.19 amu
              $(12.01) y \text{ amu} + 40.15 \text{ amu}$ = 88.19 amu
                     $(12.01) y \text{ amu}$ = 48.04 amu
                                y = 4

9.17 $\dfrac{1 \text{ atom C}}{1 \text{ molecule}} \times \dfrac{12.01 \text{ amu C}}{1 \text{ atom C}} \times \dfrac{100 \text{ amu molecule}}{74.8 \text{ amu C}} = 16.056149733$ (calc)

                                                                   = 16.1 amu molecule (corr)

9.19 $\dfrac{2 \text{ atoms N}}{1 \text{ molecule}} \times \dfrac{14.01 \text{ amu N}}{1 \text{ atom N}} \times \dfrac{100 \text{ amu molecule}}{63.6 \text{ amu N}} = 44.0566038$ (calc)

                                                                   = 44.1 amu molecule (corr)

9.21 Percent O = $100 - 63.6 = 36.4\%$ O

       $44.1 \text{ amu molecule} \times \dfrac{36.4 \text{ amu O}}{100 \text{ amu molecule}} \times \dfrac{1 \text{ atom O}}{16.00 \text{ amu O}} = 1.003275$ (calc)

                                                                          = 1 atom O (corr) and formula is N_2O

Percent Composition (Sec. 9.4)

9.23 a) $H_3BO_3 = 3(1.01 \text{ amu}) + 10.81 \text{ amu} + 3(16.00 \text{ amu}) = 61.84$ amu total

         %H = $\dfrac{3.03 \text{ amu H}}{61.84 \text{ amu total}} \times 100 = 4.89974127$ (calc) = 4.90% H (corr)

         %B = $\dfrac{10.81 \text{ amu B}}{61.84 \text{ amu total}} \times 100 = 17.48059508$ (calc) = 17.48% B (corr)

         %O = $\dfrac{48.00 \text{ amu O}}{61.84 \text{ amu total}} \times 100 = 77.61966365$ (calc) = 77.62% O (corr)

b) $C_7H_{16} = 7(12.01 \text{ amu}) + 16(1.01 \text{ amu}) = 100.23$ amu total

         %C = $\dfrac{84.07 \text{ amu C}}{100.23 \text{ amu total}} \times 100 = 83.877083$ (calc) = 83.88% C (corr)

         %H = $\dfrac{16.16 \text{ amu H}}{100.23 \text{ amu total}} \times 100 = 16.122917$ (calc) = 16.12% H (corr)

c) $C_{12}H_{11}NO_2$ = 12(12.01 amu) + 11(1.01 amu) + 14.01 amu + 2(16.00 amu) = 201.24 amu total

$$\%C = \frac{144.12 \text{ amu C}}{201.24 \text{ amu total}} \times 100 = 71.61598092 \text{ (calc)} = 71.62\% \text{ C (corr)}$$

$$\%H = \frac{11.11 \text{ amu H}}{201.24 \text{ amu total}} \times 100 = 5.5207712184 \text{ (calc)} = 5.521\% \text{ H (corr)}$$

$$\%N = \frac{14.01 \text{ amu H}}{201.24 \text{ amu total}} \times 100 = 6.961836613 \text{ (calc)} = 6.962\% \text{ N (corr)}$$

$$\%O = \frac{32.00 \text{ amu O}}{201.24 \text{ amu total}} \times 100 = 15.90141125 \text{ (calc)} = 15.90\% \text{ O (corr)}$$

d) $C_{22}H_{30}ClNO_2$ = 22(12.01 amu) + 30(1.01 amu) + 35.45 amu + 14.01 amu + 2(16.00 amu)

$\qquad$ = 375.98 amu total

$$\%C = \frac{264.22 \text{ amu C}}{375.98 \text{ amu total}} \times 100 = 70.2750146 \text{ (calc)} = 70.275\% \text{ C (corr)}$$

$$\%H = \frac{30.30 \text{ amu H}}{375.98 \text{ amu total}} \times 100 = 8.0589393 \text{ (calc)} = 8.059\% \text{ H (corr)}$$

$$\%Cl = \frac{35.45 \text{ amu Cl}}{375.98 \text{ amu total}} \times 100 = 9.42869302 \text{ (calc)} = 9.429\% \text{ Cl (corr)}$$

$$\%N = \frac{14.01 \text{ amu N}}{375.98 \text{ amu total}} \times 100 = 3.7262620 \text{ (calc)} = 3.726\% \text{ N (corr)}$$

$$\%O = \frac{32.00 \text{ amu O}}{375.98 \text{ amu total}} \times 100 = 8.511091 \text{ (calc)} = 8.512\% \text{ O (corr)}$$

9.25 a) 1.271 g Cu + 0.320 g O = 1.591 g compound

$$\%Cu = \frac{1.271 \text{ g Cu}}{1.591 \text{ g compound}} \times 100 = 79.886864 \text{ (calc)} = 79.89\% \text{ Cu (corr)}$$

$$\%O = \frac{0.320 \text{ g O}}{1.591 \text{ g compound}} \times 100 = 20.113136 \text{ (calc)} = 20.1\% \text{ O (corr)}$$

b) total sample = 49.31 g

$$\%Na = \frac{15.96 \text{ g Na}}{49.31 \text{ g total}} \times 100 = 32.366660 \text{ (calc)} = 32.37\% \text{ Na (corr)}$$

$$\%S = \frac{11.13 \text{ g S}}{49.31 \text{ g total}} \times 100 = 22.571487 \text{ (calc)} = 22.57\% \text{ S (corr)}$$

$$\%O = \frac{22.22 \text{ g O}}{49.31 \text{ g total}} \times 100 = 45.061854 \text{ (calc)} = 45.06\% \text{ O (corr)}$$

c) 25.00 g compound − 6.48 g N = 18.52 g O

$$\%N = \frac{6.48 \text{ g N}}{25.00 \text{ g total}} \times 100 = 25.92 \text{ (calc)} = 25.9\% \text{ N (corr)}$$

$$\%O = \frac{18.52 \text{ g O}}{25.00 \text{ g total}} \times 100 = 74.08\% \text{ O (calc and corr)}$$

Chemical Calculations: The Mole Concept and Chemical Formulas

d) mass compound = 10.00 g; mass O reacted = 10.00 g (initial) − 3.34 g (left over) = 6.66 g O

$$\%S = \frac{3.34 \text{ g S}}{10.00 \text{ g compound}} \times 100 = 33.4\% \text{ S (calc and corr)}$$

$$\%O = \frac{6.66 \text{ g O}}{10.00 \text{ g compound}} \times 100 = 66.6\% \text{ O (calc and corr)}$$

9.27 a) $\%H \text{ in } NH_3 = \dfrac{3.03 \text{ g H}}{17.04 \text{ g } NH_3} \times 100 = 17.78169014 \text{ (calc)} = 17.8\% \text{ H (corr)}$

$\%H \text{ in } CH_4 = \dfrac{4.04 \text{ g H}}{16.05 \text{ g } CH_4} \times 100 = 25.17133956 \text{ (calc)} = 25.2\% \text{ H (corr)}$

No. The mass percent of H in each compound is not less than 10%.

b) $\%H \text{ in } LiH = \dfrac{1.01 \text{ g H}}{7.95 \text{ g LiH}} \times 100 = 12.704402516 \text{ (calc)} = 12.7\% \text{ H (corr)}$

$\%H \text{ in } BeH_2 = \dfrac{2.02 \text{ g H}}{11.03 \text{ g } BeH_2} \times 100 = 18.3136899 \text{ (calc)} = 18.3\% \text{ H (corr)}$

No. The mass percent of H in each compound is not less than 10%.

c) $\%H \text{ in } H_2SO_4 = \dfrac{2.02 \text{ g H}}{98.09 \text{ g } H_2SO_4} \times 100 = 2.05933327 \text{ (calc)} = 2.06\% \text{ H (corr)}$

$\%H \text{ in } HNO_3 = \dfrac{1.01 \text{ g H}}{63.02 \text{ g } HNO_3} \times 100 = 1.60266582 \text{ (calc)} = 1.60\% \text{ H (corr)}$

Yes. The mass percent of H in each compound is less than 10%.

d) $\%H \text{ in } Mg(OH)_2 = \dfrac{2.02 \text{ g H}}{58.33 \text{ g } Mg(OH)_2} \times 100 = 3.46305504 \text{ (calc)} = 3.46\% \text{ H (corr)}$

$\%H \text{ in } NaOH = \dfrac{1.01 \text{ g H}}{40.00 \text{ g NaOH}} \times 100 = 2.525 \text{ (calc)} = 2.52\% \text{ H (corr)}$

Yes. The mass percent of H in each compound is less than 10%.

9.29 $\%C \text{ in } CS_2 = \dfrac{12.01 \text{ g C}}{76.15 \text{ g } CS_2} \times 100 = 15.77150361 \text{ (calc)} = 15.78\% \text{ C (corr)}$

$\%S \text{ in } CS_2 = 100 - 15.78 = 84.23\% \text{ S}$, therefore the percent C and S match.

9.31 For C_2H_2, $\%C = \dfrac{2(12.01) \text{ g C}}{2(12.01) + 2(1.01) \text{ g total}} \times 100 = 92.242704 \text{ (calc)} = 92.24\% \text{ C (corr)}$

$\%H = \dfrac{2(1.01) \text{ g H}}{2(12.01) + 2(1.01) \text{ g total}} \times 100 = 7.757296 \text{ (calc)} = 7.76\% \text{ H (corr)}$

for C_6H_6, $\%C = \dfrac{6(12.01) \text{ g C}}{6(12.01) + 2(1.01) \text{ g total}} \times 100 = 92.242704 \text{ (calc)} = 92.24\% \text{ C (corr)}$

$\%H = \dfrac{6(1.01) \text{ g H}}{6(12.01) + 2(1.01) \text{ g total}} \times 100 = 7.757296 \text{ (calc)} = 7.76\% \text{ H (corr)}$

Both compounds have the same ratio of carbon to hydrogen. Thus, the %C and %H will be the same in each.

9.33 $\%N = 100 - 12.6 = 87.4\% \text{ N}$ $52.34 \text{ g H} \times \dfrac{87.4 \text{ g N}}{12.6 \text{ g H}} = 363.0568254 \text{ (calc)} = 364 \text{ g N (corr)}$

The Mole: The Chemist's Counting Unit (Sec. 9.5)

9.35 a) $1.00 \text{ mole Ag} \times \dfrac{6.022 \times 10^{23} \text{ Ag atoms}}{1 \text{ mole Ag}} = 6.022 \times 10^{23}$ (calc) $= 6.02 \times 10^{23}$ Ag atoms (corr)

b) $1.00 \text{ mole H}_2\text{O} \times \dfrac{6.022 \times 10^{23} \text{ H}_2\text{O molecules}}{1 \text{ mole H}_2\text{O}} = 6.022 \times 10^{23}$ (calc)
$= 6.02 \times 10^{23}$ H$_2$O molecules (corr)

c) $1.00 \text{ mole NaNO}_3 \times \dfrac{6.022 \times 10^{23} \text{ NaNO}_3 \text{ formula units}}{1 \text{ mole NaNO}_3} = 6.022 \times 10^{23}$ (calc)
$= 6.02 \times 10^{23}$ NaNO$_3$ formula units (corr)

d) $1.00 \text{ mole SO}_4^{2-} \text{ ions} \times \dfrac{6.022 \times 10^{23} \text{ SO}_4^{2-} \text{ ions}}{1 \text{ mole SO}_4^{2-} \text{ ions}} = 6.022 \times 10^{23}$ (calc)
$= 6.02 \times 10^{23}$ SO$_4^{2-}$ ions (corr)

9.37 a) $3.20 \text{ moles Si} \times \dfrac{6.022 \times 10^{23} \text{ Si atoms}}{1 \text{ mole Si}} = 1.92704 \times 10^{24}$ (calc) $= 1.93 \times 10^{24}$ Si atoms (corr)

b) $0.36 \text{ mole Si} \times \dfrac{6.022 \times 10^{23} \text{ Si atoms}}{1 \text{ mole Si}} = 2.16792 \times 10^{23}$ (calc) $= 2.2 \times 10^{23}$ Si atoms (corr)

c) $1.21 \text{ moles Si} \times \dfrac{6.022 \times 10^{23} \text{ Si atoms}}{1 \text{ mole Si}} = 7.28662 \times 10^{23}$ (calc) $= 7.29 \times 10^{23}$ Si atoms (corr)

d) $16.7 \text{ moles Si} \times \dfrac{6.022 \times 10^{23} \text{ Si atoms}}{1 \text{ mole Si}} = 1.005674 \times 10^{25}$ (calc) $= 1.01 \times 10^{25}$ C atoms (corr)

9.39 a) $1.50 \text{ moles CO}_2 \times \dfrac{6.022 \times 10^{23} \text{ molecules CO}_2}{1 \text{ mole CO}_2} = 9.033 \times 10^{23}$ (calc)
$= 9.03 \times 10^{23}$ molecules CO$_2$ (corr)

b) $0.500 \text{ mole NH}_3 \times \dfrac{6.022 \times 10^{23} \text{ molecules NH}_3}{1 \text{ mole NH}_3} = 3.011 \times 10^{23}$ (calc)
$= 3.01 \times 10^{23}$ molecules NH$_3$ (corr)

c) $2.33 \text{ moles PF}_3 \times \dfrac{6.022 \times 10^{23} \text{ molecules PF}_3}{1 \text{ mole PF}_3} = 1.40313 \times 10^{24}$ (calc)
$= 1.40 \times 10^{24}$ molecules PF$_3$ (corr)

d) $1.115 \text{ moles N}_2\text{H}_4 \times \dfrac{6.022 \times 10^{23} \text{ molecules N}_2\text{H}_4}{1 \text{ mole N}_2\text{H}_4} = 6.71453 \times 10^{23}$ (calc)
$= 6.715 \times 10^{23}$ molecules N$_2$H$_4$ (corr)

9.41 $719{,}000{,}000 \text{ people} \times \dfrac{1 \text{ mole people}}{6.022 \times 10^{23} \text{ people}} = 1.1939554 \times 10^{-15}$ (calc)
$= 1.19 \times 10^{-15}$ mole people (corr)

Chemical Calculations: The Mole Concept and Chemical Formulas

Molar Mass (Sec. 9.6)

9.43 a) $1.000 \text{ mole Ca} \times \dfrac{40.08 \text{ g Ca}}{1 \text{ mole Ca}} = 40.08 \text{ g Ca}$ (calc and corr)

b) $1.000 \text{ mole Si} \times \dfrac{28.09 \text{ g Si}}{1 \text{ mole Si}} = 28.09 \text{ g Si}$ (calc and corr)

c) $1.000 \text{ mole Co} \times \dfrac{58.93 \text{ g Co}}{1 \text{ mole Co}} = 58.93 \text{ g Co}$ (calc and corr)

d) $1.000 \text{ mole Ag} \times \dfrac{107.87 \text{ g Ag}}{1 \text{ mole Ag}} = 107.87$ (calc) $= 107.9 \text{ g Ag}$ (corr)

9.45 a) 22.99 amu + 35.45 amu = 58.44 amu (calc and corr)
mass of 1 NaCl formula unit = 58.44 amu
molar mass of NaCl formula units = 58.44 g/mole

b) 2(22.99 amu) + 32.07 amu = 78.05 amu (calc and corr)
mass 1 Na_2S formula unit = 78.05 amu
molar mass of Na_2S formula units = 78.05 g/mole

c) 22.99 amu + 14.01 amu + 3(16.00 amu) = 85.00 amu (calc and corr)
mass of 1 $NaNO_3$ formula unit = 85.00 amu
molar mass of $NaNO_3$ formula units = 85.00 g/mole

d) 3(22.99 amu) + 30.97 amu + 4(16.00 amu) = 163.94 amu (calc and corr)
mass of 1 Na_3PO_4 formula unit = 163.94 amu
molar mass of Na_3PO_4 formula units = 163.94 g/mole

9.47 a) $1.000 \text{ moles NaCl} \times \dfrac{58.44 \text{ g NaCl}}{1 \text{ mole NaCl}} = 58.44 \text{ g NaCl}$ (calc and corr)

b) $1.000 \text{ moles Na}_2\text{S} \times \dfrac{78.05 \text{ g Na}_2\text{S}}{1 \text{ mole Na}_2\text{S}} = 78.05 \text{ g Na}_2\text{S}$ (calc and corr)

c) $1.000 \text{ moles NaNO}_3 \times \dfrac{85.00 \text{ g NaNO}_3}{1 \text{ mole NaNO}_3} = 85$ (calc) $= 85.00 \text{ g NaNO}_3$ (corr)

d) $1.000 \text{ moles Na}_3\text{PO}_4 \times \dfrac{163.94 \text{ g Na}_3\text{PO}_4}{1 \text{ mole Na}_3\text{PO}_4} = 163.94 \text{ g Na}_3\text{PO}_4$ (calc and corr)

9.49 a) $2.314 \text{ moles Al(OH)}_3 \times \dfrac{78.01 \text{ g Al(OH)}_3}{1 \text{ mole Al(OH)}_3} = 180.51514$ (calc) $= 180.5 \text{ g Al(OH)}_3$ (corr)

b) $2.314 \text{ mole Mg}_3\text{N}_2 \times \dfrac{100.95 \text{ g Mg}_3\text{N}_2}{1 \text{ mole Mg}_3\text{N}_2} = 233.5983$ (calc) $= 233.6 \text{ g Mg}_3\text{N}_2$ (corr)

c) $2.314 \text{ moles Cu(NO}_3)_2 \times \dfrac{187.57 \text{ g Cu(NO}_3)_2}{1 \text{ mole Cu(NO}_3)_2} = 434.03698$ (calc) $= 434.0 \text{ g Cu(NO}_3)_2$ (corr)

d) $2.314 \text{ moles La}_2\text{O}_3 \times \dfrac{325.82 \text{ g La}_2\text{O}_3}{1 \text{ mole La}_2\text{O}_3} = 753.94748$ (calc) $= 753.9 \text{ g La}_2\text{O}_3$ (corr)

9.51 a) 1 mole CO_2 = 12.01 g C + 2(16.00) g O = $\dfrac{44.01 \text{ g } CO_2}{1 \text{ mole } CO_2}$

b) 1 mole N_2O = 2(14.01) g N + 16.00 g O = $\dfrac{44.02 \text{ g } N_2O}{1 \text{ mole } N_2O}$

c) 1 mole C_3H_8 = 3(12.01) g C + 8(1.01) g H = $\dfrac{44.11 \text{ g } C_3H_8}{1 \text{ mole } C_3H_8}$

9.53 a) 2.00 moles Cu × $\dfrac{63.55 \text{ g Cu}}{1 \text{ mole Cu}}$ = 127.1 (calc) = 127 g Cu (corr)

2.00 moles O × $\dfrac{16.00 \text{ g O}}{1 \text{ mole O}}$ = 32.00 (calc) = 32.0 g O (corr) ∴ 2.00 mole Cu greater

b) 1.00 mole Br × $\dfrac{79.90 \text{ g Br}}{1 \text{ mole Br}}$ = 79.90 (calc) = 79.9 g Br (corr)

5.00 moles Be × $\dfrac{9.01 \text{ g Be}}{1 \text{ mole Be}}$ = 45.05 (calc) = 45.0 g Be (corr) ∴ 1.00 mole Br greater

c) 2.00 moles CO × $\dfrac{28.01 \text{ g CO}}{1 \text{ mole CO}}$ = 56.02 (calc) = 56.0 g CO (corr)

1.50 moles N_2O × $\dfrac{44.02 \text{ g } N_2O}{1 \text{ mole } N_2O}$ = 66.03 (calc) = 66.0 g N_2O (corr) ∴ 1.50 mole N_2O greater

d) 4.87 moles B_2H_6 × $\dfrac{27.68 \text{ g } B_2H_6}{1 \text{ mole } B_2H_6}$ = 134.802 (calc) = 135 g B_2H_6 (corr)

0.35 mole U × $\dfrac{238.03 \text{ g U}}{1 \text{ mole U}}$ = 83.31 (calc) = 83 g U (corr) ∴ 4.87 moles B_2H_6 greater

9.55 Molar mass has units of $\dfrac{g}{mole}$, so take the number of grams and divide by the number of moles.

a) $\dfrac{11.23 \text{ g}}{0.6232 \text{ mole}}$ = 18.0198973 (calc) = 18.02 $\dfrac{g}{mole}$ (corr)

b) $\dfrac{20.14 \text{ g}}{0.5111 \text{ mole}}$ = 39.40520446 (calc) = 39.41 $\dfrac{g}{mole}$ (corr)

c) $\dfrac{352.6 \text{ g}}{2.357 \text{ mole}}$ = 149.5969453 (calc) = 149.6 $\dfrac{g}{mole}$ (corr)

d) $\dfrac{100.0 \text{ g}}{1.253 \text{ mole}}$ = 79.8084597 (calc) = 79.81 $\dfrac{g}{mole}$ (corr)

9.57 Compound A = $\dfrac{3.62 \text{ g}}{0.0521 \text{ mole}}$ = 69.48176583 (calc) = 69.5 $\dfrac{g}{mole}$ (corr)

Compound B = $\dfrac{83.64 \text{ g}}{1.23 \text{ mole}}$ = 68 (calc) = 68.0 $\dfrac{g}{mole}$ (corr)

Compound A has the greater molar mass.

Chemical Calculations: The Mole Concept and Chemical Formulas

Relationship Between Atomic Mass Units and Gram Units (Sec. 9.8)

9.59 a) $19.00 \text{ amu} \times \dfrac{1.000 \text{ g}}{6.022 \times 10^{23} \text{ amu}} = 3.155098 \times 10^{-23}$ (calc) $= 3.156 \times 10^{-23}$ g (corr)

b) $52.00 \text{ amu} \times \dfrac{1.000 \text{ g}}{6.022 \times 10^{23} \text{ amu}} = 8.635004982 \times 10^{-23}$ (calc) $= 8.635 \times 10^{-23}$ g (corr)

c) $118.71 \text{ amu} \times \dfrac{1.000 \text{ g}}{6.022 \times 10^{23} \text{ amu}} = 1.971272003 \times 10^{-22}$ (calc) $= 1.971 \times 10^{-23}$ g (corr)

d) $196.97 \text{ amu} \times \dfrac{1.000 \text{ g}}{6.022 \times 10^{23} \text{ amu}} = 3.270840252 \times 10^{-22}$ (calc) $= 3.271 \times 10^{-23}$ g (corr)

9.61 $5.143 \times 10^{-23} \text{ g} \times \dfrac{6.022 \times 10^{23} \text{ amu}}{1.000 \text{ g}} = 30.971146$ (calc) $= 30.97$ amu (corr)

The element with an atomic mass of 30.97 amu is phosphorus.

9.63 $26.98 \text{ g Al} \times \dfrac{6.022 \times 10^{23} \text{ amu}}{1.000 \text{ g}} \times \dfrac{1 \text{ atom Al}}{26.98 \text{ amu}} = 6.022 \times 10^{23}$ atoms Al (calc and corr)

The Mole and Chemical Formulas (Sec. 9.9)

9.65 $\dfrac{3 \text{ moles Na}}{1 \text{ mole Na}_3\text{PO}_4}$ $\dfrac{1 \text{ mole P}}{1 \text{ mole Na}_3\text{PO}_4}$ $\dfrac{4 \text{ moles O}}{1 \text{ mole Na}_3\text{PO}}$

$\dfrac{3 \text{ moles Na}}{1 \text{ mole P}}$ $\dfrac{3 \text{ moles Na}}{4 \text{ moles O}}$ $\dfrac{1 \text{ mole P}}{4 \text{ moles O}}$

9.67 a) $\dfrac{4 \text{ moles N}}{1 \text{ mole S}_4\text{N}_4\text{Cl}_2}$ b) $\dfrac{2 \text{ moles Cl}}{1 \text{ mole S}_4\text{N}_4\text{Cl}_2}$ c) $\dfrac{10 \text{ moles atoms}}{1 \text{ mole S}_4\text{N}_4\text{Cl}_2}$ d) $\dfrac{2 \text{ moles Cl}}{4 \text{ moles S}}$

9.69 a) $1.0 \text{ mole Na}_2\text{SO}_4 \times \dfrac{1 \text{ mole S}}{1 \text{ mole Na}_2\text{SO}_4} = 1.0$ mole S (calc and corr);

$0.50 \text{ mole Na}_2\text{S}_2\text{O}_3 \times \dfrac{2 \text{ moles S}}{1 \text{ mole Na}_2\text{S}_2\text{O}_3} = 1.0$ mole (calc and corr). Same moles of S.

b) $2.00 \text{ moles S}_3\text{Cl}_2 \times \dfrac{3 \text{ moles S}}{1 \text{ mole S}_3\text{Cl}_2} = 6.00$ moles S (calc and corr);

$1.50 \text{ moles S}_2\text{O} \times \dfrac{2 \text{ moles S}}{1 \text{ mole S}_2\text{O}} = 3.00$ moles (calc and corr). Not the same.

c) $3.00 \text{ moles H}_2\text{S}_2\text{O}_5 \times \dfrac{2 \text{ moles S}}{1 \text{ mole H}_2\text{S}_2\text{O}_5} = 6.00$ moles S (calc and corr);

$6.00 \text{ moles H}_2\text{SO}_4 \times \dfrac{1 \text{ mole S}}{1 \text{ mole H}_2\text{SO}_4} = 6.00$ moles S (calc and corr). Same moles of S.

d) $1.00 \text{ mole Na}_3\text{Ag}(\text{S}_2\text{O}_3)_2 \times \dfrac{4 \text{ moles S}}{1 \text{ mole Na}_3\text{Ag}(\text{S}_2\text{O}_3)_2} = 4.00$ moles S (calc and corr);

$2.00 \text{ moles S}_2\text{F}_{10} \times \dfrac{2 \text{ moles S}}{1 \text{ mole S}_2\text{F}_{10}} = 4.00$ moles S (calc and corr). Same moles of S.

9.71 a) 2.00 moles NaAuBr$_4$ × $\dfrac{6 \text{ moles total atoms}}{1 \text{ mole NaAuBr}_4}$ = 12.0 moles atoms (calc and corr)

2.00 moles Au$_2$Te$_3$ × $\dfrac{5 \text{ moles total atoms}}{1 \text{ mole Au}_2\text{Te}_3}$ = 10.0 moles atoms (calc and corr)

∴ 2.00 moles NaAuBr$_4$

b) 1.00 mole C$_2$H$_2$Cl$_4$ × $\dfrac{8 \text{ moles total atoms}}{1 \text{ mole C}_2\text{H}_2\text{Cl}_4}$ = 8.00 moles atoms (calc and corr)

1.00 mole CCl$_4$ × $\dfrac{5 \text{ moles total atoms}}{1 \text{ mole CCl}_4}$ = 5.00 moles atoms (calc and corr)   ∴ 1.00 mole C$_2$H$_2$Cl$_4$

c) 3.00 moles Ba(NO$_3$)$_2$ × $\dfrac{9 \text{ moles total atoms}}{1 \text{ mole Ba(NO}_3)_2}$ = 27.0 moles atoms (calc and corr)

3.00 moles BaSO$_4$ × $\dfrac{6 \text{ moles total atoms}}{1 \text{ mole BaSO}_4}$ = 18.0 moles atoms (calc and corr)

∴ 3.00 moles Ba(NO$_3$)$_2$

d) 1.20 moles NH$_4$CN × $\dfrac{7 \text{ moles total atoms}}{1 \text{ mole NH}_4\text{CN}}$ = 8.40 moles atoms (calc and corr)

1.30 moles NH$_4$Cl × $\dfrac{6 \text{ moles total atoms}}{1 \text{ mole NH}_4\text{Cl}}$ = 7.80 moles atoms (calc and corr)

∴ 1.20 moles NH$_4$CN

The Mole and Chemical Calculations (Sec. 9.10)

9.73 a) 23.0 g Be × $\dfrac{1 \text{ mole Be}}{9.01 \text{ g Be}}$ × $\dfrac{6.022 \times 10^{23} \text{ atoms Be}}{1 \text{ mole Be}}$

= 1.537247503 × 10^{24} (calc) = 1.54 × 10^{24} atoms Be (corr)

b) 23.0 g Mg × $\dfrac{1 \text{ mole Mg}}{24.31 \text{ g Mg}}$ × $\dfrac{6.022 \times 10^{23} \text{ atoms Mg}}{1 \text{ mole Mg}}$

= 5.697490745 × 10^{23} (calc) = 5.70 × 10^{23} atoms Mg (corr)

c) 23.0 g Ca × $\dfrac{1 \text{ mole Ca}}{48.08 \text{ g Ca}}$ × $\dfrac{6.022 \times 10^{23} \text{ atoms Ca}}{1 \text{ mole Ca}}$

= 3.455738523 × 10^{23} (calc) = 3.46 × 10^{23} atoms Ca (corr)

d) 23.0 g Sr × $\dfrac{1 \text{ mole Sr}}{87.62 \text{ g Sr}}$ × $\dfrac{6.022 \times 10^{23} \text{ atoms Sr}}{1 \text{ mole Sr}}$

= 1.580757818 × 10^{23} (calc) = 1.58 × 10^{23} atoms Sr (corr)

9.75 a) 25.0 g NO × $\dfrac{1 \text{ mole NO}}{30.01 \text{ g NO}}$ × $\dfrac{6.022 \times 10^{23} \text{ molecules NO}}{1 \text{ mole NO}}$

= 5.016661113 × 10^{23} (calc) = 5.02 × 10^{23} molecules NO (corr)

b) 25.0 g N$_2$O × $\dfrac{1 \text{ mole N}_2\text{O}}{44.02 \text{ g N}_2\text{O}}$ × $\dfrac{6.022 \times 10^{23} \text{ molecules N}_2\text{O}}{1 \text{ mole N}_2\text{O}}$

= 3.420036347 × 10^{23} (calc) = 3.42 × 10^{23} molecules N$_2$O (corr)

Chemical Calculations: The Mole Concept and Chemical Formulas 69

c) $25.0 \text{ g N}_2\text{O}_3 \times \dfrac{1 \text{ mole N}_2\text{O}_3}{76.02 \text{ g N}_2\text{O}_3} \times \dfrac{6.022 \times 10^{23} \text{ molecules N}_2\text{O}_3}{1 \text{ mole N}_2\text{O}_3}$
$= 1.980399895 \times 10^{23} \text{ (calc)} = 1.98 \times 10^{23} \text{ molecules N}_2\text{O}_3 \text{ (corr)}$

d) $25.0 \text{ g N}_2\text{O}_5 \times \dfrac{1 \text{ mole N}_2\text{O}_5}{108.02 \text{ g N}_2\text{O}_5} \times \dfrac{6.022 \times 10^{23} \text{ molecules N}_2\text{O}_5}{1 \text{ mole N}_2\text{O}_5}$
$= 1.393723385 \times 10^{23} \text{ (calc)} = 1.39 \times 10^{23} \text{ molecules N}_2\text{O}_5 \text{ (corr)}$

9.77 a) less than 1 mole of molecules   b) less than 1 mole of molecules
c) less than 1 mole of molecules   d) less than 1 mole of molecules

9.79 a) $3.333 \times 10^{23} \text{ atoms P} \times \dfrac{1 \text{ mole P}}{6.022 \times 10^{23} \text{ atoms P}} \times \dfrac{30.97 \text{ g P}}{1 \text{ mole P}} = 17.14098472 \text{ (calc)}$
$= 17.13 \text{ g P (corr)}$

b) $3.333 \times 10^{23} \text{ molecules PH}_3 \times \dfrac{1 \text{ mole PH}_3}{6.022 \times 10^{23} \text{ atoms PH}_3} \times \dfrac{34.00 \text{ g PH}_3}{1 \text{ mole PH}_3} = 18.81800066 \text{ (calc)}$
$= 18.82 \text{ g PH}_3 \text{ (corr)}$

c) $2431 \text{ atoms P} \times \dfrac{1 \text{ mole P}}{6.022 \times 10^{23} \text{ molecules P}} \times \dfrac{30.97 \text{ g P}}{1 \text{ mole P}}$
$= 1.250217038 \times 10^{-19} \text{ (calc)} = 1.250 \times 10^{-19} \text{ g P (corr)}$

d) $2431 \text{ molecules PH}_3 \times \dfrac{1 \text{ mole PH}_3}{6.022 \times 10^{23} \text{ atoms PH}_3} \times \dfrac{34.00 \text{ g PH}_3}{1 \text{ mole PH}_3}$
$= 1.372534042 \times 10^{-19} \text{ (calc)} = 1.373 \times 10^{-19} \text{ g PH}_3 \text{ (corr)}$

9.81 a) $1 \text{ atom Na} \times \dfrac{22.99 \text{ amu}}{1 \text{ atom Na}} \times \dfrac{1 \text{ g}}{6.022 \times 10^{23} \text{ amu}} = 3.8176685 \times 10^{-23} \text{ (calc)}$
$= 3.818 \times 10^{-23} \text{ g Na (corr)}$

b) $1 \text{ atom Mg} \times \dfrac{24.31 \text{ amu}}{1 \text{ atom Mg}} \times \dfrac{1 \text{ g}}{6.022 \times 10^{23} \text{ amu}} = 4.0368648 \times 10^{-23} \text{ (calc)}$
$= 4.037 \times 10^{-23} \text{ g Mg (corr)}$

c) $1 \text{ molecule C}_4\text{H}_{10} \times \dfrac{[4(12.01) + 10(1.01)] \text{ amu}}{1 \text{ molecule C}_4\text{H}_{10}} \times \dfrac{1 \text{ g}}{6.022 \times 10^{23} \text{ amu}}$
$= 9.6545998 \times 10^{-23} \text{ (calc)} = 9.655 \times 10^{-23} \text{ g C}_4\text{H}_{10} \text{ (corr)}$

d) $1 \text{ molecule C}_6\text{H}_6 \times \dfrac{[6(12.01) + 6(1.01)] \text{ amu}}{1 \text{ molecule C}_6\text{H}_6} \times \dfrac{1 \text{ g}}{6.022 \times 10^{23} \text{ amu}}$
$= 1.2972434 \times 10^{-22} \text{ (calc)} = 1.297 \times 10^{-22} \text{ g C}_6\text{H}_6 \text{ (corr)}$

9.83 a) $100.0 \text{ g NaCl} \times \dfrac{1 \text{ mole NaCl}}{58.44 \text{ g NaCl}} \times \dfrac{1 \text{ mole Cl}}{1 \text{ mole NaCl}} \times \dfrac{35.45 \text{ g Cl}}{1 \text{ mole Cl}} = 60.660506 \text{ (calc)}$
$= 60.66 \text{ g Cl (corr)}$

b) $980.0 \text{ g CCl}_4 \times \dfrac{1 \text{ mole CCl}_4}{153.81 \text{ g CCl}_4} \times \dfrac{4 \text{ moles Cl}}{1 \text{ mole CCl}_4} \times \dfrac{35.45 \text{ g Cl}}{1 \text{ mole Cl}} = 903.47832 \text{ (calc)}$
$= 903.5 \text{ g Cl (corr)}$

70	Chemical Calculations: The Mole Concept and Chemical Formulas

c) $10.0 \text{ g HCl} \times \dfrac{1 \text{ mole HCl}}{36.46 \text{ g HCl}} \times \dfrac{1 \text{ mole Cl}}{1 \text{ mole HCl}} \times \dfrac{35.45 \text{ g Cl}}{1 \text{ mole Cl}} = 9.7229841$ (calc)

$= 9.72 \text{ g Cl}$ (corr)

d) $50.0 \text{ g BaCl}_2 \times \dfrac{1 \text{ mole BaCl}_2}{208.23 \text{ g BaCl}_2} \times \dfrac{2 \text{ moles Cl}}{1 \text{ mole BaCl}_2} \times \dfrac{35.45 \text{ g Cl}}{1 \text{ mole Cl}} = 17.024444$ (calc)

$= 17.0 \text{ g Cl}$ (corr)

9.85 a) $25.0 \text{ g PF}_3 \times \dfrac{1 \text{ mole PF}_3}{87.97 \text{ g PF}_3} \times \dfrac{1 \text{ mole P}}{1 \text{ mole PF}_3} \times \dfrac{6.022 \times 10^{23} \text{ atoms P}}{1 \text{ mole P}}$

$= 1.7113789 \times 10^{23}$ (calc) $= 1.71 \times 10^{23}$ atoms P (corr)

b) $25.0 \text{ g Be}_3\text{P}_2 \times \dfrac{1 \text{ mole Be}_3\text{P}_2}{88.97 \text{ g Be}_3\text{P}_2} \times \dfrac{2 \text{ moles P}}{1 \text{ mole Be}_3\text{P}_2} \times \dfrac{6.022 \times 10^{23} \text{ atoms P}}{1 \text{ mole P}}$

$= 3.3842868 \times 10^{23}$ (calc) $= 3.38 \times 10^{23}$ atoms P (corr)

c) $25.0 \text{ g POCl}_3 \times \dfrac{1 \text{ mole POCl}_3}{153.32 \text{ g POCl}_3} \times \dfrac{1 \text{ mole P}}{1 \text{ mole POCl}_3} \times \dfrac{6.022 \times 10^{23} \text{ atoms P}}{1 \text{ mole P}}$

$= 9.8193321 \times 10^{22}$ (calc) $= 9.82 \times 10^{22}$ atoms P (corr)

d) $25.0 \text{ g Na}_5\text{P}_3\text{O}_{10} \times \dfrac{1 \text{ mole Na}_5\text{P}_3\text{O}_{10}}{367.86 \text{ g Na}_5\text{P}_3\text{O}_{10}} \times \dfrac{3 \text{ moles P}}{1 \text{ mole Na}_5\text{P}_3\text{O}_{10}} \times \dfrac{6.022 \times 10^{23} \text{ atoms P}}{1 \text{ mole P}}$

$= 1.2277769 \times 10^{23}$ (calc) $= 1.23 \times 10^{23}$ atoms P (corr)

9.87 a) $4.7 \times 10^{24} \text{ molecules XeO}_3 \times \dfrac{1 \text{ mole XeO}_3}{6.022 \times 10^{23} \text{ molecules XeO}_3} \times \dfrac{3 \text{ moles O}}{1 \text{ mole XeO}_3} \times \dfrac{16.00 \text{ g O}}{1 \text{ mole O}}$

$= 374.62637$ (calc) $= 3.7 \times 10^2$ g O (corr)

b) $55.00 \text{ g SO}_2\text{Cl}_2 \times \dfrac{1 \text{ mole SO}_2\text{Cl}_2}{134.97 \text{ g SO}_2\text{Cl}_2} \times \dfrac{2 \text{ moles O}}{1 \text{ mole SO}_2\text{Cl}_2} \times \dfrac{16.00 \text{ g O}}{1 \text{ mole O}} = 13.0399348$ (calc)

$= 13.04 \text{ g O}$ (corr)

c) $0.30 \text{ moles C}_6\text{H}_{12}\text{O}_6 \times \dfrac{6 \text{ moles O}}{1 \text{ mole C}_6\text{H}_{12}\text{O}_6} \times \dfrac{16.00 \text{ g O}}{1 \text{ mole O}} = 28.8$ (calc) $= 29$ g O (corr)

d) $475 \text{ g Na}_2\text{CO}_3 \times \dfrac{1 \text{ mole Na}_2\text{CO}_3}{105.99 \text{ g Na}_2\text{CO}_3} \times \dfrac{3 \text{ moles O}}{1 \text{ mole Na}_2\text{CO}_3} \times \dfrac{16.00 \text{ g O}}{1 \text{ mole O}} = 215.1146335$ (calc)

$= 215 \text{ g O}$ (corr)

9.89 a) $1.000 \text{ g B} \times \dfrac{1 \text{ mole B}}{10.81 \text{ g B}} \times \dfrac{1 \text{ mole B(OH)}_3}{1 \text{ mole B}} = 0.09250694$ (calc) $= 0.09251$ mole B(OH)$_3$ (corr)

b) $1.000 \text{ g B} \times \dfrac{1 \text{ mole B}}{10.81 \text{ g B}} \times \dfrac{1 \text{ mole B}_4\text{H}_{10}}{4 \text{ moles B}} \times \dfrac{6.022 \times 10^{23} \text{ molecules B}_4\text{H}_{10}}{1 \text{ mole B}_4\text{H}_{10}}$

$= 1.392691952 \times 10^{22}$ (calc) $= 1.393 \times 10^{22}$ molecules B$_4$H$_{10}$ (corr)

c) $1.000 \text{ g B} \times \dfrac{1 \text{ mole B}}{10.81 \text{ g B}} \times \dfrac{1 \text{ mole C}_2\text{B}_4\text{H}_6}{4 \text{ moles B}} \times \dfrac{73.32 \text{ g C}_2\text{B}_4\text{H}_6}{1 \text{ mole C}_2\text{B}_4\text{H}_6} = 1.695652174$ (calc)

$= 1.696 \text{ g C}_2\text{B}_4\text{H}_6$ (corr)

d) $1.000 \text{ g B} \times \dfrac{1 \text{ mole B}}{10.81 \text{ g B}} \times \dfrac{6.022 \times 10^{23} \text{ atoms B}}{1 \text{ mole B}} = 5.5707678076 \times 10^{22}$ (calc)

$= 5.571 \times 10^{22}$ atoms B (corr)

Chemical Calculations: The Mole Concept and Chemical Formulas

9.91 a) 2.70×10^{23} atoms N $\times \dfrac{1 \text{ mole N}}{6.022 \times 10^{23} \text{ atoms N}} \times \dfrac{1 \text{ mole NH}_3}{1 \text{ mole N}} \times \dfrac{17.04 \text{ g NH}_3}{1 \text{ mole NH}_3}$

$= 7.639986715$ (calc) $= 7.64$ g NH_3 (corr)

b) 2.70×10^{23} atoms H $\times \dfrac{1 \text{ mole H}}{6.022 \times 10^{23} \text{ atoms H}} \times \dfrac{1 \text{ mole NH}_3}{3 \text{ moles H}} \times \dfrac{17.04 \text{ g NH}_3}{1 \text{ mole NH}_3}$

$= 2.546662238$ (calc) $= 2.55$ g NH_3 (corr)

c) There are 4 atoms in 1 molecule of NH_3 (1 atom N + 3 atoms H)

2.70×10^{23} atoms total $\times \dfrac{1 \text{ molecule NH}_3}{4 \text{ atoms}} \times \dfrac{1 \text{ mole NH}_3}{6.022 \times 10^{23} \text{ molecules NH}_3} \times \dfrac{17.04 \text{ g NH}_3}{1 \text{ mole NH}_3}$

$= 1.909996679$ (calc) $= 1.91$ g NH_3 (corr)

d) 2.70×10^{25} molecules $NH_3 \times \dfrac{1 \text{ mole NH}_3}{6.022 \times 10^{23} \text{ molecules NH}_3} \times \dfrac{17.04 \text{ g NH}_3}{1 \text{ mole NH}_3}$

$= 7.639986715$ (calc) $= 7.64$ g NH_3 (corr)

9.93 a) 20.0 g Si $\times \dfrac{1 \text{ mole Si}}{28.09 \text{ g Si}} \times \dfrac{1 \text{ mole SiH}_4}{1 \text{ mole Si}} \times \dfrac{32.13 \text{ g SiH}_4}{1 \text{ mole SiH}_4} = 22.87646849$ (calc)

$= 22.9$ g SiH_4 (corr)

b) 5.75 moles $SiH_4 \times \dfrac{32.13 \text{ g SiH}_4}{1 \text{ mole SiH}_4} = 184.7475$ (calc) $= 185$ g SiH_4 (corr)

c) 5.75 moles atoms $\times \dfrac{1 \text{ mole SiH}_4}{5 \text{ moles atoms}} \times \dfrac{32.13 \text{ g SiH}_4}{1 \text{ mole SiH}_4} = 36.9495$ (calc) $= 36.9$ g SiH_4 (corr)

d) 25.0 grams of Si and H: since SiH_4 contains only Si and H, this is the mass of SiH_4 present.

9.95 The molar mass of $C_{21}H_{30}O_2$ is $21(12.01) + 30(1.01) + 2(16.00) = 314.51$ g/mole (corr)

a) 25.00 g $C_{21}H_{30}O_2 \times \dfrac{1 \text{ mole } C_{21}H_{30}O_2}{314.51 \text{ g } C_{21}H_{30}O_2} \times \dfrac{(21 + 30 + 2) \text{ moles atoms}}{1 \text{ mole } C_{21}H_{30}O_2}$

$= 4.2129026$ (calc) $= 4.213$ moles atoms (corr)

b) 25.00 g $C_{21}H_{30}O_2 \times \dfrac{1 \text{ mole } C_{21}H_{30}O_2}{314.51 \text{ g } C_{21}H_{30}O_2} \times \dfrac{21 \text{ moles C}}{1 \text{ mole } C_{21}H_{30}O_2} \times \dfrac{6.022 \times 10^{23} \text{ atoms C}}{1 \text{ mole C}}$

$= 1.0052304 \times 10^{24}$ (calc) $= 1.005 \times 10^{24}$ atoms C (corr)

c) 25.00 g $C_{21}H_{30}O_2 \times \dfrac{1 \text{ mole } C_{21}H_{30}O_2}{314.51 \text{ g } C_{21}H_{30}O_2} \times \dfrac{2 \text{ moles O}}{1 \text{ mole } C_{21}H_{30}O_2} \times \dfrac{16.00 \text{ g O}}{1 \text{ mole O}}$

$= 2.5436393$ (calc) $= 2.544$ g O (corr)

d) 25.00 g $C_{21}H_{30}O_2 \times \dfrac{1 \text{ mole } C_{21}H_{30}O_2}{314.51 \text{ g } C_{21}H_{30}O_2} \times \dfrac{6.022 \times 10^{23} \text{ molecules } C_{21}H_{30}O_2}{1 \text{ mole } C_{21}H_{30}O_2}$

$= 4.7868112 \times 10^{22}$ (calc) $= 4.787 \times 10^{22}$ molecules $C_{21}H_{30}O_2$ (corr)

9.97 Molecule is triatomic and has a molar mass of 120.0 g

a) $60.0 \text{ g of compound} \times \dfrac{1 \text{ mole compound}}{120.0 \text{ g compound}} \times \dfrac{6.022 \times 10^{23} \text{ molecules}}{1 \text{ mole compound}} \times$

$\dfrac{3 \text{ atoms}}{1 \text{ unit of compound}} = 9.033 \times 10^{23}$ (calc) $= 9.03 \times 10^{23}$ total atoms (corr)

b) $90.0 \text{ g compound} \times \dfrac{1 \text{ mole compound}}{120.0 \text{ g compound}} \times \dfrac{6.022 \times 10^{23} \text{ molecules}}{1 \text{ mole compound}} = 4.5165 \times 10^{23}$ (calc)

$= 4.52 \times 10^{23}$ molecules (corr)

Purity of Samples (Sec. 9.11)

9.99 a) $325 \text{ g sample Fe}_2\text{S}_3 \times \dfrac{95.4 \text{ g Fe}_2\text{S}_3}{100 \text{ g sample Fe}_2\text{S}_3} = 3.1005 \times 10^2$ (calc) $= 3.10 \times 10^2$ g Fe_2S_3 (corr)

b) $325 \text{ g sample Fe}_2\text{S}_3 - 310. \text{ g Fe}_2\text{S}_3 = 15$ g impurities in sample (calc and corr)

9.101 Percent impurity $= \dfrac{0.23 \text{ g impurity}}{32.21 \text{ g sample CaCO}_3} \times 100 = 0.7140639553$ (calc) $= 0.71\%$ impurity

9.103 For every $100 - 85.0$ g of $\text{Cu}_2\text{S} = 15.0$ g are impurities in sample

$3.00 \text{ g impurities} \times \dfrac{100 \text{ g sample}}{15.0 \text{ g impurities}} = 20$ (calc) $= 20.0$ g sample (corr)

9.105 $35.00 \text{ g sample} \times \dfrac{92.35 \text{ g Cu}}{100 \text{ g sample}} \times \dfrac{1 \text{ mole Cu}}{63.55 \text{ g Cu}} \times \dfrac{6.022 \times 10^{23} \text{ atoms Cu}}{1 \text{ mole Cu}}$

$= 3.062881117 \times 10^{23}$ (calc) $= 3.063 \times 10^{23}$ atoms Cu (corr)

9.107 $25.00 \text{ g sample} \times \dfrac{64.0 \text{ g Cr}_2\text{O}_3}{100 \text{ g sample}} \times \dfrac{104.00 \text{ g Cr}}{152.00 \text{ g Cr}_2\text{O}_3} = 1.094737 \times 10^1$ (calc) $= 10.9$ g Cr (corr)

Determination of Empirical Formulas (Secs. 9.13 and 9.14)

9.109 a) CH b) NS c) C_2H_6O d) BNH_2

9.111 a) not the same b) not the same c) yes, the same (CH) d) yes the same (NO_2)

9.113 a) Molecular and empirical formula is $C_6H_8O_7$.

b) Molecular formula is $C_6H_{12}O_6$; empirical formula is CH_2O.

9.115 a) $58.91 \text{ g Na} \times \dfrac{1 \text{ mole Na}}{22.99 \text{ g Na}} = 2.5624184$ (calc) $= 2.562$ moles Na (corr)

$41.09 \text{ g S} \times \dfrac{1 \text{ mole S}}{32.07 \text{ g S}} = 1.2812597$ (calc) $= 1.281$ moles S (corr)

Na: $\dfrac{2.562}{1.281} = 2.000$ S: $\dfrac{1.281}{1.281} = 1.000$ Empirical formula is Na_2S.

Chemical Calculations: The Mole Concept and Chemical Formulas 73

b) $24.74 \text{ g K} \times \dfrac{1 \text{ mole K}}{39.10 \text{ g K}} = 0.63273657 \text{ (calc)} = 0.6327 \text{ mole K (corr)}$

$34.76 \text{ g Mn} \times \dfrac{1 \text{ mole Mn}}{54.94 \text{ g Mn}} = 0.6326902 \text{ (calc)} = 0.6327 \text{ mole Mn (corr)}$

$40.50 \text{ g O} \times \dfrac{1 \text{ mole O}}{16.00 \text{ g O}} = 2.53125 \text{ (calc)} = 2.531 \text{ moles O (corr)}$

K: $\dfrac{0.6327}{0.6327} = 1.000$ Mn: $\dfrac{0.6327}{0.6327} = 1.000$ O: $\dfrac{2.531}{0.6327} = 4.000$ Empirical formula is $KMnO_4$.

c) $2.06 \text{ g H} \times \dfrac{1 \text{ mole H}}{1.01 \text{ g H}} = 2.0396039 \text{ (calc)} = 2.04 \text{ moles H (corr)}$

$32.69 \text{ g S} \times \dfrac{1 \text{ mole S}}{32.07 \text{ g S}} = 1.0193327 \text{ (calc)} = 1.020 \text{ moles S (corr)}$

$65.25 \text{ g O} \times \dfrac{1 \text{ mole O}}{16.00 \text{ g O}} = 4.078125 \text{ (calc)} = 4.078 \text{ moles O (corr)}$

H: $\dfrac{2.04}{1.020} = 2.00$ S: $\dfrac{1.020}{1.020} = 1.000$ O: $\dfrac{4.078}{1.020} = 3.998$ Empirical formula is H_2SO_4.

d) $19.84 \text{ g C} \times \dfrac{1 \text{ mole C}}{12.01 \text{ g C}} = 1.6519567 \text{ (calc)} = 1.652 \text{ moles C (corr)}$

$2.50 \text{ g H} \times \dfrac{1 \text{ mole H}}{1.01 \text{ g H}} = 2.4752475 \text{ (calc)} = 2.48 \text{ moles H (corr)}$

$66.08 \text{ g O} \times \dfrac{1 \text{ mole O}}{16.00 \text{ g O}} = 4.13 \text{ (calc)} = 4.130 \text{ moles O (corr)}$

$11.57 \text{ g N} \times \dfrac{1 \text{ mole N}}{14.01 \text{ g N}} = 0.82583868 \text{ (calc)} = 0.8258 \text{ mole N (corr)}$

C: $\dfrac{1.652}{0.8258} = 2.000$ H: $\dfrac{2.48}{0.8258} = 3.00$ O: $\dfrac{4.130}{0.8258} = 5.001$ N: $\dfrac{0.8258}{0.8258} = 1.000$

Empirical formula is $C_2H_3O_5N$.

9.117 a) $1.00 \times 3 = 3.00$, $1.67 \times 3 = 5.01$ therefore 3 to 5
b) $1.00 \times 2 = 2.00$, $1.50 \times 2 = 3.00$ therefore 2 to 3
c) $2.00 \times 3 = 6.00$, $2.33 \times 3 = 6.99$ therefore 6 to 7
d) $1.33 \times 3 = 3.99$, $2.33 \times 3 = 6.99$, $2.00 \times 3 = 6.00$ therefore 4 to 7 to 6

9.119 a) $43.64 \text{ g P} \times \dfrac{1 \text{ mole P}}{30.97 \text{ g P}} = 1.4091055 \text{ (calc)} = 1.409 \text{ moles P (corr)}$

$56.36 \text{ g O} \times \dfrac{1 \text{ mole O}}{16.00 \text{ g O}} = 3.5225 \text{ (calc)} = 3.522 \text{ moles O (corr)}$

P: $\dfrac{1.409}{1.409} = 1.000$ O: $\dfrac{3.502}{1.409} = 2.500$

O: $2.500 \times 2 = 5.000$ P: $1.000 \times 2 = 2.000$ Empirical formula is P_2O_5.

b) $72.24 \text{ g Mg} \times \dfrac{1 \text{ mole Mg}}{24.30 \text{ g Mg}} = 2.9716166$ (calc) $= 2.972$ moles Mg (corr)

$27.76 \text{ g N} \times \dfrac{1 \text{ mole N}}{14.01 \text{ g N}} = 1.9814418$ (calc) $= 1.981$ moles N (corr)

Mg: $\dfrac{2.972}{1.981} = 1.500$ N: $\dfrac{1.981}{1.981} = 1.000$

Mg: $1.500 \times 2 = 3.000$ N: $1.000 \times 2 = 2.000$ Empirical formula is Mg_3N_2.

c) $29.08 \text{ g Na} \times \dfrac{1 \text{ mole Na}}{22.99 \text{ g Na}} = 1.2648977$ (calc) $= 1.265$ moles Na (corr)

$40.56 \text{ g S} \times \dfrac{1 \text{ mole S}}{32.07 \text{ g S}} = 1.2647334$ (calc) $= 1.265$ moles S (corr)

$30.36 \text{ g O} \times \dfrac{1 \text{ mole O}}{16.00 \text{ g O}} = 1.8975$ (calc) $= 1.898$ moles O (corr)

Na: $\dfrac{1.265}{1.265} = 1.000$ S: $\dfrac{1.265}{1.265} = 1.000$ O: $\dfrac{1.898}{1.265} = 1.500$

Na: $1.000 \times 2 = 2.000$ S: $1.000 \times 2 = 2.000$ O: $1.500 \times 2 = 3.000$

Empirical formula is $Na_2S_2O_3$.

d) $21.85 \text{ g Mg} \times \dfrac{1 \text{ mole Mg}}{24.31 \text{ g Mg}} = 0.89880708$ (calc) $= 0.8988$ mole Mg (corr)

$27.83 \text{ g P} \times \dfrac{1 \text{ mole P}}{30.97 \text{ g P}} = 0.89861155$ (calc) $= 0.8986$ mole P (corr)

$50.32 \text{ g O} \times \dfrac{1 \text{ mole O}}{16.00 \text{ g O}} = 3.145$ moles O (calc and corr)

Mg: $\dfrac{0.8988}{0.8986} = 1.000$ P: $\dfrac{0.8986}{0.8986} = 1.000$ O: $\dfrac{3.145}{0.8986} = 3.500$

Mg: $1.000 \times 2 = 2.000$ P: $1.000 \times 2 = 2.000$ O: $3.500 \times 2 = 7.000$

Empirical formula is $Mg_2P_2O_7$.

9.121 Compound A: Take 100 g compound. %Cu $= 100 - 33.54 = 66.46\%$ Cu

$66.46 \text{ g Cu} \times \dfrac{1 \text{ mole Cu}}{63.55 \text{ g Cu}} = 1.045790716$ (calc) $= 1.046$ mole Cu (corr)

$33.54 \text{ g S} \times \dfrac{1 \text{ mole S}}{32.07 \text{ g S}} = 1.04583723$ (calc) $= 1.046$ moles S (corr)

Cu: $\dfrac{1.046}{1.046} = 1.000$ S: $\dfrac{1.046}{1.046} = 1.000$ Empirical formula is CuS.

Compound B: Take 100 g compound. %Cu $= 100 - 20.15 = 79.85\%$ Cu

$79.85 \text{ g Cu} \times \dfrac{1 \text{ mole Cu}}{63.55 \text{ g Cu}} = 1.25649095$ (calc) $= 1.256$ mole Cu (corr)

$20.15 \text{ g S} \times \dfrac{1 \text{ mole S}}{32.07 \text{ g S}} = 0.628313065$ (calc) $= 0.6283$ moles S (corr)

Cu: $\dfrac{1.256}{0.6283} = 1.999$ S: $\dfrac{0.6283}{0.6283} = 1.000$ Empirical formula is Cu_2S.

Chemical Calculations: The Mole Concept and Chemical Formulas 75

9.123 $5.798 \text{ g C} \times \dfrac{1 \text{ mole C}}{12.01 \text{ g C}} = 0.48276436 \text{ (calc)} = 0.4828 \text{ mole C (corr)}$

$1.46 \text{ g H} \times \dfrac{1 \text{ mole H}}{1.01 \text{ g H}} = 1.4455445 \text{ (calc)} = 1.45 \text{ moles H (corr)}$

$7.740 \text{ g S} \times \dfrac{1 \text{ mole S}}{32.07 \text{ g S}} = 0.24134705 \text{ (calc)} = 0.2413 \text{ mole S (corr)}$

C: $\dfrac{0.4828}{0.2413} = 2.001$ H: $\dfrac{1.45}{0.2413} = 6.01$ S: $\dfrac{0.2413}{0.2413} = 1.000$ Empirical formula is C_2H_6S.

9.125 Mass of oxygen in compound $= 5.55 \text{ g} - 2.00 \text{ g} = 3.55 \text{ g oxygen}$

$2.00 \text{ g Be} \times \dfrac{1 \text{ mole Be}}{9.01 \text{ g Be}} = 0.22197558 \text{ (calc)} = 0.222 \text{ moles Be (corr)}$

$3.55 \text{ g O} \times \dfrac{1 \text{ mole O}}{16.00 \text{ g O}} = 0.221875 \text{ (calc)} = 0.222 \text{ moles O (corr)}$

Be: $\dfrac{0.222}{0.222} = 1.00$ O: $\dfrac{0.222}{0.222} = 1.00$ Empirical formula is BeO.

Formula Determination Using Combustion Analysis (Sec. 9.13)

9.127 Since this compound has only C and H, find moles C and moles H from data:

a) $0.338 \text{ g CO}_2 \times \dfrac{1 \text{ mole CO}_2}{44.01 \text{ g CO}_2} \times \dfrac{1 \text{ mole C}}{1 \text{ mole CO}_2} = 0.007680072 \text{ (calc)} = 0.00768 \text{ moles C (corr)}$

$0.277 \text{ g H}_2\text{O} \times \dfrac{1 \text{ mole H}_2\text{O}}{18.02 \text{ g H}_2\text{O}} \times \dfrac{2 \text{ moles H}}{1 \text{ mole H}_2\text{O}} = 0.0307436 \text{ (calc)} = .0307 \text{ moles H (corr)}$

H: $\dfrac{0.0307}{0.00768} = 4.00$ C: $\dfrac{0.00768}{0.00768} = 1.00$ Empirical formula is CH_4.

b) $0.303 \text{ g CO}_2 \times \dfrac{1 \text{ mole CO}_2}{44.01 \text{ g CO}_2} \times \dfrac{1 \text{ mole C}}{1 \text{ mole CO}_2} = 0.006884799 \text{ (calc)} = 0.00688 \text{ moles C (corr)}$

$0.0621 \text{ g H}_2\text{O} \times \dfrac{1 \text{ mole H}_2\text{O}}{18.02 \text{ g H}_2\text{O}} \times \dfrac{2 \text{ mole H}}{1 \text{ mole H}_2\text{O}} = 0.0068923 \text{ (calc)} = 0.00689 \text{ moles H (corr)}$

H: $\dfrac{0.00689}{0.00688} = 1.00$ C: $\dfrac{0.00688}{0.00688} = 1.00$ Empirical formula is CH.

c) $0.225 \text{ g CO}_2 \times \dfrac{1 \text{ mole CO}_2}{44.01 \text{ g CO}_2} \times \dfrac{1 \text{ mole C}}{1 \text{ mole CO}_2} = 0.00511247 \text{ (calc)} = 0.00511 \text{ moles C (corr)}$

$0.115 \text{ g H}_2\text{O} \times \dfrac{1 \text{ mole H}_2\text{O}}{18.02 \text{ g H}_2\text{O}} \times \dfrac{2 \text{ mole H}}{1 \text{ mole H}_2\text{O}} = 0.0127635 \text{ (calc)} = 0.0128 \text{ moles H (corr)}$

H: $\dfrac{0.0128}{0.00511} = 2.50$ C: $\dfrac{0.00511}{0.00511} = 1.00$ Multiplying by 2 gives the empirical formula C_2H_5.

d) $0.314 \text{ g CO}_2 \times \dfrac{1 \text{ mole CO}_2}{44.01 \text{ g CO}_2} \times \dfrac{1 \text{ mole C}}{1 \text{ mole CO}_2} = 0.00713474 \text{ (calc)} = 0.00713 \text{ mole C}$

$0.192 \text{ g H}_2\text{O} \times \dfrac{1 \text{ mole H}_2\text{O}}{18.02 \text{ g H}_2\text{O}} \times \dfrac{2 \text{ moles H}}{1 \text{ mole H}_2\text{O}} = 0.0213096 \text{ (calc)} = 0.0213 \text{ moles H}$

H: $\dfrac{0.0213}{0.00713} = 2.99$ C: $\dfrac{0.00713}{0.00713} = 1.00$ Empirical formula is CH_3.

9.129 $2.328 \text{ mg CO}_2 \times \dfrac{12.01 \text{ mg C}}{44.01 \text{ mg CO}_2} = 0.6352937 \text{ (calc)} = 0.6353 \text{ mg C (corr)}$

$0.7429 \text{ mg sample} - 0.6353 \text{ mg C} = 0.1076 \text{ mg H}$

$0.6353 \text{ mg C} \times \dfrac{1 \text{ g C}}{10^3 \text{ mg C}} \times \dfrac{1 \text{ mole C}}{12.01 \text{ g C}} = 5.2897587 \times 10^{-5} \text{ (calc)} = 5.290 \times 10^{-5} \text{ moles C (corr)}$

$0.1076 \text{ mg H} \times \dfrac{1 \text{ g C}}{10^3 \text{ mg C}} \times \dfrac{1 \text{ mole H}}{1.01 \text{ g H}} = 1.0653465 \times 10^{-4} \text{ (calc)} = 1.065 \times 10^{-4} \text{ moles H (corr)}$

C: $\dfrac{5.290 \times 10^{-5}}{5.290 \times 10^{-5}} = 1.000$ H: $\dfrac{1.065 \times 10^{-4}}{5.290 \times 10^{-5}} = 2.013$ Empirical formula is CH_2.

9.131 The grams of O and the moles of C, H, and O are needed.

C: $10.05 \text{ g CO}_2 \times \dfrac{1 \text{ mole CO}_2}{44.01 \text{ g CO}_2} \times \dfrac{1 \text{ mole C}}{1 \text{ mole CO}_2} = 0.22835719 \text{ (calc)} = 0.2284 \text{ mole C (corr)}$

$0.2284 \text{ mole C} \times \dfrac{12.01 \text{ g C}}{1 \text{ mole C}} = 2.743084 \text{ (calc)} = 2.743 \text{ g C (corr)}$

H: $2.47 \text{ g H}_2\text{O} \times \dfrac{1 \text{ mole H}_2\text{O}}{18.02 \text{ g H}_2\text{O}} \times \dfrac{2 \text{ moles H}}{1 \text{ mole H}_2\text{O}} = 0.27413984 \text{ (calc)} = 0.274 \text{ mole H (corr)}$

$0.274 \text{ mole H} \times \dfrac{1.01 \text{ g H}}{1 \text{ mole H}} = 0.27674 \text{ (calc)} = 0.277 \text{ g H (corr)}$

O: $3.750 \text{ g compound} - 2.743 \text{ g C} - 0.277 \text{ g H} = 0.73 \text{ (calc)} = 0.730 \text{ g O (corr)}$

$0.730 \text{ g O} \times \dfrac{1 \text{ mole O}}{16.00 \text{ g O}} = 0.045625 \text{ (calc)} = 0.0456 \text{ mole O (corr)}$

C: $\dfrac{0.2284}{0.0456} = 5.01$ H: $\dfrac{0.274}{0.0456} = 6.01$ O: $\dfrac{0.0456}{0.0456} = 1.00$ Empirical formula is C_5H_6O.

Determination of Molecular Formulas (Sec. 9.14)

9.133 a) formula mass $CH_2 = 12.01$ amu C $+ 2(1.01)$ amu H $= 14.03$ amu

$\dfrac{\text{molecular formula mass}}{\text{empirical formula mass}} = \dfrac{42.08 \text{ amu}}{14.03 \text{ amu}} = 2.9992872 \text{ (calc)} = 3$

molecular formula $= (CH_2)_3 = C_3H_6$

b) formula mass $NaS_2O_3 = 22.99$ amu Na $+ 2(32.07)$ amu S $+ 3(16.00) = 135.13$ amu

$\dfrac{\text{molecular formula mass}}{\text{empirical formula mass}} = \dfrac{270.26 \text{ amu}}{135.13 \text{ amu}} = 2 \text{ (calc)} = 2$

molecular formula $= (NaS_2O_3)_2 = Na_2S_4O_6$

c) formula mass $C_3H_6O_2 = 3(12.01)$ amu C $+ 6(1.01)$ amu H $+ 2(16.00)$ amu O $= 74.09$ amu
empirical formula mass $=$ molecular formula mass

molecular formula $= (C_3H_6O_2)_1 = C_3H_6O_2$

d) formula mass CHN $= 12.01$ amu C $+ 1.01$ amu H $+ 14.01$ amu N $= 27.02$ amu

$\dfrac{\text{molecular formula mass}}{\text{empirical formula mass}} = \dfrac{135.15 \text{ amu}}{27.03 \text{ amu}} = 5 \text{(calc)} = 5$

molecular formula $= (CHN)_5 = C_5H_5N_5$

Chemical Calculations: The Mole Concept and Chemical Formulas 77

9.135 a) To have 3 X atoms in our molecular formula, the empirical formula must be multiplied by 3. $(XY_3)_3 = X_3Y_9$
b) To have 3 Y atoms in our molecule, the empirical formula will be the same as the molecular formula.
c) The total number of atoms in one molecule of the empirical formula = 1 X + 3 Y = 4 atoms. To have 8 atoms in the molecular formula, we multiply the empirical formula by 2. $(XY_3)_2 = X_2Y_6$
d) The total number of atoms in one molecule of the empirical formula = 1 X + 3 Y = 4 atoms. Tetratomic is defined as 4 atoms in our molecular formula, thus the empirical formula is the same as the molecular formula.

9.137 a) For the molar mass of the molecular compound to equal twice the molar mass of the empirical formula, the molecular formula must be 2x that of the empirical formula: $(C_2H_3O)_2 = C_4H_6O_2$
b) One molecule of the empirical formula, C_2H_3O, contains 2 C + 3 H + 1 O = 6 atoms. To have 18 atoms in one molecule, we multiply the empirical formula by 3. $(C_2H_3O)_3 = C_6H_9O_3$
c) One molecule of the empirical formula, C_2H_3O, contains 2 C + 1 O = 3 C and O atoms. To have 18 C and O atoms, we multiply our empirical formula by 6. $(C_2H_3O)_6 = C_{12}H_{18}O_6$
d) The mass of 0.010 mole of the empirical formula

$$0.010 \text{ mole of empirical formula} \times \frac{43.05 \text{ g empirical formula}}{1 \text{ mole empirical formula}} = 0.4305 \text{ g}$$

To determine the molecular formula, we divide the molar mass of 0.010 mole molecular formula by the molar mass of 0.010 mole of the empirical formula.

$$\text{The molecular formula} = \frac{0.86 \text{ g molecular formula}}{0.4305 \text{ g empirical formula}} = 2 \text{ (calc)} = 2$$

$$\text{molecular formula} = (C_2H_3O)_2 = C_4H_6O_2$$

9.139 a) $100.0 \text{ g compound} \times \frac{70.57 \text{ g C}}{100 \text{ g compound}} \times \frac{1 \text{ mole C}}{12.01 \text{ g C}} = 5.8759367 \text{ (calc)}$
$= 5.876 \text{ moles C (corr)}$

$100.0 \text{ g compound} \times \frac{5.93 \text{ g H}}{100 \text{ g compound}} \times \frac{1 \text{ mole H}}{1.01 \text{ g H}} = 5.87129 \text{ (calc)} = 5.87 \text{ moles H (corr)}$

$100.0 \text{ g compound} \times \frac{23.49 \text{ g O}}{100 \text{ g compound}} \times \frac{1 \text{ mole O}}{16.00 \text{ g O}} = 1.468125 \text{ (calc)}$
$= 1.468 \text{ moles O (corr)}$

C: $\frac{5.876}{1.468} = 4.003$ H: $\frac{5.87}{1.468} = 4.00$ (corr) O: $\frac{1.468}{1.468} = 1.00$ Empirical formula = C_4H_4O

Empirical formula mass = 4(12.01) amu C + 4(1.01) amu H + 1(16.00) amu O = 68.08 amu

$$\frac{\text{molecular formula mass}}{\text{empirical formula mass}} = \frac{136 \text{ amu}}{68.08 \text{ amu}} = 1.9976498 \text{ (calc)} = 2$$

$$\text{molecular formula} = (C_4H_4O)_2 = C_8H_8O_2$$

b) $1.00 \text{ mole compound} \times \frac{136 \text{ g compound}}{1 \text{ mole compound}} \times \frac{70.57 \text{ g C}}{100 \text{ g compound}} \times \frac{1 \text{ mole C}}{12.01 \text{ g C}}$
$= 7.991274 \text{ (calc)} = 8.00 \text{ moles C (corr)}$

$1.00 \text{ mole compound} \times \frac{136 \text{ g compound}}{1 \text{ mole compound}} \times \frac{5.93 \text{ g H}}{100 \text{ g compound}} \times \frac{1 \text{ mole H}}{1.01 \text{ g H}}$
$= 7.9849505 \text{ (calc)} = 7.98 \text{ moles H (corr)}$

$1.00 \text{ mole compound} \times \frac{136 \text{ g compound}}{1 \text{ mole compound}} \times \frac{23.49 \text{ g O}}{100 \text{ g compound}} \times \frac{1 \text{ mole O}}{16.00 \text{ g O}}$
$= 1.99665 \text{ (calc)} = 2 \text{ moles O and molecular formula is } C_8H_8O_2$

9.141 a) $100.0 \text{ g compound} \times \dfrac{40.0 \text{ g C}}{100 \text{ g compound}} \times \dfrac{1 \text{ mole C}}{12.01 \text{ g C}} = 3.3305579 \text{ (calc)} = 3.33 \text{ moles C (corr)}$

$100.0 \text{ g compound} \times \dfrac{6.71 \text{ g H}}{100 \text{ g compound}} \times \dfrac{1 \text{ mole H}}{1.01 \text{ g H}} = 6.6435644 \text{ (calc)} = 6.64 \text{ moles H (corr)}$

$100.0 \text{ g compound} \times \dfrac{53.3 \text{ g O}}{100 \text{ g compound}} \times \dfrac{1 \text{ mole O}}{16.00 \text{ g O}} = 3.33125 \text{ (calc)} = 3.33 \text{ moles O (corr)}$

H: $\dfrac{6.64}{3.33} = 1.99399399 \text{ (calc)} = 2$ C: $\dfrac{3.33}{3.33} = 1 \text{ (calc)} = 1$ O: $\dfrac{3.33}{3.33} = 1.00 \text{ (calc)} = 1$

Empirical formula is CH_2O.

Empirical formula mass = 12.01 amu C + 2(1.01) amu H + 16.00 amu O = 30.03 amu

$\dfrac{\text{molecular formula mass}}{\text{empirical formula mass}} = \dfrac{90.0 \text{ amu}}{30.03 \text{ amu}} = 2.997003 \text{ (calc)} = 3$

molecular formula = $(CH_2O)_3 = C_3H_6O_3$

b) $1.00 \text{ mole compound} \times \dfrac{90.0 \text{ g compound}}{1 \text{ mole compound}} \times \dfrac{40.0 \text{ g C}}{100 \text{ g compound}} \times \dfrac{1 \text{ mole C}}{12.01 \text{ g C}}$

$= 2.9975021 \text{ (calc)} = 3.00 \text{ moles (corr)}$

$1.00 \text{ mole compound} \times \dfrac{90.0 \text{ g compound}}{1 \text{ mole compound}} \times \dfrac{6.71 \text{ g H}}{100 \text{ g compound}} \times \dfrac{1 \text{ mole H}}{1.01 \text{ g H}}$

$= 5.979201 \text{ (calc)} = 5.98 \text{ moles H (corr)}$

$1.00 \text{ mole compound} \times \dfrac{90.0 \text{ g compound}}{1 \text{ mole compound}} \times \dfrac{53.3 \text{ g O}}{100 \text{ g compound}} \times \dfrac{1 \text{ mole O}}{16.00 \text{ g O}}$

$= 2.99813 \text{ (calc)} = 3.00 \text{ moles O (corr)}$

molecular formula = $C_3H_6O_3$

Additional Problems

9.143 a) gold   b) sulfur   c) Cl_2 molecules   d) Ne

9.145 a) P_4 b) 1.00 mol Na   c) Cu   d) Be

9.147 a) false   b) true   c) false   d) false

9.149 Method 1: Molecular mass of K_2S = 2(39.10) + 32.07 = 110.27 amu

$4.000 \text{ g } K_2S \times \dfrac{1 \text{ mole } K_2S}{110.27 \text{ g } K_2S} \times \dfrac{2 \text{ moles K}}{1 \text{ mole } K_2S} \times \dfrac{39.10 \text{ g K}}{1 \text{ mole K}} = 2.8366736 \text{ (calc)} = 2.837 \text{ g K (corr)}$

$4.000 \text{ g } K_2S \times \dfrac{1 \text{ mole } K_2S}{110.27 \text{ g } K_2S} \times \dfrac{1 \text{ mole S}}{1 \text{ mole } K_2S} \times \dfrac{32.07 \text{ g S}}{1 \text{ mole S}} = 1.1633264 \text{ (calc)} = 1.163 \text{ g S (corr)}$

Method 2: %K in $K_2S = \dfrac{2(39.10) \text{ amu}}{110.27 \text{ amu}} \times 100 = 70.91684 \text{ (calc)} = 70.92\% \text{ K (corr)}$

$\%S = \dfrac{32.07 \text{ amu}}{110.27 \text{ amu}} \times 100 = 29.08315952 \text{ (calc)} = 29.08\% \text{ S (corr)}$

$4.000 \text{ g } K_2S \times \dfrac{70.92 \text{ g K}}{100 \text{ g } K_2S} = 2.8368 \text{ (calc)} = 2.837 \text{ g K (corr)}$

$4.000 \text{ g } K_2S \times \dfrac{29.08 \text{ g S}}{100 \text{ g } K_2S} = 1.1632 \text{ (calc)} = 1.163 \text{ g S (corr)}$

Chemical Calculations: The Mole Concept and Chemical Formulas

9.151 3.50 moles B contain the same number of atoms as 3.50 moles Xe

$$3.50 \text{ moles B} \times \frac{10.81 \text{ g B}}{\text{mole B}} = 37.835 \text{ (calc)} = 37.8 \text{ g B (corr)}$$

9.153 The same mass of carbon contains the same moles of C atoms.

$$3.44 \text{ g C}_2\text{H}_6\text{O} \times \frac{1 \text{ mole C}_2\text{H}_6\text{O}}{46.08 \text{ g C}_2\text{H}_6\text{O}} \times \frac{2 \text{ moles C}}{1 \text{ mole C}_2\text{H}_6\text{O}} \times \frac{1 \text{ mole C}_6\text{H}_{12}\text{O}_6}{6 \text{ moles C}} \times \frac{180.18 \text{ g C}_6\text{H}_{12}\text{O}_6}{1 \text{ mole C}_6\text{H}_{12}\text{O}_6}$$

$$= 4.4836458 \text{ (calc)} = 4.48 \text{ g C}_6\text{H}_{12}\text{O}_6 \text{ (corr)}$$

9.155 From Al_2O_3, for every 2 atoms of Al, we have 3 atoms of O.

$$7.23 \times 10^{24} \text{ atoms Al} \times \frac{3 \text{ atoms O}}{2 \text{ atoms Al}} \times \frac{1 \text{ mole O}}{6.022 \times 10^{23} \text{ atoms O}} \times \frac{16.00 \text{ g O}}{1 \text{ mole O}}$$

$$= 288.1434739 \text{ (calc)} = 288 \text{ g O (corr)}$$

9.157 a) $3(12.01)$ amu C + $9(1.01)$ amu H + 28.09 amu Si + 35.45 amu Cl = 108.66 amu

$$\text{mass \% H} = \frac{9.09 \text{ amu H}}{108.66 \text{ amu}} \times 100 = 8.3655439 \text{ (calc)} = 8.37\% \text{ H (corr)}$$

b) $1 \text{ molecule } (CH_3)_3SiCl \times \dfrac{14 \text{ atoms total}}{1 \text{ molecule } (CH_3)_3SiCl} = 14 \text{ atoms total (calc and corr)}$

$1 \text{ molecule } (CH_3)_3SiCl \times \dfrac{9 \text{ H atoms}}{1 \text{ molecule } (CH_3)_3SiCl} = 9 \text{ H atoms (calc and corr)}$

$\text{atom \% H} = \dfrac{9 \text{ H atoms}}{14 \text{ total atoms}} \times 100 = 64.285714 \text{ (calc)} = 64.29\% \text{ H (corr)}$

c) $1 \text{ mole } (CH_3)_3SiCl \times \dfrac{14 \text{ moles atoms}}{1 \text{ mole } (CH_3)_3SiCl} = 14 \text{ moles atoms (calc and corr)}$

$1 \text{ mole } (CH_3)_3SiCl \times \dfrac{9 \text{ moles H atoms}}{1 \text{ mole } (CH_3)_3 SiCl} = 9 \text{ moles H atoms (calc and corr)}$

$\text{mole \% H} = \dfrac{9 \text{ moles H atoms}}{14 \text{ total moles atoms}} \times 100 = 64.285714 \text{ (calc)} = 64.29\% \text{ H (corr)}$

9.159 a) Since the ratio of atoms equals the ratio of moles of atoms, the atom ratio can be used instead of the mole ratio to determine the empirical formula.

Na: $\dfrac{9.0 \times 10^{23}}{3.0 \times 10^{23}} = 3.0$ Al: $\dfrac{3.0 \times 10^{23}}{3.0 \times 10^{23}} = 1.0$ F: $\dfrac{1.8 \times 10^{24}}{3.0 \times 10^{23}} = 6.0$

Empirical formula is Na_3AlF_6.

b) $3.2 \text{ g S} \times \dfrac{1 \text{ mole S}}{32.07 \text{ g S}} = 0.09978172747 \text{ (calc)} = 0.10 \text{ mole S (corr)}$

$1.20 \times 10^{23} \text{ atoms O} \times \dfrac{1 \text{ mole O}}{6.022 \times 10^{23} \text{ atoms O}} = 0.1992693457 \text{ mole O (calc)}$

$= 0.199 \text{ mole O (corr)}$

S: $\dfrac{0.10}{0.10} = 1.0$ O: $\dfrac{0.199}{0.10} = 2.0$ Empirical formula is SO_2.

c) 0.36 mole Ba, 0.36 mole C $17.2 \text{ g O} \times \dfrac{1 \text{ mole O}}{16.00 \text{ g O}} = 1.075 \text{ (calc)} = 1.08 \text{ moles O (corr)}$

Ba: $\dfrac{0.36}{0.36} = 1.0$ C: $\dfrac{0.36}{0.36} = 1.0$ O: $\dfrac{1.08}{0.36} = 3.0$ Empirical formula is $BaCO_3$.

d) 1.81×10^{23} atoms H $\times \dfrac{1 \text{ mole H}}{6.022 \times 10^{23} \text{ atoms H}} = 0.300564697$ mole H (calc)

$= 0.301$ mole H (corr)

$10.65 \text{ g Cl} \times \dfrac{1 \text{ mole}}{35.45 \text{ g Cl}} = 0.30042313 \text{ (calc)} = 0.3004$ mole Cl (corr)

0.30 mole O (from problem)

H: $\dfrac{0.301}{0.30} = 1.0$ O: $\dfrac{0.30}{0.30} = 1.0$ Cl: $\dfrac{0.3004}{0.30} = 1.0$ Empirical formula is HClO.

9.161   0.331 g sample H_xO_y contains 0.311 g O: 0.331 g H_xO_y − 0.311 g O = 0.20 g H

$0.311 \text{ g O} \times \dfrac{1 \text{ mole O}}{16.00 \text{ g O}} = 0.0194375 \text{ (calc)} = 0.0194$ mole O (corr)

$0.020 \text{ g H} \times \dfrac{1 \text{ mole H}}{1.01 \text{ g H}} = 0.01980198 \text{ (calc)} = 0.020$ mole H (corr)

Ratio of O to H $\dfrac{0.194}{0.19} = 1.0210526$ (calc) = 1.0 and empirical formula is HO.

Molecular mass HO: 1.01 amu H + 16.00 amu O = 17.01 amu HO

$\dfrac{\text{molecular formula mass}}{\text{empirical formula mass}} = \dfrac{34.02 \text{ amu}}{17.01 \text{ amu}} = 2$ molecular formula = $(HO)_2 = H_2O_2$

9.163 $5.25 \text{ g compound} \times \dfrac{92.26 \text{ g C}}{100 \text{ g compound}} \times \dfrac{1 \text{ mole C}}{12.01 \text{ g C}} \times \dfrac{6.022 \times 10^{23} \text{ atoms C}}{1 \text{ mole C}}$

$= 2.4286811 \times 10^{23}$ (calc) $= 2.43 \times 10^{23}$ atoms C (corr)

9.165   Since the compound contains only C and H, the mass of C in 13.75 g CO_2 plus the mass of H in 11.25 g H_2O equals the mass of the compound.

mass C = $13.75 \text{ g CO}_2 \times \dfrac{1 \text{ mole CO}_2}{44.01 \text{ g CO}_2} \times \dfrac{1 \text{ mole C}}{1 \text{ mole CO}_2} \times \dfrac{12.01 \text{ g C}}{1 \text{ mole C}}$

$= 3.75227221$ (calc) = 3.752 g C (corr)

mass H = $11.25 \text{ g H}_2\text{O} \times \dfrac{1 \text{ mole H}_2\text{O}}{18.02 \text{ g H}_2\text{O}} \times \dfrac{2 \text{ moles H}}{1 \text{ mole H}_2\text{O}} \times \dfrac{1.01 \text{ g H}}{1 \text{ mole H}}$

$= 1.26109878$ (calc) = 1.261 g H (corr)

mass compound = 3.752 g C + 1.261 g H = 5.013 g compound

9.167   Both the moles and grams of C, H, and S are needed.

Carbon: $6.60 \text{ g CO}_2 \times \dfrac{1 \text{ mole CO}_2}{44.01 \text{ g CO}_2} \times \dfrac{1 \text{ mole C}}{1 \text{ mole CO}_2} = 0.1499659$ (calc) = 0.150 mole C (corr)

$0.150 \text{ mole C} \times \dfrac{12.01 \text{ g C}}{1 \text{ mole C}} = 1.8015$ (calc) = 1.80 g C (corr)

Chemical Calculations: The Mole Concept and Chemical Formulas

Hydrogen: $5.41 \text{ g H}_2\text{O} \times \dfrac{1 \text{ mole H}_2\text{O}}{18.02 \text{ g H}_2\text{O}} \times \dfrac{2 \text{ moles H}}{1 \text{ mole H}_2\text{O}} = 0.600444 \text{ (calc)} = 0.600 \text{ mole H (corr)}$

$0.600 \text{ mole H} \times \dfrac{1.01 \text{ g H}}{1 \text{ mole H}} = 0.606 \text{ g H (calc and corr)}$

Sulfur: $9.61 \text{ g SO}_2 \times \dfrac{1 \text{ mole SO}_2}{64.07 \text{ g SO}_2} \times \dfrac{1 \text{ mole S}}{1 \text{ mole SO}_2} = 0.149992196 \text{ (calc)} = 0.150 \text{ mole S (corr)}$

$0.150 \text{ mole S} \times \dfrac{32.07 \text{ g S}}{1 \text{ mole S}} = 4.8105 \text{ (calc)} = 4.81 \text{ g S (corr)}$

a) C: $\dfrac{0.150}{0.150} = 1.00$ H: $\dfrac{0.600}{0.150} = 4.00$ S: $\dfrac{0.150}{0.150} = 1.00$ Empirical formula is CH_4S.

b) Sample mass = 1.80 g C + 0.606 g H + 4.81 g S = 7.216 (calc) = 7.22 g sample burned (corr)

9.169 Basis: 100.0 g sample: Since all the F is in NaF,

mass NaF = $100.0 \text{ g sample} \times \dfrac{18.1 \text{ g F}}{100 \text{ g sample}} \times \dfrac{1 \text{ mole F}}{19.00 \text{ g F}} \times \dfrac{1 \text{ mole NaF}}{1 \text{ mole F}} \times \dfrac{41.99 \text{ g NaF}}{1 \text{ mole NaF}}$
$= 40.001 \text{ (calc)} = 40.0 \text{ g NaF (corr)}$

%NaF = $\dfrac{40.0 \text{ g NaF}}{100.0 \text{ g sample}} \times 100 = 40.0\%$ NaF, since all N atoms are in $NaNO_3$, mass $NaNO_3$ =

$100.0 \text{ g sample} \times \dfrac{6.60 \text{ g N}}{100 \text{ g sample}} \times \dfrac{1 \text{ mole N}}{14.01 \text{ g N}} \times \dfrac{1 \text{ mole NaNO}_3}{1 \text{ mole N}} \times \dfrac{85.00 \text{ g NaNO}_3}{1 \text{ mole NaNO}_3}$
$= 40.042827 \text{ (calc)} = 40.0 \text{ g NaNO}_3 \text{ (corr)}$

%$NaNO_3$ = $\dfrac{40.0 \text{ g NaNO}_3}{100.0 \text{ g sample}} \times 100 = 40.0\%$ $NaNO_3$

Mass Na_2SO_4 = mass sample − mass NaF − mass $NaNO_3$ = 100.0 − 40.0 − 40.0 = 20.0 g Na_2SO_4

%Na_2SO_4 = $\dfrac{20.0 \text{ g Na}_2\text{SO}_4}{100.0 \text{ g sample}} \times 100 = 20.0\%$ Na_2SO_4

9.171 100 g mixture − (5.000 g CO_2 + 10.00 g N_2O) = 85.00 g H_2O. Mass of each element in mixture:

$5.000 \text{ g CO}_2 \times \dfrac{12.01 \text{ g C}}{44.01 \text{ g CO}_2} = 1.364462622 \text{ (calc)} = 1.364 \text{ g C (corr)}$

$5.000 \text{ g CO}_2 \times \dfrac{32.00 \text{ g O}}{44.01 \text{ g CO}_2} = 3.63553737 \text{ (calc)} = 3.636 \text{ g O (corr)}$

$10.00 \text{ g N}_2\text{O} \times \dfrac{28.02 \text{ g N}}{44.02 \text{ g N}_2\text{O}} = 6.365288505 \text{ (calc)} = 6.365 \text{ g N (corr)}$

$10.00 \text{ g N}_2\text{O} \times \dfrac{16.00 \text{ g O}}{44.02 \text{ g N}_2\text{O}} = 3.634711495 \text{ (calc)} = 3.635 \text{ g O (corr)}$

$85.00 \text{ g H}_2\text{O} \times \dfrac{2.02 \text{ g H}}{18.02 \text{ g H}_2\text{O}} = 9.528301887 \text{ (calc)} = 9.528 \text{ g H (corr)}$

$85.00 \text{ g H}_2\text{O} \times \dfrac{16.00 \text{ g O}}{18.02 \text{ g H}_2\text{O}} = 75.47169811 \text{ (calc)} = 75.47 \text{ g O (corr)}$

Percent C in mixture: $\dfrac{1.346 \text{ g C}}{100. \text{ g mixture}} \times 100 = 1.346\%$ C (calc and corr)

Percent N in mixture: $\dfrac{6.365 \text{ g N}}{100. \text{ g mixture}} \times 100 = 6.365\%$ N (calc and corr)

Percent H in mixture: $\dfrac{9.528 \text{ g H}}{100. \text{ g mixture}} \times 100 = 9.528\%$ H (calc and corr)

Percent O: = 3.636 g (CO_2) + 3.635 g (N_2O) + 75.47 g (H_2O) = 82.741 (calc) = 82.74 g O (corr)

$\dfrac{82.47 \text{ g O}}{100. \text{ g mixture}} \times 100 = 82.47\%$ (calc and corr)

Confirming answer: 1.346% C + 6.365% N + 9.528% H + 82.74% O = 99.979 (100%)

9.173 28.9% of the formula mass must be from the chlorine atom. 0.289 × (formula mass) = 35.45 amu
formula mass = 122.66435 amu (calc) = 123 amu (corr)
formula mass = at. mass K + at. mass Cl + X(at. mass O)
123 amu   = 39.10 amu + 35.45 amu + X(16.00 amu)
3.028125 = X   The value of X is the integer 3.

9.175 mass O = total mass − mass M = 10.498 g − 7.503 g = 2.995 g O (calc and corr)

$2.995 \text{ g O} \times \dfrac{1 \text{ mole O}}{16.00 \text{ g O}} \times \dfrac{1 \text{ mole M}}{1 \text{ mole O}} = 0.1871875$ (calc) = 0.1872 mole M (corr)

molar mass = g of M divided by the moles of M. $\dfrac{7.503 \text{ g}}{0.1872 \text{ mole}}$

= 40.08013 (calc) = 40.08 $\dfrac{\text{g}}{\text{mole}}$ (corr)

9.177 Basis: 100.0 g of compound.

Carbon: 100.0 g compound × $\dfrac{47.4 \text{ g C}}{100 \text{ g compound}}$ = 47.4 g C (calc and corr)

47.4 g C × $\dfrac{1 \text{ mole C}}{12.01 \text{ g C}}$ = 3.9467111 (calc) = 3.95 moles C (corr)

OH_4 units: mass OH_4 = mass compound − mass C = 100.0 g − 47.4 g = 52.6 g (calc and corr)

52.6 g OH_4 × $\dfrac{1 \text{ mole } OH_4}{20.04 \text{ g } OH_4}$ = 2.6247505 (calc) = 2.62 moles OH_4 (corr)

C: $\dfrac{3.95}{2.62}$ = 1.5076336 (calc) = 1.51 (corr) OH_4: $\dfrac{2.62}{2.62}$ = 1.00 (calc and corr)

multiply by 2 to get a whole number ratio:
C: 1.51 × 2 = 3.02 OH_4: 1.00 × 2 = 2.00

Empirical formula is $C_3(OH_4)_2$ or $C_3H_8O_2$.

Cumulative Problems

9.179 Add the exact atomic masses and round appropriately:

3(1.00794) amu H + 74.9216 amu As + 4(15.9994) amu O = 141.94302 (calc) = 141.9430 (corr)

a) 142   b) 141.9   c) 141.94   d) 141.943

9.181 Molar mass to 4 significant figures:

a) 1 mole $MgSO_4$ = 24.31 g Mg + 32.07 g S + 4(16.00) g O = 120.38 (calc)
   = 120.4 g $MgSO_4$ (corr)

b) 1 mole K_3P = 3(39.10) g K + 30.97 g P = 148.27 (calc) = 148.3 g K_3P (corr)

c) 1 mole $CuBr_2$ = 63.55 g Cu + 2(79.90) g Br = 223.35 (calc) = 223.4 g $CuBr_2$ (corr)

d) 1 mole $(NH_4)_3PO_4$ = 3(14.01) g N + 12(1.01) g H + 30.97 g P + 4(16.00) g O
   = 149.12 (calc) = 149.1 g $(NH_4)_3PO_4$ (corr)

9.183 Assume a 1 mole sample of each compound.

a) 1 mole S_2O: 2(32.07)g S + 16.00 g O = 80.14 g S_2O

$$\%S = \frac{64.14 \text{ g S}}{80.14 \text{ g } S_2O} \times 100 = 80.0349388 \text{ (calc)} = 80.03\% \text{ S in } S_2O \text{ (corr)}$$

b) 1 mole H_2SO_3: 2(1.01) g H + 32.07 g S + 3(16.00) g O = 82.09 g H_2SO_3

$$\%S = \frac{32.07 \text{ g S}}{82.09 \text{ g } H_2SO_3} \times 100 = 39.06687782 \text{ (calc)} = 39.07\% \text{ S in } H_2SO_3 \text{ (corr)}$$

c) 1 mole $(NH_4)_2SO_4$ = 2(14.01) g N + 8(1.01) g H + 32.07 g S + 4(16.00) g O
   = 132.17 g $(NH_4)_2SO_4$

$$\%S = \frac{32.07 \text{ g S}}{132.17 \text{ g } (NH_4)_2SO_4} \times 100 = 24.264205 \text{ (calc)} = 24.26\% \text{ S in } (NH_4)_2SO_4 \text{ (calc and corr)}$$

d) 1 mole H_2S = 2(1.01) g H + 32.07 g S = 34.09 g H_2S

$$\%S = \frac{32.07 \text{ g S}}{34.09 \text{ g } H_2S} \times 100 = 94.0745086 \text{ (calc)} = 94.07\% \text{ S in } H_2S \text{ (corr)}$$

9.185 $0.422 \text{ mole Pb} \times \dfrac{207.2 \text{ g Pb}}{1 \text{ mole Pb}} = 87.4384$ (calc) = 87.4 g Pb (corr) $d = \dfrac{87.4 \text{ g}}{7.74 \text{ cm}^3}$
= 11.291989 (calc) = 11.3 g/cm³ (corr)

9.187 $3.752 \text{ moles } CO_2 \times \dfrac{44.01 \text{ g } CO_2}{1 \text{ mole } CO_2} \times \dfrac{1 \text{ L } CO_2 \text{ gas}}{1.96 \text{ g } CO_2 \text{ gas}} = 84.24771429$ (calc) = 84.2 L CO_2 (corr)

9.189 $225 \text{ mL } C_2H_6O \times \dfrac{0.789 \text{ g } C_2H_6O}{1 \text{ mL } C_2H_6O} \times \dfrac{1 \text{ mole } C_2H_6O}{46.08 \text{ g } C_2H_6O} \times \dfrac{6.022 \times 10^{23} \text{ molecules } C_2H_6O}{1 \text{ mole } C_2H_6O}$

= 2.319999×10^{24} (calc) = 2.32×10^{24} molecules C_2H_6O (corr) = 2.32×10^{24} molecules H_2O

2.32×10^{24} molecules $H_2O \times \dfrac{1 \text{ mole } H_2O}{6.022 \times 10^{23} \text{ molecules } H_2O} \times \dfrac{18.02 \text{ g } H_2O}{1 \text{ mole } H_2O} \times \dfrac{1 \text{ mL } H_2O}{1.00 \text{ g } H_2O}$

= 69.422783 (calc) = 69.4 mL H_2O (corr)

9.191 Assume that the measurement, one cup, has one significant figure.

$1 \text{ cup} \times \dfrac{371 \text{ mg K}}{1 \text{ cup}} \times \dfrac{10^{-3} \text{ g K}}{1 \text{ mg K}} \times \dfrac{1 \text{ mole K}}{39.10 \text{ g K}} \times \dfrac{6.022 \times 10^{23} \text{ atoms K}}{1 \text{ mole K}} \times \dfrac{0.012 \text{ atoms } ^{40}K}{100 \text{ atoms K}}$

= 6.8567632×10^{17} (calc) = 6.9×10^{17} atoms ^{40}K (corr)

9.193 a) $2.33 \text{ moles Al}_2(\text{SO}_4)_3 \times \dfrac{6.022 \times 10^{23} \text{ formula units Al}_2(\text{SO}_4)_3}{1 \text{ mole Al}_2(\text{SO}_4)_3}$

$= 1.403126 \times 10^{24}$ (calc) $= 1.40 \times 10^{24}$ formula units $\text{Al}_2(\text{SO}_4)_3$ (corr)

b) $2.33 \text{ moles Al}_2(\text{SO}_4)_3 \times \dfrac{2 \text{ moles Al}^{3+} \text{ ions}}{1 \text{ mole Al}_2(\text{SO}_4)_3} \times \dfrac{6.022 \times 10^{23} \text{ Al}^{3+} \text{ ions}}{1 \text{ mole Al}^{3+} \text{ ions}}$

$= 2.806252 \times 10^{24}$ (calc) $= 2.81 \times 10^{24}$ Al^{3+} ions (corr)

c) $2.33 \text{ moles Al}_2(\text{SO}_4)_3 \times \dfrac{3 \text{ moles SO}_4^{2-} \text{ ions}}{1 \text{ mole Al}_2(\text{SO}_4)_3} \times \dfrac{6.022 \times 10^{23} \text{ SO}_4^{2-} \text{ ions}}{1 \text{ mole SO}_4^{2-} \text{ ions}}$

$= 4.209378 \times 10^{24}$ (calc) $= 4.21 \times 10^{24}$ SO_4^{2-} ions (corr)

d) $2.33 \text{ moles Al}_2(\text{SO}_4)_3 \times \dfrac{5 \text{ moles ions}}{1 \text{ mole Al}_2(\text{SO}_4)_3} \times \dfrac{6.022 \times 10^{23} \text{ ions}}{1 \text{ mole ions}}$

$= 7.01563 \times 10^{24}$ (calc) $= 7.02 \times 10^{24}$ ions (corr)

9.195 $32.5 \text{ g CaCl}_2 \times \dfrac{1 \text{ mole CaCl}_2}{110.98 \text{ g CaCl}_2} \times \dfrac{3 \text{ moles ions}}{1 \text{ mole CaCl}_2} = 0.87853667$ (calc) $= 0.879$ moles ions (corr)

$0.879 \text{ moles ions} \times \dfrac{1 \text{ mole KCl}}{2 \text{ moles ions}} \times \dfrac{74.55 \text{ g KCl}}{1 \text{ mole KCl}} = 32.764725$ (calc) $= 32.8$ g KCl (corr)

9.197 $425 \text{ g mixture} \times \dfrac{42.0 \text{ g NaCl}}{100 \text{ g mixture}} \times \dfrac{1 \text{ mole NaCl}}{58.44 \text{ g NaCl}} \times \dfrac{1 \text{ mole Cl}^- \text{ ions}}{1 \text{ mole NaCl}} \times \dfrac{6.022 \times 10^{23} \text{ Cl}^- \text{ ions}}{1 \text{ mole Cl}^- \text{ ions}}$

$= 1.8393686 \times 10^{24}$ (calc) $= 1.84 \times 10^{24}$ Cl^- ions (corr)

$425 \text{ g mixture} \times \dfrac{58.0 \text{ g CaCl}_2}{100 \text{ g mixture}} \times \dfrac{1 \text{ mole CaCl}_2}{110.98 \text{ g CaCl}_2} \times \dfrac{2 \text{ moles Cl}^- \text{ ions}}{1 \text{ mole CaCl}_2} \times$

$\dfrac{6.022 \times 10^{23} \text{ Cl}^- \text{ ions}}{1 \text{ mole Cl}^- \text{ ions}} = 2.675118 \times 10^{24}$ (calc) $= 2.68 \times 10^{24}$ Cl^- ions (corr)

Total Cl^- ions $= 1.84 \times 10^{24} + 2.68 \times 10^{24}$ Cl^- ions $= 4.52 \times 10^{24}$ Cl^- ions (calc and corr)

9.199 $V = 10.0 \text{ cm} \times 30.0 \text{ cm} \times 62.0 \text{ cm} = 18{,}600 \text{ cm}^3$ (calc and corr)

$18{,}600 \text{ cm}^3 \times \dfrac{8.31 \text{ g alloy}}{1 \text{ cm}^3} \times \dfrac{43.0 \text{ g Ni}}{100 \text{ g alloy}} \times \dfrac{1 \text{ mole Ni}}{58.69 \text{ g Ni}} \times \dfrac{6.022 \times 10^{23} \text{ atoms Ni}}{1 \text{ mole Ni}}$

$= 6.8196026 \times 10^{26}$ (calc) $= 6.82 \times 10^{26}$ atoms Ni (corr)

9.201 $0.275 \text{ mole NaOH} \times \dfrac{40.00 \text{ g NaOH}}{1 \text{ mole NaOH}} \times \dfrac{100 \text{ g solution}}{12.0 \text{ g NaOH}} \times \dfrac{1 \text{ mL solution}}{1.131 \text{ g solution}}$

$= 81.049218$ (calc) $= 81.0$ mL solution (corr)

Answers to Multiple-Choice Practice Test

9.203	e	9.204	c	9.205	d	9.206	d	9.207	c	9.208	c
9.209	b	9.210	d	9.211	e	9.212	d	9.213	d	9.214	c
9.215	a	9.216	c	9.217	d	9.218	e	9.219	d	9.220	e
9.221	a	9.222	c								

CHAPTER TEN
Chemical Calculations Involving Chemical Equations

Practice Problems

The Law of Conservation of Mass (Sec. 10.1)

10.1 The total mass of products = total mass of reactants
 a) 127.10 g Cu + 34.00 g O_2 = 161.10 g reactants, 159.10 g CuO products (calc and corr)
  This is inconsistent with the law of conservation of mass.
 b) 76.15 g CS_2 + 96.00 g O_2 = 172.15 g reactants (calc and corr)
  44.01 g CO_2 + 128.14 g SO_2 = 172.15 g products (calc and corr)
  This is consistent with the law of conservation of mass.

10.3 The total mass of products = total mass of reactants
 (16.05 g CH_4) + (64.00 g O_2) = (x g CO_2) + (36.04 g H_2O)
 Mass of reactants = (16.05 g CH_4) + (64.00 g O_2) = 80.05 g reactants (calc and corr)
 Mass of products = (x g CO_2) + (36.04 g H_2O)
 (x g CO_2) = 80.05 g reactants − (36.04 g H_2O) = 44.01 g CO_2 (calc and corr)

10.5 The law of conservation of mass requires the mass of the reactants to equal the mass of the products
 Mass of reactants:
  4.2 g sodium hydrogen carbonate + 10.0 g acetic acid = 14.2 g reactants (calc and corr)
  Mass of contents in vessel = 12.0 g (products)
  Mass of missing carbon dioxide = (14.2 − 12.0) g = 2.2 g carbon dioxide (calc and corr)

10.7 Diagrams I and III are consistent with law of conservation of mass.
 Diagram I = 4 units of 2 empty paired circles (8) + 6 solid unpaired circles = 14 total
 Diagram III = 6 units with a solid and empty circle (12) + 2 extra paired solid circles = 14 total
 The type and number of circles remains the same in III.

Chemical Equation Notation (Secs. 10.2 and 10.4)

10.9 Answers "a," "b," and "c" contain symbols correctly written for gases in an equation and therefore are appropriate. However, in "d" Br is not correct; it should be Br_2, and therefore this is not appropriate.

10.11 a) (s) means solid, (l) means liquid, (aq) means water (aqueous) solution, and (g) means gas.
 b) (s) means solid, (aq) means water (aqueous) solution, (s) means solid, (g) means gas, and (l) means liquid.

Balancing Chemical Equations (Sec. 10.4)

10.13 a) balanced b) unbalanced c) balanced d) unbalanced

10.15

	Reaction	Reactants	Products
a)	$2\ Cu + O_2 \rightarrow 2\ CuO$	2 Cu, 2 O	2 Cu, 2 O
b)	$2\ H_2O \rightarrow 2\ H_2 + O_2$	4 H, 2 O	4 H, 2 O
c)	$BaCl_2 + Na_2S \rightarrow BaS + 2\ NaCl$	1 Ba, 2 Cl, 2 Na, 1 S	1 Ba, 1 S, 2 Na, 2 Cl
d)	$2\ C_3H_8O + 9\ O_2 \rightarrow 6\ CO_2 + 8\ H_2O$	6 C, 16 H, 20 O	6 C, 20 O, 16 H

10.17   a) $2\ N_2 + 3\ O_2 \rightarrow 2\ N_2O_3$ b) $4\ NH_3 + 6\ NO \rightarrow 5\ N_2 + 6\ H_2O$
   b) $CS_2 + 3\ O_2 \rightarrow CO_2 + 2\ SO_2$ d) $Mg + 2\ HBr \rightarrow MgBr_2 + H_2$

10.19   a) $3\ PbO + 2\ NH_3 \rightarrow 3\ Pb + N_2 + 3\ H_2O$
   b) $2\ NaHCO_3 + H_2SO_4 \rightarrow Na_2SO_4 + 2\ CO_2 + 2\ H_2O$
   c) $TiO_2 + C + 2\ Cl_2 \rightarrow TiCl_4 + CO_2$
   d) $2\ NBr_3 + 3\ NaOH \rightarrow N_2 + 3\ NaBr + 3\ HBrO$

10.21   a) $C_5H_{12} + 8\ O_2 \rightarrow 5\ CO_2 + 6\ H_2O$ b) $2\ C_5H_{10} + 15\ O_2 \rightarrow 10\ CO_2 + 10\ H_2O$
   c) $C_5H_8 + 7\ O_2 \rightarrow 5\ CO_2 + 4\ H_2O$ d) $C_5H_{10}O + 7\ O_2 \rightarrow 5\ CO_2 + 5\ H_2O$

10.23   a) $Ca(OH)_2 + 2\ HNO_3 \rightarrow Ca(NO_3)_2 + 2\ H_2O$
   b) $BaCl_2 + (NH_4)_2SO_4 \rightarrow BaSO_4 + 2\ NH_4Cl$
   c) $2\ Fe(OH)_3 + 3\ H_2SO_4 \rightarrow Fe_2(SO_4)_3 + 6\ H_2O$
   d) $Na_3PO_4 + 3\ AgNO_3 \rightarrow 3\ NaNO_3 + Ag_3PO_4$

10.25   a) divide each coefficient by 3: $AgNO_3 + KCl \rightarrow AgCl + KNO_3$
   b) divide each coefficient by 2: $CS_2 + 3\ O_2 \rightarrow CO_2 + 2\ SO_2$
   c) multiply each coefficient by 2: $2\ H_2 + O_2 \rightarrow 2\ H_2O$
   d) multiply each coefficient by 2: $2\ Ag_2CO_3 \rightarrow 4\ Ag + 2\ CO_2 + O_2$

10.27   a) $6\ A_2 + 2\ B_2 \rightarrow 4\ A_3B$ b) $3\ A_2 + 3\ B_2 \rightarrow 6\ AB$

10.29   Diagram III is consistent with Diagram I.

Classes of Chemical Reactions (Sec. 10.6)

10.31   a) synthesis   b) synthesis   c) double replacement   d) single replacement

10.33   a) synthesis   b) synthesis

10.35   a) The correct formula for zinc nitrate is $Zn(NO_3)_2$ and the symbol for copper is Cu;
      $Zn + Cu(NO_3)_2 \rightarrow Zn(NO_3)_2 + Cu$
   b) The correct formula for calcium oxide is CaO; $2\ Ca + O_2 \rightarrow 2\ CaO$.
   c) The correct formula for barium sulfate is $BaSO_4$, and for potassium nitrate it is KNO_3;
      $K_2SO_4 + Ba(NO_3)_2 \rightarrow BaSO_4 + 2\ KNO_3$.
   d) The correct formula for oxygen is O_2, the symbol for silver is Ag; $2\ Ag_2O \rightarrow 4\ Ag + O_2$.

10.37   a) $Rb_2CO_3 \rightarrow Rb_2O + CO_2$ b) $SrCO_3 \rightarrow SrO + CO_2$
   c) $Al_2(CO_3)_3 \rightarrow Al_2O_3 + 3\ CO_2$ d) $Cu_2CO_3 \rightarrow Cu_2O + CO_2$

10.39   a) $2\ C_3H_6 + 9\ O_2 \rightarrow 6\ CO_2 + 6\ H_2O$ b) $C_2H_4 + 3\ O_2 \rightarrow 2\ CO_2 + 2\ H_2O$
   c) $C_7H_{16} + 11\ O_2 \rightarrow 7\ CO_2 + 8\ H_2O$ d) $C_8H_{16} + 12\ O_2 \rightarrow 8\ CO_2 + 8\ H_2O$

Chemical Calculations Involving Chemical Equations 87

10.41 a) $CH_2O + O_2 \rightarrow CO_2 + H_2O$ b) $2\,C_5H_{12}O + 15\,O_2 \rightarrow 10\,CO_2 + 12\,H_2O$
  c) $C_5H_{10}O + 7\,O_2 \rightarrow 5\,CO_2 + 5\,H_2O$ d) $2\,C_5H_{10}O_2 + 13\,O_2 \rightarrow 10\,CO_2 + 10\,H_2O$

10.43 a) $4\,C_2H_7N + 19\,O_2 \rightarrow 8\,CO_2 + 14\,H_2O + 4\,NO_2$
  b) $CH_4S + 3\,O_2 \rightarrow CO_2 + 2\,H_2O + SO_2$

10.45 a) single-replacement, combustion, and synthesis
  b) single-replacement and decomposition
  c) synthesis, decomposition, single-replacement, double-replacement, and combustion
  d) synthesis, decomposition, single-replacement, double replacement, and combustion

Chemical Equations and the Mole Concept (Sec. 10.7)

10.47 a) $1.0 \text{ mole A} \times \dfrac{2 \text{ moles B}}{4 \text{ moles A}} = 0.5 \text{ (calc)} = 0.50 \text{ moles B (corr)}$

  b) $2.0 \text{ moles A} \times \dfrac{2 \text{ moles B}}{4 \text{ moles A}} = 1 \text{ (calc)} = 1.0 \text{ mole B (corr)}$

  c) $5.0 \text{ moles A} \times \dfrac{2 \text{ moles B}}{4 \text{ moles A}} = 2.5 \text{ moles B (calc and corr)}$

  d) $7.0 \text{ moles A} \times \dfrac{2 \text{ moles B}}{4 \text{ moles A}} = 3.5 \text{ moles B (calc and corr)}$

10.49 $4\,NH_3 + 3\,O_2 \rightarrow 2\,N_2 + 6\,H_2O$

$\dfrac{4 \text{ moles } NH_3}{3 \text{ moles } O_2}$ $\dfrac{3 \text{ moles } O_2}{4 \text{ moles } NH_3}$ $\dfrac{4 \text{ moles } NH_3}{2 \text{ moles } N_2}$ $\dfrac{2 \text{ moles } N_2}{4 \text{ moles } NH_3}$

$\dfrac{4 \text{ moles } NH_3}{6 \text{ moles } H_2O}$ $\dfrac{6 \text{ moles } H_2O}{4 \text{ moles } NH_3}$ $\dfrac{3 \text{ moles } O_2}{2 \text{ moles } N_2}$ $\dfrac{2 \text{ moles } N_2}{3 \text{ moles } O_2}$

$\dfrac{3 \text{ moles } O_2}{6 \text{ moles } H_2O}$ $\dfrac{6 \text{ moles } H_2O}{3 \text{ moles } O_2}$ $\dfrac{2 \text{ moles } N_2}{6 \text{ moles } H_2O}$ $\dfrac{6 \text{ moles } H_2O}{2 \text{ moles } N_2}$

10.51 $6\,ClO_2 + 3\,H_2O \rightarrow 5\,HClO_3 + HCl$

  a) $\dfrac{3 \text{ moles } H_2O}{6 \text{ moles } ClO_2}$ b) $\dfrac{1 \text{ mole } HCl}{5 \text{ moles } HClO_3}$ c) $\dfrac{1 \text{ mole } HCl}{3 \text{ moles } H_2O}$ d) $\dfrac{6 \text{ moles } ClO_2}{5 \text{ moles } HClO_3}$

10.53 a) $3.00 \text{ moles } N_2 \times \dfrac{2 \text{ moles } NaN_3}{3 \text{ moles } N_2} = 2 \text{ (calc)} = 2.00 \text{ moles } NaN_3 \text{ (corr)}$

  b) $3.00 \text{ moles } N_2 \times \dfrac{3 \text{ moles } CO}{1 \text{ mole } N_2} = 9 \text{ (calc)} = 9.00 \text{ moles CO (corr)}$

  c) $3.00 \text{ moles } N_2 \times \dfrac{2 \text{ moles } NH_2Cl}{1 \text{ mole } N_2} = 6 \text{ (calc)} = 6.00 \text{ moles } NH_2Cl \text{ (corr)}$

  d) $3.00 \text{ moles } N_2 \times \dfrac{4 \text{ moles } C_3H_5O_9N_3}{6 \text{ moles } N_2} = 2 \text{ (calc)} = 2.00 \text{ moles } C_3H_5O_9N_3 \text{ (corr)}$

10.55 a) $1.42 \text{ moles H}_2\text{SO}_2 \times \dfrac{1 \text{ mole H}_2\text{O}_2}{1 \text{ mole H}_2\text{S}} = 1.42 \text{ mole H}_2\text{O}_2$ (calc and corr)

b) $1.42 \text{ moles O}_2 \times \dfrac{1 \text{ mole CS}_2}{3 \text{ moles O}_2} = 0.4733333 \text{ (calc)} = 0.473 \text{ moles CS}_2$ (corr)

c) $1.42 \text{ moles HCl} \times \dfrac{1 \text{ mole Mg}}{2 \text{ moles HCl}} = 0.71 \text{ (corr)} = 0.710 \text{ moles Mg}$ (corr)

d) $1.42 \text{ moles Al} \times \dfrac{6 \text{ moles HCl}}{2 \text{ moles Al}} = 4.26 \text{ moles HCl}$ (calc and corr)

10.57 a) $1.75 \text{ moles NH}_4\text{NO}_3 \times \dfrac{7 \text{ moles products}}{2 \text{ moles NH}_4\text{NO}_3} = 6.125 \text{ (calc)} = 6.12 \text{ moles products}$ (corr)

b) $1.75 \text{ moles NaClO}_3 \times \dfrac{5 \text{ moles products}}{2 \text{ moles NaClO}_3} = 4.375 \text{ (calc)} = 4.38 \text{ moles products}$ (corr)

c) $1.75 \text{ moles KNO}_3 \times \dfrac{3 \text{ moles products}}{2 \text{ moles KNO}_3} = 2.625 \text{ (calc)} = 2.62 \text{ moles products}$ (corr)

d) $1.75 \text{ moles I}_4\text{O}_9 \times \dfrac{11 \text{ moles products}}{4 \text{ moles I}_4\text{O}_9} = 4.8125 \text{ (calc)} = 4.81 \text{ moles products}$ (corr)

10.59 a) $6.0 \text{ moles Ag} \times \dfrac{2 \text{ moles Ag}_2\text{CO}_3}{4 \text{ moles Ag}} = 3 \text{ (calc)} = 3.0 \text{ moles Ag}_2\text{CO}_3$ (corr)

b) $8.0 \text{ moles CO}_2 \times \dfrac{2 \text{ moles Ag}_2\text{CO}_3}{2 \text{ moles CO}_2} = 8 \text{ (calc)} = 8.0 \text{ moles Ag}_2\text{CO}_3$ (corr)

c) $2.4 \text{ moles O}_2 \times \dfrac{2 \text{ moles Ag}_2\text{CO}_3}{1 \text{ mole O}_2} = 4.8 \text{ moles Ag}_2\text{CO}_3$ (calc and corr)

d) $15.0 \text{ moles of products} \times \dfrac{2 \text{ moles Ag}_2\text{CO}_3}{7 \text{ moles product}} = 4.2857142857 \text{ (calc)} = 4.29 \text{ moles Ag}_2\text{CO}_3$ (corr)

Balanced Chemical Equations and the Law of Conservation of Mass (Sec. 10.8)

10.61 Law of Conservation of Mass states mass reactants = mass products
 a) 4.00 g A + 1.67 g B = 5.67 g C
 b) 3.76 g A + x g B = 7.02 g C   ∴ 7.02 g C − 3.76 g A = 3.26 g B

10.63 a) $C_7H_{16} + 11 \text{ O}_2 \rightarrow 7 \text{ CO}_2 + 8 \text{ H}_2\text{O}$
  100.23 g C_7H_{16} + 11(32.00) g O_2 = 7(44.01) g CO_2 + 8(18.02) g H_2O
                      452.23 g reactants = 452.23 g products

 b) $2 \text{ HCl} + \text{CaCO}_3 \rightarrow \text{CaCl}_2 + \text{CO}_2 + \text{H}_2\text{O}$
  2(36.46) g HCl + 100.09 g $CaCO_3$ = 110.98 g $CaCl_2$ + 44.01 g CO_2 + 18.02 g H_2O
                      173.01 g reactants = 173.01 g products

 c) $4 \text{ Na}_2\text{SO}_4 + 2 \text{ C} \rightarrow \text{Na}_2\text{S} + 2 \text{ CO}_2$
  142.05 g Na_2SO_4 + 2(12.01) g C = 78.09 g Na_2S + 2(44.01) g CO_2
                      166.07 g reactants = 166.07 g products

 d) $4 \text{ Na}_2\text{CO}_3 + \text{Fe}_3\text{Br}_8 \rightarrow 8 \text{ NaBr} + 4 \text{ CO}_2 + \text{Fe}_3\text{O}_4$
  4(105.99) g Na_2CO_3 + 806.75 g Fe_3Br_8 = 8(102.89) g NaBr + 4(44.01) g CO_2 + 231.55 g Fe_3O_4
                      1230.71 g reactants = 1230.71 g products

Chemical Calculations Involving Chemical Equations 89

Calculations Based on Chemical Equations (Sec. 10.9)

10.65 a) $7.00 \text{ moles KClO}_3 \times \dfrac{3 \text{ moles O}_2}{2 \text{ moles KClO}_3} \times \dfrac{32.00 \text{ g O}_2}{1 \text{ mole O}_2} = 336 \text{ g O}_2$ (calc and corr)

b) $7.00 \text{ moles CuO} \times \dfrac{1 \text{ mole O}_2}{2 \text{ moles CuO}} \times \dfrac{32.00 \text{ g O}_2}{1 \text{ mole O}_2} = 112 \text{ g O}_2$ (calc and corr)

c) $7.00 \text{ moles NaNO}_3 \times \dfrac{1 \text{ mole O}_2}{2 \text{ moles NaNO}_3} \times \dfrac{32.00 \text{ g O}_2}{1 \text{ mole O}_2} = 112 \text{ g O}_2$ (calc and corr)

d) $7.00 \text{ moles HNO}_3 \times \dfrac{1 \text{ mole O}_2}{4 \text{ moles HNO}_3} \times \dfrac{32.00 \text{ g O}_2}{1 \text{ mole O}_2} = 56 \text{ (calc)} = 56.0 \text{ g O}_2$ (corr)

10.67 a) $1.00 \text{ mole H}_2\text{O} \times \dfrac{4 \text{ moles HNO}_3}{2 \text{ moles H}_2\text{O}} \times \dfrac{63.02 \text{ g HNO}_3}{1 \text{ mole HNO}_3} = 126.04 \text{ (calc)} = 126 \text{ g HNO}_3$ (corr)

b) $1.00 \text{ mole H}_2\text{O} \times \dfrac{6 \text{ moles HNO}_3}{3 \text{ moles H}_2\text{O}} \times \dfrac{63.02 \text{ g HNO}_3}{1 \text{ mole HNO}_3} = 126.04 \text{ (calc)} = 126 \text{ g HNO}_3$ (corr)

c) $1.00 \text{ mole H}_2\text{O} \times \dfrac{1 \text{ mole HNO}_3}{2 \text{ moles H}_2\text{O}} \times \dfrac{63.02 \text{ g HNO}_3}{1 \text{ mole HNO}_3} = 31.51 \text{ (calc)} = 31.5 \text{ g HNO}_3$ (corr)

d) $1.00 \text{ mole H}_2\text{O} \times \dfrac{10 \text{ moles HNO}_3}{3 \text{ moles H}_2\text{O}} \times \dfrac{63.02 \text{ g HNO}_3}{1 \text{ mole HNO}_3} = 210.06666 \text{ (calc)} = 210. \text{ g HNO}_3$ (corr)

10.69 a) $1.772 \text{ g CaO} \times \dfrac{1 \text{ mole CaO}}{56.08 \text{ g CaO}} \times \dfrac{2 \text{ moles HNO}_3}{1 \text{ mole CaO}} \times \dfrac{63.02 \text{ g HNO}_3}{1 \text{ mole HNO}_3} = 3.98257632$ (calc)

$= 3.983 \text{ g HNO}_3$ (corr)

b) $1.772 \text{ g PCl}_5 \times \dfrac{1 \text{ mole PCl}_5}{208.22 \text{ g PCl}_5} \times \dfrac{4 \text{ moles H}_2\text{O}}{1 \text{ mole PCl}_5} \times \dfrac{18.02 \text{ g H}_2\text{O}}{1 \text{ mole H}_2\text{O}} = 0.6134173$ (calc)

$= 0.6134 \text{ g H}_2\text{O}$ (corr)

c) $1.772 \text{ g CuCl}_2 \times \dfrac{1 \text{ mole CuCl}_2}{134.45 \text{ g CuCl}_2} \times \dfrac{2 \text{ moles NaOH}}{1 \text{ mole CuCl}_2} \times \dfrac{40.00 \text{ g NaOH}}{1 \text{ mole NaOH}} = 1.05436965$ (calc)

$= 1.054 \text{ g NaOH}$ (corr)

d) $1.772 \text{ g Na}_2\text{S} \times \dfrac{1 \text{ mole Na}_2\text{S}}{78.05 \text{ g Na}_2\text{S}} \times \dfrac{2 \text{ moles AgC}_2\text{H}_3\text{O}_2}{1 \text{ mole Na}_2\text{S}} \times \dfrac{166.92 \text{ g AgC}_2\text{H}_3\text{O}_2}{1 \text{ mole AgC}_2\text{H}_3\text{O}_2} = 7.5793014$ (calc)

$= 7.579 \text{ g AgC}_2\text{H}_3\text{O}_2$ (corr)

10.71 a) $1.50 \text{ moles C} \times \dfrac{1 \text{ mole SiO}_2}{3 \text{ moles C}} \times \dfrac{60.09 \text{ g SiO}_2}{1 \text{ mole SiO}_2} = 30.045 \text{ (calc)} = 30.0 \text{ g SiO}_2$ (corr)

b) $1.37 \text{ moles SiO}_2 \times \dfrac{2 \text{ moles CO}}{1 \text{ mole SiO}_2} \times \dfrac{28.01 \text{ g CO}}{1 \text{ mole CO}} = 76.7474 \text{ (calc)} = 76.7 \text{ g CO}$ (corr)

c) $3.33 \text{ moles CO} \times \dfrac{1 \text{ mole SiC}}{2 \text{ moles CO}} \times \dfrac{40.10 \text{ g SiC}}{1 \text{ mole SiC}} = 66.7665 \text{ (calc)} = 66.8 \text{ g SiC}$ (corr)

d) $0.575 \text{ mole SiC} \times \dfrac{3 \text{ moles C}}{1 \text{ mole SiC}} \times \dfrac{12.01 \text{ g C}}{1 \text{ mole C}} = 20.71725 \text{ (calc)} = 20.7 \text{ g C}$ (corr)

10.73 a) $4.50 \text{ moles } CO_2 \times \dfrac{2 \text{ moles LiOH}}{1 \text{ mole } CO_2} \times \dfrac{23.95 \text{ g LiOH}}{1 \text{ mole LiOH}} = 215.55 \text{ (calc)} = 216 \text{ g LiOH (corr)}$

b) $3.00 \times 10^{24} \text{ molecules } CO_2 \times \dfrac{1 \text{ mole } CO_2}{6.022 \times 10^{23} \text{ molecules } CO_2} \times \dfrac{2 \text{ moles LiOH}}{1 \text{ mole } CO_2} \times \dfrac{23.95 \text{ g LiOH}}{1 \text{ mole LiOH}}$

$= 238.62504 \text{ (calc)} = 239 \text{ g LiOH (corr)}$

c) $10.0 \text{ g } H_2O \times \dfrac{1 \text{ mole } H_2O}{18.02 \text{ g } H_2O} \times \dfrac{2 \text{ moles LiOH}}{1 \text{ mole } H_2O} \times \dfrac{23.95 \text{ g LiOH}}{1 \text{ mole LiOH}} = 26.581576 \text{ (calc)}$

$= 26.6 \text{ g LiOH (corr)}$

d) $10.0 \text{ g } Li_2CO_3 \times \dfrac{1 \text{ mole } Li_2CO_3}{73.89 \text{ g } Li_2CO_3} \times \dfrac{2 \text{ moles LiOH}}{1 \text{ mole } Li_2CO_3} \times \dfrac{23.95 \text{ g LiOH}}{1 \text{ mole LiOH}} = 6.4826093 \text{ (calc)}$

$= 6.48 \text{ g LiOH (corr)}$

10.75 a) $25.00 \text{ g NaF} \times \dfrac{1 \text{ mole NaF}}{41.99 \text{ g NaF}} \times \dfrac{1 \text{ mole } Na_2SiO_3}{2 \text{ moles NaF}}$

$= 0.29768992 \text{ (calc)} = 0.2977 \text{ mole } Na_2SiO_3 \text{ (corr)}$

b) $27.00 \text{ g } H_2O \times \dfrac{1 \text{ mole } H_2O}{18.02 \text{ g } H_2O} \times \dfrac{8 \text{ moles HF}}{3 \text{ moles } H_2O} \times \dfrac{20.01 \text{ g HF}}{1 \text{ mole HF}}$

$= 79.951165 \text{ (calc)} = 79.95 \text{ g HF (corr)}$

c) $2.000 \text{ g } Na_2SiO_3 \times \dfrac{1 \text{ mole } Na_2SiO_3}{122.07 \text{ g } Na_2SiO_3} \times \dfrac{1 \text{ mole } H_2SiF_6}{1 \text{ mole } Na_2SiO_3} \times \dfrac{6.022 \times 10^{23} \text{ molecules } H_2SiF_6}{1 \text{ mole } H_2SiF_6}$

$= 9.8667401 \times 10^{21} \text{ (calc)} = 9.867 \times 10^{21} \text{ molecules } H_2SiF_6 \text{ (corr)}$

d) $50.00 \text{ g } Na_2SiO_3 \times \dfrac{1 \text{ mole } Na_2SiO_3}{122.07 \text{ g } Na_2SiO_3} \times \dfrac{8 \text{ moles HF}}{1 \text{ mole } Na_2SiO_3} \times \dfrac{20.01 \text{ g HF}}{1 \text{ mole HF}} = 65.568936 \text{ (calc)}$

$= 65.57 \text{ g HF (corr)}$

10.77 Balanced equation: $4 \text{ Al} + 3 \text{ O}_2 \rightarrow 2 \text{ Al}_2O_3$

$23.7 \text{ g } O_2 \times \dfrac{1 \text{ mole } O_2}{32.00 \text{ g } O_2} \times \dfrac{4 \text{ moles Al}}{3 \text{ moles } O_2} \times \dfrac{26.98 \text{ g Al}}{1 \text{ mole Al}} = 26.64275 \text{ (calc)} = 26.6 \text{ g Al (corr)}$

10.79 mass of chromium: $200.0 \text{ g } CrCl_3 \times \dfrac{1 \text{ mole } CrCl_3}{158.35 \text{ g } CrCl_3} \times \dfrac{2 \text{ moles Cr}}{2 \text{ moles } CrCl_3} \times \dfrac{52.00 \text{ g Cr}}{1 \text{ mole Cr}}$

$= 65.677297 \text{ (calc)} = 65.68 \text{ g Cr (corr)}$

mass of chlorine: $200.0 \text{ g } CrCl_3 \times \dfrac{1 \text{ mole } CrCl_3}{158.35 \text{ g } CrCl_3} \times \dfrac{3 \text{ moles } Cl_2}{2 \text{ moles } CrCl_3} \times \dfrac{70.90 \text{ g } Cl_2}{1 \text{ mole } Cl_2}$

$= 134.3227 \text{ (calc)} = 134.3 \text{ g } Cl_2 \text{ (corr)}$

Note: The mass of Cl_2 could also be obtained by subtracting 65.68 g Cr from 200.0 g $CrCl_3$.

Chemical Calculations Involving Chemical Equations

Limiting Reactant Calculations (Sec. 10.10)

10.81 $216 \text{ nuts} \times \dfrac{1 \text{ combination}}{3 \text{ nuts}} = 72 \text{ combinations} \qquad 284 \text{ bolts} \times \dfrac{1 \text{ combination}}{4 \text{ bolts}} = 71 \text{ combinations}$

The limiting reactant is the 284 bolts.

10.83 $426 \text{ wings} \times \dfrac{1 \text{ kit}}{2 \text{ wings}} = 213 \text{ kits (calc and corr)}$

$224 \text{ fuselages} \times \dfrac{1 \text{ kit}}{1 \text{ fuselage}} = 224 \text{ kits (calc and corr)}$

$860 \text{ engines} \times \dfrac{1 \text{ kit}}{4 \text{ engines}} = 215 \text{ kits (calc and corr)}$

$1578 \text{ wheels} \times \dfrac{1 \text{ kit}}{6 \text{ wheels}} = 263 \text{ kits (calc and corr)}$ $\qquad \therefore$ 213 kits can be produced.

10.85 a) $2.00 \text{ moles Be} \times \dfrac{3 \text{ moles } H_2}{3 \text{ moles Be}} = 2.00 \text{ moles } H_2 \text{ (calc and corr)}$

$0.500 \text{ moles } NH_3 \times \dfrac{3 \text{ moles } H_2}{2 \text{ moles } NH_3} = 0.75 \text{ (calc)} = 0.750 \text{ moles } H_2 \text{ (corr)}$

The 0.500 moles NH_3 is the limiting reactant.

b) $3.00 \text{ moles Be} \times \dfrac{3 \text{ moles } H_2}{3 \text{ moles Be}} = 3.00 \text{ moles } H_2 \text{ (calc and corr)}$

$3.00 \text{ moles } NH_3 \times \dfrac{3 \text{ moles } H_2}{2 \text{ moles } NH_3} = 4.5 \text{ (calc)} = 4.50 \text{ moles } NH_3 \text{ (corr)}$

The 3.00 moles of Be is the limiting reactant.

c) $3.00 \text{ g Be} \times \dfrac{1 \text{ mole Be}}{9.01 \text{ g Be}} \times \dfrac{3 \text{ moles } H_2}{3 \text{ moles Be}} = 0.33296337 \text{ (calc)} = 0.333 \text{ moles } H_2 \text{ (corr)}$

$0.100 \text{ moles } NH_3 \times \dfrac{3 \text{ moles } H_2}{2 \text{ moles } NH_3} = 0.15 \text{ (calc)} = 0.150 \text{ moles } NH_3 \text{ (corr)}$

The 0.100 moles of NH_3 is the limiting reactant.

d) $20.00 \text{ g Be} \times \dfrac{1 \text{ mole Be}}{9.01 \text{ g Be}} \times \dfrac{3 \text{ moles } H_2}{3 \text{ moles Be}} = 2.219755827 \text{ (calc)} = 2.22 \text{ moles } H_2 \text{ (corr)}$

$60.00 \text{ g } NH_3 \times \dfrac{1 \text{ mole } NH_3}{17.04 \text{ g } NH_3} \times \dfrac{3 \text{ moles } H_2}{2 \text{ moles } NH_3} = 5.281690141 \text{ (calc)} = 5.282 \text{ moles } NH_3 \text{ (corr)}$

The 20.00 g of Be is the limiting reactant.

10.87 Determine the limiting reactant in each. $N_2 + 3 H_2 \rightarrow 2 NH_3$

a) $3.0 \text{ g } N_2 \times \dfrac{1 \text{ mole } N_2}{28.02 \text{ g } N_2} \times \dfrac{2 \text{ moles } NH_3}{1 \text{ mole } N_2} \times \dfrac{17.04 \text{ g } NH_3}{1 \text{ mole } NH_3} = 3.6488223 \text{ (calc)} = 3.6 \text{ g } NH_3 \text{ (corr)}$

$5.0 \text{ g } H_2 \times \dfrac{1 \text{ mole } H_2}{2.02 \text{ g } H_2} \times \dfrac{2 \text{ moles } NH_3}{3 \text{ moles } H_2} \times \dfrac{17.04 \text{ g } NH_3}{1 \text{ mole } NH_3} = 28.11881188 \text{ (calc)} = 28 \text{ g } NH_3 \text{ (corr)}$

N_2 is the limiting reactant, and 3.6 g of NH_3 can be produced.

b) $30.0 \text{ g N}_2 \times \dfrac{1 \text{ mole N}_2}{28.02 \text{ g N}_2} \times \dfrac{2 \text{ moles NH}_3}{1 \text{ mole N}_2} \times \dfrac{17.04 \text{ g NH}_3}{1 \text{ mole NH}_3} = 36.488223 \text{ (calc)} = 36.5 \text{ g NH}_3 \text{ (corr)}$

$10.0 \text{ g H}_2 \times \dfrac{1 \text{ mole H}_2}{2.02 \text{ g H}_2} \times \dfrac{2 \text{ moles NH}_3}{3 \text{ moles H}_2} \times \dfrac{17.04 \text{ g NH}_3}{1 \text{ mole NH}_3} = 56.2376237 \text{ (calc)} = 56.2 \text{ g NH}_3 \text{ (corr)}$

N_2 is the limiting reactant, and 36.5 g of NH_3 can be produced.

c) $50.0 \text{ g N}_2 \times \dfrac{1 \text{ mole N}_2}{28.02 \text{ g N}_2} \times \dfrac{2 \text{ moles NH}_3}{1 \text{ mole N}_2} \times \dfrac{17.04 \text{ g NH}_3}{1 \text{ mole NH}_3} = 60.8137045 \text{ (calc)} = 60.8 \text{ g NH}_3 \text{ (corr)}$

$8.00 \text{ g H}_2 \times \dfrac{1 \text{ mole H}_2}{2.02 \text{ g H}_2} \times \dfrac{2 \text{ moles NH}_3}{3 \text{ moles H}_2} \times \dfrac{17.04 \text{ g NH}_3}{1 \text{ mole NH}_3} = 44.990099 \text{ (calc)} = 45.0 \text{ g NH}_3 \text{ (corr)}$

H_2 is the limiting reactant, and 45.0 g of NH_3 can be produced.

d) $56 \text{ g N}_2 \times \dfrac{1 \text{ mole N}_2}{28.02 \text{ g N}_2} \times \dfrac{2 \text{ moles NH}_3}{1 \text{ mole N}_2} \times \dfrac{17.04 \text{ g NH}_3}{1 \text{ mole NH}_3} = 68.111349 \text{ (calc)} = 68 \text{ g NH}_3 \text{ (corr)}$

$12 \text{ g H}_2 \times \dfrac{1 \text{ mole H}_2}{2.02 \text{ g H}_2} \times \dfrac{2 \text{ moles NH}_3}{3 \text{ moles H}_2} \times \dfrac{17.04 \text{ g NH}_3}{1 \text{ mole NH}_3} = 67.4851485 \text{ (calc)} = 67 \text{ g NH}_3 \text{ (corr)}$

H_2 is the limiting reactant, and 67 g of NH_3 can be produced.

10.89 Determine the amount of excess reactant in each mixture. $N_2 + 3 H_2 \rightarrow 2 NH_3$

a) N_2 is the limiting reactant, and 3.6 g of NH_3 can be produced from 3.0 g N_2 and 5.0 g H_2.

$3.0 \text{ g N}_2 \times \dfrac{1 \text{ mole N}_2}{28.02 \text{ g N}_2} \times \dfrac{3 \text{ moles H}_2}{1 \text{ mole N}_2} \times \dfrac{2.02 \text{ g H}_2}{1 \text{ mole H}_2} = 0.64882227 \text{ (calc)} = 0.65 \text{ g H}_2 \text{ used (corr)}$

5.0 g H_2 present − 0.65 g H_2 used = 4.35 (calc) = 4.4 g excess H_2 (corr)

b) N_2 is the limiting reactant, and 36.5 g of NH_3 can be produced from 30.0 g N_2 and 10.0 g H_2.

$30.0 \text{ g N}_2 \times \dfrac{1 \text{ mole N}_2}{28.02 \text{ g N}_2} \times \dfrac{3 \text{ moles H}_2}{1 \text{ mole N}_2} \times \dfrac{2.02 \text{ g H}_2}{1 \text{ mole H}_2} = 6.4882227 \text{ (calc)} = 6.50 \text{ g H}_2 \text{ used (corr)}$

10.0 g H_2 present − 6.50 g H_2 used = 3.5 g excess H_2 (calc and corr)

c) H_2 is the limiting reactant, and 45 g of NH_3 can be produced from 50.0 g N_2 and 8.0 g H_2.

$8.0 \text{ g H}_2 \times \dfrac{1 \text{ mole H}_2}{2.02 \text{ g H}_2} \times \dfrac{1 \text{ mole N}_2}{3 \text{ moles H}_2} \times \dfrac{28.02 \text{ g N}_2}{1 \text{ mole N}_2} = 37.0693069 \text{ (calc)} = 37.0 \text{ g N}_2 \text{ used (corr)}$

50.0 g N_2 present − 37.0 g N_2 used = 13.0 g excess N_2 (calc and corr)

d) H_2 is the limiting reactant, and 67.5 g of NH_3 can be produced from 56 g N_2 and 12 g H_2.

$12.0 \text{ g H}_2 \times \dfrac{1 \text{ mole H}_2}{2.02 \text{ g H}_2} \times \dfrac{1 \text{ mole N}_2}{3 \text{ moles H}_2} \times \dfrac{28.02 \text{ g N}_2}{1 \text{ mole N}_2} = 55.485149 \text{ (calc)} = 55 \text{ g N}_2 \text{ used (corr)}$

56.0 g N_2 present − 55 g N_2 used = 1 g excess N_2 (calc and corr)

10.91 a) Balanced reaction: $6 A_2 + 2 B_2 \rightarrow 4 BA_3$ we have 6 A_2 and 4 B_2

$6 A_2 \times \dfrac{4 \text{ BA}_3}{6 A_2} = 4 \text{ BA}_3$ and $4 B_2 \times \dfrac{4 \text{ BA}_3}{2 B_2} = 8 \text{ BA}_3$ therefore A_2 is limiting.

b) Balanced reaction: $3 A_2 + 3 B_2 \rightarrow 6 AB$ we have 3 A_2 and 6 B_2

$3 A_2 \times \dfrac{6 \text{ AB}}{3 A_2} = 6 \text{ AB}$ and $6 B_2 \times \dfrac{6 \text{ AB}}{3 B_2} = 9 \text{ AB}$ therefore A_2 is limiting.

Chemical Calculations Involving Chemical Equations

10.93 $525 \text{ Co atoms} \times \dfrac{2 \text{ formula units CoCl}_3}{2 \text{ atoms Co}} = 525 \text{ formula units CoCl}_3 \text{ (calc and corr)}$

$525 \text{ HCl molecules} \times \dfrac{2 \text{ formula units CoCl}_3}{6 \text{ molecules HCl}} = 175 \text{ formula units (calc and corr)}$

The HCl is the limiting reactant. 175 CoCl$_3$ formula units can be made.

10.95   The limiting reactant will all react. Use the limiting reactant to determine the mass of the other reactant that reacts, and then find the unreacted by difference.

$70.0 \text{ g Fe}_3\text{O}_4 \times \dfrac{1 \text{ mole Fe}_3\text{O}_4}{231.55 \text{ g Fe}_3\text{O}_4} \times \dfrac{6 \text{ moles Fe}_2\text{O}_3}{4 \text{ moles Fe}_3\text{O}_4} = 0.453465774 \text{ (calc)} = 0.453 \text{ mole Fe}_2\text{O}_3 \text{ (corr)}$

$12.0 \text{ g O}_2 \times \dfrac{1 \text{ mole O}_2}{32.00 \text{ g O}_2} \times \dfrac{6 \text{ moles Fe}_2\text{O}_3}{1 \text{ mole O}_2} = 2.25 \text{ moles Fe}_2\text{O}_3 \text{ (calc and corr)}$

The Fe$_3$O$_4$ is the limiting reactant. There will be none left upon completion.
Calculate the mass of O$_2$ reacted:

$0.454 \text{ mole Fe}_2\text{O}_3 \times \dfrac{1 \text{ mole O}_2}{6 \text{ moles Fe}_2\text{O}_3} \times \dfrac{32.00 \text{ g O}_2}{1 \text{ mole O}_2} = 2.4213333 \text{ (calc)} = 2.42 \text{ g O}_2 \text{ reacted (corr)}$

Unreacted: Fe$_3$O$_4$ = 0
O$_2$ = 12.0 g − 2.42 g = 9.58 (calc) = 9.6 g O$_2$ unreacted (corr)

10.97 $8.00 \text{ g SCl}_2 \times \dfrac{1 \text{ mole SCl}_2}{102.97 \text{ g SCl}_2} \times \dfrac{1 \text{ mole SF}_4}{3 \text{ moles SCl}_2} = 0.0258975106 \text{ (calc)} = 0.0259 \text{ mole SF}_4 \text{ (corr)}$

$4.00 \text{ g NaF} \times \dfrac{1 \text{ mole NaF}}{41.99 \text{ g NaF}} \times \dfrac{1 \text{ mole SF}_4}{4 \text{ moles NaF}} = 0.023815194 \text{ (calc)} = 0.0238 \text{ mole SF}_4 \text{ (corr)}$

NaF is the limiting reactant.

$0.0238 \text{ mole SF}_4 \times \dfrac{108.07 \text{ g SF}_4}{1 \text{ mole SF}_4} = 2.572066 \text{ (calc)} = 2.57 \text{ g SF}_4 \text{ (corr)}$

$0.0238 \text{ mole SF}_4 \times \dfrac{1 \text{ mole S}_2\text{Cl}_2}{1 \text{ mole SF}_4} \times \dfrac{135.04 \text{ g S}_2\text{Cl}_2}{1 \text{ mole S}_2\text{Cl}_2} = 3.213952 \text{ (calc)} = 3.21 \text{ g S}_2\text{Cl}_2 \text{ (corr)}$

$0.0238 \text{ mole SF}_4 \times \dfrac{4 \text{ moles NaCl}}{1 \text{ mole SF}_4} \times \dfrac{58.44 \text{ g NaCl}}{1 \text{ mole NaCl}} = 5.563488 \text{ (calc)} = 5.56 \text{ g NaCl (corr)}$

Theoretical Yield and Percent Yield (Sec. 10.11)

10.99 $\% \text{ yield} = \dfrac{16.0 \text{ g (actual)}}{52.0 \text{ g (theoretical)}} \times 100 = 30.76923077 \text{ (calc)} = 30.8\% \text{ (corr)}$

10.101   2 Mg + O$_2$ → 2 MgO Given 125.6 g MgO isolated out of possible 172.2 g MgO
a) Theoretical yield = 172.2 g MgO
b) Actual yield = 125.6 g MgO
c) Percent yield = $\dfrac{\text{actual yield}}{\text{theoretical yield}} \times 100 = \dfrac{125.6 \text{ g MgO}}{172.2 \text{ g MgO}} \times 100 = 72.93844367 \text{(calc)}$

$= 72.94\% \text{ yield (corr)}$

94   Chemical Calculations Involving Chemical Equations

10.103   Diagram III

10.105   Theoretical yield of HCl:

$$2.13 \text{ g H}_2 \times \frac{1 \text{ mole H}_2}{2.02 \text{ g H}_2} \times \frac{2 \text{ moles HCl}}{1 \text{ mole H}_2} \times \frac{36.46 \text{ g HCl}}{1 \text{ mole HCl}} = 76.890891 \text{ (calc)} = 76.9 \text{ g HCl (corr)}$$

$$\% \text{ yield} = \frac{74.30 \text{ g HCl}}{76.9 \text{ g HCl}} \times 100 = 96.6189857 \text{ (calc)} = 96.6\% \text{ (corr)}$$

10.107   Theoretical yield of Al_2S_3:

$$55.0 \text{ g Al} \times \frac{1 \text{ mole Al}}{26.98 \text{ g Al}} \times \frac{1 \text{ mole Al}_2S_3}{2 \text{ moles Al}} \times \frac{150.17 \text{ g Al}_2S_3}{1 \text{ mole Al}_2S_3} = 153.0643069 \text{ (calc)}$$

$$= 153 \text{ g Al}_2S_3 \text{ (corr)}$$

Actual yield = theoretical yield x% yield

$$= 153 \text{ g Al}_2S_3 \text{ (theor)} \times \frac{85.6 \text{ g Al}_2S_3 \text{ (actual)}}{100 \text{ g Al}_2S_3 \text{ (theor)}} = 130.968 \text{ (calc)} = 131 \text{ g Al}_2S_3 \text{ (corr)}$$

10.109   Find the limiting reactant.

$$35.0 \text{ g CO} \times \frac{1 \text{ mole CO}}{28.01 \text{ g CO}} \times \frac{2 \text{ moles CO}_2}{2 \text{ moles CO}} = 1.2495537 \text{ (calc)} = 1.25 \text{ moles CO}_2 \text{ (calc and corr)}$$

$$35.0 \text{ g O}_2 \times \frac{1 \text{ mole O}_2}{32.00 \text{ g O}_2} \times \frac{2 \text{ moles CO}_2}{1 \text{ mole O}_2} = 2.1875 \text{ (calc)} = 2.19 \text{ moles CO}_2 \text{ (corr)}$$

The CO is the limiting reactant.

$$\text{Theoretical yield} = 1.25 \text{ moles CO}_2 \times \frac{44.01 \text{ g CO}_2}{1 \text{ mole CO}_2} = 55.0125 \text{ (calc)} = 55.0 \text{ g CO}_2 \text{ (corr)}$$

$$\text{Actual yield} = 55.0 \text{ g CO}_2 \text{ (theor)} \times \frac{57.8 \text{ g CO}_2 \text{ (actual)}}{100 \text{ g CO}_2 \text{ (theor)}} = 31.79 \text{ (calc)} = 31.8 \text{ g CO}_2 \text{ (corr)}$$

10.111   3 sequential reactions, A → B → C → D. The percent yield for this synthesis is the product of the percent yields of each step in the multi-step process.

$$100. \text{ g A} \times \frac{92 \text{ g B}}{100. \text{ g A}} \times \frac{87 \text{ g C}}{100. \text{ g B}} \times \frac{43 \text{ g D}}{100. \text{ g C}} = 34.4172 \text{ (calc)} = 34 \text{ g D (corr)}$$

$$\frac{34 \text{ g D}}{100. \text{ g A}} \times 100 = 34\% \text{ yield}$$

Simultaneous Chemical Reactions (Sec. 10.12)

10.113 $82.5 \text{ g mixture} \times \dfrac{60.0 \text{ g ZnS}}{100 \text{ g mixture}} \times \dfrac{1 \text{ mole ZnS}}{97.45 \text{ g ZnS}} \times \dfrac{2 \text{ moles SO}_2}{2 \text{ moles ZnS}} \times \dfrac{64.07 \text{ g SO}_2}{1 \text{ mole SO}_2}$

$$= 32.544536 \text{ (calc)} = 32.5 \text{ g SO}_2 \text{ (corr)}$$

$82.5 \text{ g mixture} \times \dfrac{40.0 \text{ g CuS}}{100 \text{ g mixture}} \times \dfrac{1 \text{ mole CuS}}{95.62 \text{ g CuS}} \times \dfrac{2 \text{ moles SO}_2}{2 \text{ moles CuS}} \times \dfrac{64.07 \text{ g SO}_2}{1 \text{ mole SO}_2}$

$$= 22.111588 \text{ (calc)} = 22.1 \text{ g SO}_2 \text{ (corr)}$$

Total SO_2 = 32.5 g + 22.1 g = 54.6 g (calc and corr)

Chemical Calculations Involving Chemical Equations

10.115 $75.0 \text{ g mixture} \times \dfrac{70.0 \text{ g CH}_4}{100 \text{ g mixture}} \times \dfrac{1 \text{ mole CH}_4}{16.05 \text{ g CH}_4} \times \dfrac{2 \text{ moles O}_2}{1 \text{ mole CH}_4} \times \dfrac{32.00 \text{ g O}_2}{1 \text{ mole O}_2}$

$= 209.34579 \text{ (calc)} = 209 \text{ g O}_2 \text{ (corr)}$

$75.0 \text{ g mixture} \times \dfrac{30.0 \text{ g C}_2\text{H}_6}{100 \text{ g mixture}} \times \dfrac{1 \text{ mole C}_2\text{H}_6}{30.08 \text{ g C}_2\text{H}_6} \times \dfrac{7 \text{ moles O}_2}{2 \text{ moles C}_2\text{H}_6} \times \dfrac{32.00 \text{ g O}_2}{1 \text{ mole O}_2}$

$= 83.776596 \text{ (calc)} = 83.8 \text{ g O}_2 \text{ (corr)}$

Total = 209 g O$_2$ + 83.8 g O$_2$ = 292.8 (calc) = 293 g O$_2$ (corr)

Sequential Chemical Reactions (Sec. 10.11)

10.117 a) $6.00 \text{ moles SO}_2 \times \dfrac{1 \text{ mole O}_2}{1 \text{ mole SO}_2} \times \dfrac{2 \text{ moles NaClO}_3}{3 \text{ moles O}_2} = 49 \text{ (calc)} = 4.00 \text{ moles NaClO}_3 \text{ (corr)}$

b) $20.0 \text{ g SO}_2 \times \dfrac{1 \text{ mole SO}_2}{64.07 \text{ g SO}_2} \times \dfrac{1 \text{ mole O}_2}{1 \text{ mole SO}_2} \times \dfrac{2 \text{ moles NaClO}_3}{3 \text{ moles O}_2} \times \dfrac{106.44 \text{ g NaClO}_3}{1 \text{ mole NaClO}_3}$

$= 22.150773 \text{ (calc)} = 22.2 \text{ g NaClO}_3 \text{ (corr)}$

10.119 a) $2.00 \text{ moles N}_2 \times \dfrac{2 \text{ moles NO}}{1 \text{ mole N}_2} \times \dfrac{2 \text{ moles NO}_2}{2 \text{ moles NO}} \times \dfrac{4 \text{ moles HNO}_3}{4 \text{ moles NO}_2} = 4 \text{ (calc)}$

$= 4.00 \text{ moles HNO}_3 \text{ (corr)}$

b) $2.00 \text{ g N}_2 \times \dfrac{1 \text{ mole N}_2}{28.02 \text{ g N}_2} \times \dfrac{2 \text{ moles NO}}{1 \text{ mole N}_2} \times \dfrac{2 \text{ moles NO}_2}{2 \text{ moles NO}} \times \dfrac{4 \text{ moles HNO}_3}{4 \text{ moles NO}_2} \times \dfrac{63.02 \text{ g HNO}_3}{1 \text{ mole HNO}_3}$

$= 8.9964311 \text{ (calc)} = 9.00 \text{ g HNO}_3 \text{ (corr)}$

10.121 $5.00 \text{ g I}_2 \times \dfrac{1 \text{ mole I}_2}{253.80 \text{ g I}_2} \times \dfrac{2 \text{ moles FeI}_2}{2 \text{ moles I}_2} \times \dfrac{2 \text{ moles AgI}}{1 \text{ mole FeI}_2} \times \dfrac{1 \text{ mole AgNO}_3}{1 \text{ mole AgI}} \times \dfrac{169.88 \text{ g AgNO}_3}{1 \text{ mole AgNO}_3}$

$= 6.6934594 \text{ (calc)} = 6.69 \text{ g AgNO}_3 \text{ (corr)}$

10.123 2 NaClO$_3$ → 2 NaCl + 3 O$_2$
+ 3 S + 3 O$_2$ → 3 SO$_2$
----
2 NaClO$_3$ + 3 S → 2 NaCl + 2 O$_2$ + 3 SO$_2$

10.125 2 N$_2$ + 2 O$_2$ → 4 NO
+ 4 NO + 2 O$_2$ → 4 NO$_2$
  4 NO$_2$ + 2 H$_2$O + O$_2$ → 4 HNO$_3$
----
2 N$_2$ + 2 NO$_2$ + 5 O$_2$ → 4 HNO$_3$

Additional Problems

10.127 a) $3.00 \text{ moles CH}_4 \times \dfrac{2 \text{ moles O}_2}{1 \text{ mole CH}_4} = 6 \text{ (calc)} = 6.00 \text{ moles O}_2 \text{ (corr)}$

b) $3.00 \text{ moles PCl}_3 \times \dfrac{3 \text{ moles H}_2\text{O}}{1 \text{ mole PCl}_3} = 9 \text{ (calc)} = 9.00 \text{ moles H}_2\text{O (corr)}$

c) $3.00 \text{ moles NaOH} \times \dfrac{1 \text{ mole } H_3PO_4}{3 \text{ moles NaOH}} = 1 \text{ (calc)} = 1.00 \text{ moles } H_3PO_4 \text{ (corr)}$

d) $3.00 \text{ moles NaClO}_2 \times \dfrac{1 \text{ mole Cl}_2}{2 \text{ moles NaClO}_2} = 1.5 \text{ (calc)} = 1.50 \text{ moles Cl}_2 \text{ (corr)}$

10.129 $75.0 \text{ g }(NH_4)_2Cr_2O_7 \times \dfrac{1 \text{ mole }(NH_4)_2Cr_2O_7}{252.10 \text{ g }(NH_4)_2Cr_2O_7} \times \dfrac{1 \text{ mole } N_2}{1 \text{ mole }(NH_4)_2Cr_2O_7} \times \dfrac{28.02 \text{ g } N_2}{1 \text{ mole } N_2}$

$= 8.3359778 \text{ (calc)} = 8.34 \text{ g } N_2 \text{ (corr)}$

$75.0 \text{ g }(NH_4)_2Cr_2O_7 \times \dfrac{1 \text{ mole }(NH_4)_2Cr_2O_7}{252.10 \text{ g }(NH_4)_2Cr_2O_7} \times \dfrac{4 \text{ moles } H_2O}{1 \text{ mole }(NH_4)_2Cr_2O_7} \times \dfrac{18.02 \text{ g } H_2O}{1 \text{ mole } H_2O}$

$= 21.443871 \text{ (calc)} = 21.4 \text{ g } H_2O \text{ (corr)}$

$75.0 \text{ g }(NH_4)_2Cr_2O_7 \times \dfrac{1 \text{ mole }(NH_4)_2Cr_2O_7}{252.10 \text{ g }(NH_4)_2Cr_2O_7} \times \dfrac{1 \text{ mole } Cr_2O_3}{1 \text{ mole }(NH_4)_2Cr_2O_7} \times \dfrac{152.00 \text{ g } Cr_2O_3}{1 \text{ mole } Cr_2O_3}$

$= 45.220151 \text{ (calc)} = 45.2 \text{ g } Cr_2O_3 \text{ (corr)}$

10.131 Obtain a conversion factor with grams of product and grams of H_2S by finding the mass of products that could be made from 1.000 g H_2S:

$1.000 \text{ g } H_2S \times \dfrac{1 \text{ mole } H_2S}{34.09 \text{ g } H_2S} \times \dfrac{2 \text{ moles } SO_2}{2 \text{ moles } H_2S} \times \dfrac{64.07 \text{ g } SO_2}{1 \text{ mole } SO_2}$

$= 1.879436785 \text{ (calc)} = 1.879 \text{ g } SO_2 \text{ (corr)}$ and

$1.000 \text{ g } H_2S \times \dfrac{1 \text{ mole } H_2S}{34.09 \text{ g } H_2S} \times \dfrac{2 \text{ moles } H_2O}{2 \text{ moles } H_2S} \times \dfrac{18.02 \text{ g } H_2O}{1 \text{ mole } H_2O}$

$= 0.52860076 \text{ (calc)} = 0.5286 \text{ g } H_2O \text{ (corr)}$

Total: $1.879 \text{ g} + 0.5286 = 2.4076 \text{ (calc)} = 2.408 \text{ g product (corr)}$

The conversion factor is $\dfrac{2.408 \text{ g product}}{1.000 \text{ g } H_2S}$

$100.0 \text{ g product} \times \dfrac{1.000 \text{ g } H_2S}{2.408 \text{ g product}} = 41.5282392 \text{ (calc)} = 41.53 \text{ g } H_2S \text{ (corr)}$

10.133 $2.00 \text{ moles } KNO_2 \times \dfrac{4 \text{ moles NO}}{3 \text{ moles } KNO_2} = 2.6666666 \text{ (calc)} = 2.67 \text{ moles NO (corr)}$

$2.00 \text{ moles } KNO_3 \times \dfrac{4 \text{ moles NO}}{1 \text{ mole } KNO_3} = 8 \text{ (calc)} = 8.00 \text{ moles NO (corr)}$

$2.00 \text{ moles } Cr_2O_3 \times \dfrac{4 \text{ moles NO}}{1 \text{ mole } Cr_2O_3} = 8 \text{ (calc)} = 8.00 \text{ moles NO (corr)}$ KNO_2 is the limiting reactant.

$2.67 \text{ moles NO} \times \dfrac{30.01 \text{ g NO}}{1 \text{ mole NO}} = 80.1267 \text{ (calc)} = 80.1 \text{ g NO (corr)}$

10.135 $1.33 \text{ g Cu} \times \dfrac{1 \text{ mole Cu}}{63.55 \text{ g Cu}} \times \dfrac{1 \text{ mole } CuSO_4}{1 \text{ mole Cu}} \times \dfrac{159.62 \text{ g } CuSO_4}{1 \text{ mole } CuSO_4} = 3.34059166 \text{ (calc)}$

$= 3.34 \text{ g } CuSO_4 \text{ (corr)}$ and % $CuSO_4 = \dfrac{3.34 \text{ g}}{7.53 \text{ g}} \times 100 = 44.35591 \text{ (calc)} = 44.4\% \text{ (corr)}$

Chemical Calculations Involving Chemical Equations

10.137 $113.4 \text{ g } I_2O_5 \times \dfrac{1 \text{ mole } I_2O_5}{333.80 \text{ g } I_2O_5} \times \dfrac{12 \text{ moles } IF_5}{6 \text{ moles } I_2O_5} = 0.67944877 \text{ (calc)} = 0.6794 \text{ mole } IF_5 \text{ (corr)}$

$132.2 \text{ g } BrF_3 \times \dfrac{1 \text{ mole } BrF_3}{136.90 \text{ g } BrF_3} \times \dfrac{12 \text{ moles } IF_5}{20 \text{ moles } BrF_3} = 0.57940102 \text{ (calc)} = 0.5794 \text{ mole } IF_5 \text{ (corr)}$

BrF_3 is the limiting reactant. Theoretical yield of IF_5:

$0.5794 \text{ mole } IF_5 \times \dfrac{221.9 \text{ g } IF_5}{1 \text{ mole } IF_5} = 128.56886 \text{ (calc)} = 128.6 \text{ g } IF_5 \text{ (corr)}$

% yield $IF_5 = \dfrac{97.0 \text{ g}}{128.6 \text{ g}} \times 100 = 75.4276835443 \text{ (calc)} = 75.4\% \text{ (corr)}$

10.139 Mass of $NaHCO_3$ present:

$0.873 \text{ g } H_2O \times \dfrac{1 \text{ mole } H_2O}{18.02 \text{ g } H_2O} \times \dfrac{2 \text{ moles } NaHCO_3}{1 \text{ mole } H_2O} \times \dfrac{84.01 \text{ g } NaHCO_3}{1 \text{ mole } NaHCO_3}$

$= 8.139925638 \text{ (calc)} = 8.14 \text{ g } NaHCO_3 \text{ (corr)}$

Mass of $CaCO_3$ present: $(13.20 - 8.14) \text{ g } CaCO_3 = 5.06 \text{ g } CaCO_3$ (calc and corr)

% $CaCO_3$: $\dfrac{5.06 \text{ g}}{13.20 \text{ g}} \times 100 = 38.3333 \text{ (calc)} = 38.3\% \text{ (corr)}$

Cumulative Problems

10.141 a) $Zn + 2 AgNO_3 \rightarrow Zn(NO_3)_2 + 2 Ag$ b) $HCl + NaOH \rightarrow NaCl + H_2O$
c) $PCl_3 + Cl_2 \rightarrow PCl_5$ d) $2 Cu + O_2 \rightarrow 2 CuO$

10.143 Since the oxygen is balanced with 18 atoms on each side of the equation, the compound cyclopropane contains only C and H: $2 C_xH_y + 9 O_2 \rightarrow 6 CO_2 + 6 H_2O$
C balance: $2x = 6; x = 3$ H balance: $2y = 6(2); y = 6$
∴ Cyclopropane = C_3H_6

10.145 $Sn + 2 HF \rightarrow SnF_2 + H_2$ Theoretical yield:

$30.0 \text{ g Sn} \times \dfrac{1 \text{ mole Sn}}{118.71 \text{ g Sn}} \times \dfrac{1 \text{ mole } H_2}{1 \text{ mole Sn}} \times \dfrac{2.02 \text{ g } H_2}{1 \text{ mole } H_2} = 0.5104877432 \text{ (calc)} = 0.510 \text{ g } H_2 \text{ (corr)}$

$2.50 \text{ L } H_2 \times \dfrac{0.090 \text{ g } H_2}{1 \text{ L } H_2 \text{ gas}} = 0.225 \text{ (calc)} = 0.22 \text{ g } H_2 \text{ gas (corr)}$

Percent yield $= \dfrac{0.22 \text{ g } H_2 \text{ gas}}{0.510 \text{ g } H_2 \text{ gas}} \times 100 = 43.1372549 \text{ (calc)} = 43\% \text{ yield (corr)}$

10.147 Take 100.00 g of the copper compound

$100.00 \text{ g compound} \times \dfrac{88.82 \text{ g Cu}}{100.0 \text{ g compound}} \times \dfrac{1 \text{ mole Cu}}{63.55 \text{ g Cu}} = 1.397639654 \text{ (calc)}$

$= 1.398 \text{ moles Cu (corr)}$

100.00 g compound − 88.82 g Cu = 11.18 g O (calc and corr)

$11.18 \text{ g O} \times \dfrac{1 \text{ mole O}}{16.00 \text{ g O}} = 0.69875$ (calc) = 0.6988 mole O (corr)

Cu: $\dfrac{1.398}{0.6988} = 2.000$ O: $\dfrac{0.6988}{0.6988} = 1.000$ Formula = Cu_2O.

$2\ Cu_2S + 3\ O_2 \rightarrow 2\ Cu_2O + 2\ SO_2$

10.149 Empirical formula mass of CH = 13.02 amu

$\dfrac{78.12 \text{ amu}}{13.02 \text{ amu}} = 6.000$ ∴ the molecular formula = $(CH)_6 = C_6H_6$

$2\ C_6H_6 + 15\ O_2 \rightarrow 12\ CO_2 + 6\ H_2O$

10.151 Since five significant figures are wanted in part b), work part a) using 5 significant figures and then round as needed for parts b, c, and d.

a) $50.000 \text{ g Be} \times \dfrac{1 \text{ mole Be}}{9.0122 \text{ g Be}} \times \dfrac{1 \text{ mole BeF}_2}{1 \text{ mole Be}} \times \dfrac{47.012 \text{ g BeF}_2}{1 \text{ mole BeF}_2} = 260.824216179696$ (calc)

$= 261 \text{ g BeF}_2$ (corr)

b) using calculator answer in part a), rounded to 5 significant figures:

$260.82 \text{ g BeF}_2 \times \dfrac{1 \text{ kg}}{1 \times 10^3 \text{ g}} = 0.26082 \text{ kg BeF}_2$ (calc and corr)

c) using calculator answer in part a), rounded to 4 significant figures:

$260.8 \text{ g BeF}_2 \times \dfrac{1 \text{ μg}}{1 \times 10^{-6} \text{ g}} = 2.608 \times 10^8 \text{ μg BeF}_2$ (calc and corr)

d) using calculator answer in part a), rounded to 4 significant figures:

$260.8 \text{ g BeF}_2 \times \dfrac{1 \text{ lb}}{453.6 \text{ g}} = 0.5749559083$ (calc) = 0.5750 lb BeF_2 (corr)

10.153 There are five chlorine-containing product molecules or formula units: 2(KCl) + 2(ClO_2) + 1 Cl_2
Determine the limiting reactant

$100.0 \text{ g KClO}_3 \times \dfrac{1 \text{ mole KClO}_3}{122.55 \text{ g KClO}_3} \times \dfrac{2 \text{ moles H}_2O}{2 \text{ moles KClO}_3} = 0.81599347$ (calc) = 0.8160 mole H_2O (corr)

$200.0 \text{ g HCl} \times \dfrac{1 \text{ mole HCl}}{36.46 \text{ g HCl}} \times \dfrac{2 \text{ moles H}_2O}{4 \text{ moles HCl}} = 2.7427317$ (calc) = 2.743 moles H_2O (corr)

$KClO_3$ is the limiting reactant.

$0.8160 \text{ mole H}_2O \times \dfrac{5 \text{ moles chlorine-containing products}}{2 \text{ moles H}_2O} = 2.04$ (calc)

$= 2.040$ moles chlorine-containing products (corr)

10.155 $10.0 \text{ g Al} \times \dfrac{1 \text{ mole Al}}{26.98 \text{ g Al}} \times \dfrac{1 \text{ mole AlCl}_3}{3 \text{ moles Al}} \times \dfrac{[13 + 3(17)] \text{ moles electrons}}{1 \text{ mole AlCl}_3}$

$= 7.9070917$ (calc) = 7.91 moles electrons (corr)

Chemical Calculations Involving Chemical Equations

10.157 For the NaCl produced:

$$500.0 \text{ g CaCl}_2 \times \frac{1 \text{ mole CaCl}_2}{110.98 \text{ g CaCl}_2} \times \frac{2 \text{ moles NaCl}}{1 \text{ mole CaCl}_2} \times \frac{1 \text{ mole pos. ions}}{1 \text{ mole NaCl}}$$

$$\times \frac{6.022 \times 10^{23} \text{ pos. ions}}{1 \text{ mole pos. ions}} = 5.4262029 \times 10^{24} \text{ (calc)} = 5.426 \times 10^{24} \text{ pos. ions (corr)}$$

For the CaCO$_3$ produced:

$$500.0 \text{ g CaCl}_2 \times \frac{1 \text{ mole CaCl}_2}{110.98 \text{ g CaCl}_2} \times \frac{1 \text{ mole CaCO}_3}{1 \text{ mole CaCl}_2} \times \frac{1 \text{ mole pos. ions}}{1 \text{ mole CaCO}_3} \times \frac{6.022 \times 10^{23} \text{ pos. ions}}{1 \text{ mole pos. ions}}$$

$$= 2.7131015 \times 10^{24} \text{ (calc)} = 2.713 \times 10^{24} \text{ pos. ions (corr)}$$

Total = $[5.426 \times 10^{24} + 2.713 \times 10^{24}] = 8.139 \times 10^{24}$ positive ions (calc and corr)

10.159 $10.0 \text{ g AgBr} \times \dfrac{1 \text{ mole AgBr}}{187.77 \text{ g AgBr}} \times \dfrac{1 \text{ mole NaBr}}{1 \text{ mole AgBr}} \times \dfrac{102.89 \text{ g NaBr}}{1 \text{ mole NaBr}} \times \dfrac{100 \text{ g solution}}{6.00 \text{ g NaBr}}$

$$\times \frac{1 \text{ mL solution}}{1.046 \text{ g solution}} = 87.310008 \text{ (calc)} = 87.3 \text{ mL solution (corr)}$$

10.161 % yield can be expressed as either $\dfrac{\text{g (actual)}}{\text{g (theoretical)}} \times 100$ or $\dfrac{\text{moles (actual)}}{\text{moles (theoretical)}} \times 100$

$$75.0 \text{ mL NH}_3 \times \frac{10^{-3} \text{ L NH}_3}{1 \text{ mL NH}_3} \times \frac{0.695 \text{ g NH}_3}{1 \text{ L NH}_3} \times \frac{1 \text{ mole NH}_3}{17.04 \text{ g NH}_3} \times \frac{4 \text{ moles NO (theoretical)}}{4 \text{ moles NH}_3}$$

$$\times \frac{0.852 \text{ mole NO (actual)}}{1 \text{ mole NO (theoretical)}} \times \frac{2 \text{ moles NO}_2 \text{ (theoretical)}}{2 \text{ moles NO (actual)}} \times \frac{0.827 \text{ mole NO}_2 \text{ (actual)}}{1 \text{ mole NO}_2 \text{ (theoretical)}}$$

$$\times \frac{2 \text{ moles HNO}_3 \text{ (theoretical)}}{3 \text{ moles NO}_2 \text{ (actual)}} \times \frac{0.870 \text{ mole HNO}_3 \text{ (actual)}}{1 \text{ mole HNO}_3 \text{ (theoretical)}} \times \frac{63.02 \text{ g HNO}_3 \text{ (actual)}}{1 \text{ mole HNO}_3 \text{ (actual)}}$$

$$= 0.0787821764 \text{ (calc)} = 0.0788 \text{ g HNO}_3 \text{ (corr)}$$

10.163 The balanced equations are $S + O_2 \rightarrow SO_2$ and $CaO + SO_2 \rightarrow CaSO_3$.

$$1.0 \text{ ton coal} \times \frac{2000 \text{ lb coal}}{1 \text{ ton coal}} \times \frac{453.6 \text{ g coal}}{1 \text{ lb coal}} \times \frac{4.3 \text{ g S}}{100 \text{ g coal}} \times \frac{1 \text{ mole S}}{32.07 \text{ g S}}$$

$$\times \frac{1 \text{ mole CaSO}_3}{1 \text{ mole SO}_2} \times \frac{120.15 \text{ g CaSO}_3}{1 \text{ mole CaSO}_3} \times \frac{1 \text{ lb CaSO}_3}{453.6 \text{ g CaSO}_3} \times \frac{1 \text{ ton CaSO}_3}{2000 \text{ lb CaSO}_3}$$

$$= 0.1610991581 \text{ (calc)} = 0.16 \text{ ton CaSO}_3 \text{ (corr)}$$

Answers to Multiple-Choice Practice Test

10.165	c	10.166	c	10.167	e	10.168	e	10.169	d	10.170	d
10.171	b	10.172	e	10.173	a	10.174	d	10.175	a	10.176	b
10.177	d	10.178	b	10.179	c	10.180	c	10.181	d	10.182	b
10.183	e	10.184	c								

CHAPTER ELEVEN
States of Matter

Practice Problems

Physical States of Matter (Secs. 11.1 and 11.2)

11.1 a) gaseous   b) liquid   c) gaseous   d) gaseous

11.3 a) both gases   b) liquid, solid   c) both gases   d) both solids

11.5 a) 3, H_2, N_2, Ne   b) 2, Br_2, Hg

Kinetic Molecular Theory of Matter (Secs. 11.3–11.6)

11.7 a) potential   b) kinetic   c) potential   d) potential

11.9 a) direct; the average velocity increases as the temperature increases and vice versa
b) potential (attractive)
c) direct; the higher the temperature, the higher the disruptive forces
d) all three; the disruptive forces are predominant in gases and the cohesive forces are more dominant in solids

11.11 a) solid   b) liquid   c) solid   d) solid or liquid

11.13 a) The predominant cohesive forces in the solid hold the particles in essentially fixed position.
b) The gas particles are widely separated (disruptive forces). The solid and liquid particles have very little space between them (cohesive forces). The space between the particles can be decreased greatly in gases, but not in solids or liquids.
c) The cohesive forces are dominant enough that changing the temperature has only a small effect on the space between particles.
d) The disruptive forces in a gas are so dominant that each particle can act independently of the others.

11.15 a) liquid   b) liquid, gas   c) solid

Physical Changes of State (Sec. 11.8)

11.17 a) Evaporation is endothermic, freezing is exothermic, not the same thermicity.
b) Melting is endothermic, deposition is exothermic, not the same thermicity.
c) Freezing is exothermic, condensation is exothermic, same thermicity.
d) Sublimation is endothermic, evaporation is endothermic, same thermicity.

11.19 a) different, solid, liquid   b) different, liquid, solid   c) different, solid, liquid   d) same, gases

11.21 a) not opposite   b) not opposite   c) not opposite   d) not opposite

Heat Energy and Specific Heat (Sec. 11.9)

11.23 a) $2.0 \text{ calories} \times \dfrac{4.184 \text{ joules}}{1 \text{ calorie}} = 8.368 \text{ (calc)} = 8.4 \text{ joules (corr)} \quad \therefore 2.0 \text{ calories} > 2.0 \text{ joules}$

b) $1.0 \text{ kilocalorie} \times \dfrac{10^3 \text{ calories}}{1 \text{ kilocalorie}} = 1000 \text{ (calc)} = 1.0 \times 10^3 \text{ calories (corr)}$

$\therefore 1.0 \text{ kilocalorie} > 92 \text{ calories}$

c) $100 \text{ Calorie} \times \dfrac{10^3 \text{ calories}}{1 \text{ Calorie}} = 100{,}000 \text{ (calc)} = 1.0 \times 10^5 \text{ calories (corr)}$

$\therefore 100 \text{ Calories} > 100 \text{ calories}$

d) $2.3 \text{ Calories} \times \dfrac{10^3 \text{ calories}}{1 \text{ Calorie}} \times \dfrac{1 \text{ kilocalorie}}{10^3 \text{ calories}} = 2.3 \text{ kilocalories (calc and corr)}$

$\therefore 2.3 \text{ Calories} < 1000 \text{ kilocalories}$

11.25 a) $2290 \text{ kJ} \times \dfrac{10^3 \text{ J}}{1 \text{ kJ}} = 2.29 \times 10^6 \text{ J (calc and corr)}$

b) $2290 \text{ kJ} \times \dfrac{1 \text{ kcal}}{4.184 \text{ kJ}} = 547.3231358 \text{ (calc)} = 547 \text{ kcal (corr)}$

c) $2290 \text{ kJ} \times \dfrac{1 \text{ kcal}}{4.184 \text{ kJ}} \times \dfrac{10^3 \text{ cal}}{1 \text{ kcal}} = 5.473231358 \times 10^5 \text{ (calc)} = 5.47 \times 10^5 \text{ cal (corr)}$

d) $2290 \text{ kJ} \times \dfrac{1 \text{ kcal}}{4.184 \text{ kJ}} \times \dfrac{1 \text{ Cal}}{1 \text{ kcal}} = 547.3231358 \text{ (calc)} = 547 \text{ Cal (corr)}$

11.27 Use values of specific heat from Table 11.2, $\Delta T = (35.0 - 25.0)°C = 10.0°C$

a) $\dfrac{0.382 \text{ J}}{\text{g °C}} \times 40.0 \text{ g} \times 10.0°C = 152.8 \text{ (calc)} = 153 \text{ J (corr)}$

b) $\dfrac{0.444 \text{ J}}{\text{g °C}} \times 40.0 \text{ g} \times 10.0°C = 177.6 \text{ (calc)} = 178 \text{ J (corr)}$

c) $\dfrac{1.0 \text{ J}}{\text{g °C}} \times 40.0 \text{ g} \times 10.0°C = 400 \text{ (calc)} = 4.0 \times 10^2 \text{ J (corr)}$

d) $\dfrac{2.42 \text{ J}}{\text{g °C}} \times 40.0 \text{ g} \times 10.0°C = 968 \text{ J (calc and corr)}$

11.29 Using values of specific heat from Table 11.2, 7.73 g of substance, above 43.2°C.

a) $\dfrac{145 \text{ J}}{0.908 \dfrac{\text{J}}{\text{g °C}} \times 7.73 \text{ g}} = 20.6586843 \text{ (calc)} = 20.7°C \text{ (corr) above } 43.2°C = 63.9°C$

b) $\dfrac{-145 \text{ J}}{0.908 \dfrac{\text{J}}{\text{g °C}} \times 7.73 \text{ g}} = -20.6586843 \text{ (calc)} = 20.7°C \text{ (corr) below } 43.2°C = 22.5°C$

c) $\dfrac{325 \text{ J}}{0.908 \dfrac{\text{J}}{\text{g °C}} \times 7.73 \text{ g}} = 46.30394766 \text{ (calc)} = 46.3°C \text{ (corr) above } 43.2°C = 89.5°C$

d) $\dfrac{-525 \text{ J}}{0.908 \dfrac{\text{J}}{\text{g °C}} \times 7.73 \text{ g}} = -74.79868468 \text{ (calc)} = 74.8°C \text{ (corr) below } 43.2°C = -31.6°C$

States of Matter

11.31 Use values of specific heat from Table 11.2, $\Delta T = (95 - 20.0)°C = 75°C$, absorbs 371 J of energy.

a) $g = \dfrac{371 \text{ J}}{0.444 \dfrac{\text{J}}{\text{g °C}} \times 75°C} = 11.1411411$ (calc) = 11 g Fe (corr)

b) $g = \dfrac{371 \text{ J}}{0.388 \dfrac{\text{J}}{\text{g °C}} \times 75°C} = 12.74914089$ (calc) = 13 g Zn (corr)

c) $g = \dfrac{371 \text{ J}}{4.18 \dfrac{\text{J}}{\text{g °C}} \times 75°C} = 1.1834131078$ (calc) = 1.2 g water (corr)

d) $g = \dfrac{371 \text{ J}}{2.42 \dfrac{\text{J}}{\text{g °C}} \times 75°C} = 2.044077135$ (calc) = 2.0 g ethyl alcohol (corr)

11.33

a) $\dfrac{39.8 \text{ J}}{40.0 \text{ g} \times 3.0°C} = 0.3316666$ (calc) = $0.33 \dfrac{\text{J}}{\text{g °C}}$ (corr)

b) $\dfrac{46.9 \text{ J}}{40.0 \text{ g} \times 3.0°C} = 0.3908333$ (calc) = $0.39 \dfrac{\text{J}}{\text{g °C}}$ (corr)

c) $\dfrac{57.8 \text{ J}}{40.0 \text{ g} \times 3.0°C} = 0.4816666$ (calc) = $0.48 \dfrac{\text{J}}{\text{g °C}}$ (corr)

d) $\dfrac{75.0 \text{ J}}{40.0 \text{ g} \times 3.0°C} = 0.625$ (calc) = $0.62 \dfrac{\text{J}}{\text{g °C}}$ (corr)

11.35

a) $40.0 \text{ g Zn} \times \dfrac{0.388 \text{ J}}{\text{g °C}} = 15.52$ (calc) = $15.5 \dfrac{\text{J}}{°C}$ (corr)

b) $40.0 \text{ g Al} \times \dfrac{0.908 \text{ J}}{\text{g °C}} = 36.32$ (calc) = $36.3 \dfrac{\text{J}}{°C}$ (corr)

c) $40.0 \text{ g water} \times \dfrac{4.18 \text{ J}}{\text{g °C}} = 167.2$ (calc) = $167 \dfrac{\text{J}}{°C}$ (corr)

d) $40.0 \text{ g NaCl} \times \dfrac{0.88 \text{ J}}{\text{g °C}} = 35.2$ (calc) = $35 \dfrac{\text{J}}{°C}$ (corr)

11.37 Heat capacity of 40.0 g gold = 40.0 g × 0.13 J/g °C = 5.2 J/°C (calc and corr)
Heat capacity of 80.0 g copper = 80.0 g × 0.382 J/g °C = 30.56 (calc) = 30.6 J/°C (corr)
∴ heat capacity of 80.0 g copper is higher.

11.39 Specific heat = $\dfrac{28.7 \dfrac{\text{J}}{°C}}{75.1 \text{ g}} = 0.3821571238$ (calc) = $0.382 \dfrac{\text{J}}{\text{g °C}}$ (corr)

11.41 Heat gained (copper) = −Heat lost (water)
Heat lost by the water is equal in magnitude, but opposite in sign to the heat gained by the copper.

$$0.382 \frac{J}{g\,°C} \times X\,g \times 20°C = -4.18 \frac{J}{g\,°C} \times 1.0\,g \times -85°C$$

Solving for X: $X = \dfrac{4.18 \frac{J}{g\,°C} \times 1.0\,g \times 85°C}{0.382 \frac{J}{g\,°C} \times 20°C} = 46.505236\ (calc) = 47\,g\ (corr)$

Energy and Changes of State (Sec. 11.11)

11.43 a) heat of solidification   b) heat of condensation   c) heat of fusion   d) heat of vaporization

11.45 a) $50.0\,g\,Al \times \dfrac{393\,J}{g\,Al} = -19650\ (calc) = -1.96 \times 10^4\,J$ released (corr)

b) $50.0\,g\,steam \times \dfrac{2260\,J}{g\,steam} = -113{,}000\ (calc)\ or\ -1.13 \times 10^5\,J$ released (corr)

c) $50.0\,g\,Cu \times \dfrac{205\,J}{g\,Cu} = 10{,}250\ (calc) = 1.02 \times 10^4\,J$ absorbed (corr)

d) $50.0\,g\,H_2O \times \dfrac{2260\,J}{g\,H_2O} = 113{,}000\ (calc)\ or\ 1.13 \times 10^5\,J$ absorbed (corr)

11.47 The same amount, −3075 J

11.49 $\dfrac{1.00\ mole\ Na_2SO_4 \times 80.93 \frac{kJ}{mole}\ Na_2SO_4}{1.00\ mole\ NaOH \times 15.79 \frac{kJ}{mole}\ NaOH} = 5.1253958\ (calc) = 5.13$ times as much (corr)

11.51 $432\,kJ \times \dfrac{1\ mole\ methane}{10.4\,kJ} = 41.53846154\ (calc) = 41.5$ mole methane (corr)

11.53 Experiment 1 Heat of fusion = $\dfrac{1273\,J}{6.21\,g} = 204.9919847\ (calc) = 205\ \dfrac{J}{g}$ (corr)

Experiment 2 Heat of fusion = $\dfrac{914\,J}{4.46\,g} = 204.932735426\ (calc) = 205\ \dfrac{J}{g}$ (corr)

Both experiments give the same Heat of fusion, $205\ \dfrac{J}{g}$ = Heat of fusion of copper from Table 11.3.

11.55 $7.00\,g\,Na \times \dfrac{1\ mole\ Na}{22.99\,g\,Na} \times \dfrac{2.40\,kJ}{1\ mole\ Na} \times \dfrac{10^3\,J}{1\,kJ} = 730.7525\ (calc) = 7.31 \times 10^2\,J$ (corr)

States of Matter

Heat Energy Calculations (Sec. 11.12)

11.57

[Graph: Temperature (°C) vs Heat added. Y-axis marks at −100, −23, 77, 100. Curve rises from −100 to −23 (Solid), plateaus at −23 (Melting), rises to 77 (Liquid), plateaus at 77 (Evaporation), rises (Gas).]

11.59 specific heat ice = 2.09 J/g °C   heat of fusion = 334 J/g   specific heat steam = 2.03 J/g °C
specific heat water = 4.18 J/g °C   heat of vaporization = 2260 J/g

a) Step 1: Heat ice from −18°C to ice at −3°C.
   Q = 52.0 g ice × 2.09 J/g °C × 15°C = 1630.2 (calc) = 1.6 × 10³ J (corr)

b) Step 1: Heat ice from −18°C to 0°C.
   Q = 52.0 g × 2.09 J/g °C × 18°C = 1956.24 (calc) = 2.0 × 10³ J (corr)
   Step 2: Melt ice at 0°C to water at 0°C.
   Q = 52.0 g × 334 J/g = 17368 (calc) = 17.4 × 10³ J (corr)
   Step 3: Heat water from 0°C to 37°C.
   Q = 52.0 g × 4.18 J/g °C × 37°C = 8042.32 (calc) = 8.04 × 10³ J (corr)
   Q total = Q ice + Q melting + Q water
   = 2.0 × 10³ J + 17.4 × 10³ J + 8.04 × 10³ J = 27.44 × 10³ J (calc)
   = 27.4 × 10³ J (corr)

c) Step 1: Heat ice from −18°C to ice at 0°C.
   Q = 52.0 g × 2.09 J/g °C × 18°C = 1956.24 (calc) = 2.0 × 10³ J (corr)
   Step 2: Melt ice at 0°C to water at 0°C.
   Q = 52 g × 334 J/g = 17368 (calc) = 17.4 × 10³ J (corr)
   Step 3: Heat water from 0°C to 100°C.
   52.0 g H_2O × 4.18 J/g °C × 100°C = 21,736 (calc) = 21.7 × 10³ J (corr)
   Q total = Q ice + Q melting + Q water
   = 2.0 × 10³ J + 17.4 × 10³ J + 21.7 × 10³ J = 41.1 × 10³ J (calc and corr)

d) Step 1: Heat ice from −18°C to ice at 0°C.
   Q = 52.0 g × 2.09 J/g °C × 18°C = 1956.24 (calc) = 2.0 × 10³ J (corr)
   Step 2: Melt ice at 0°C to water at 0°C.
   Q = 52 g × 334 J/g = 17368 (calc) = 17.4 × 10³ J (corr)
   Step 3: Heat water from 0°C to 100°C.
   52.0 g H_2O × 4.18 J/g °C × 100°C = 21,736 (calc) = 21.7 × 10³ J (corr)
   Step 4: Vaporize water at 100°C.
   52.0 g × 2260 J/g = 117,520 (calc) = 118 × 10³ J (corr)
   Step 5: Heat steam from 100°C to 115°C.
   52.0 g × 2.03 J/g °C × 15.°C = 1583.4 (calc) = 1.6 × 10³ J (corr)
   Q total = Q ice + Q melting + Q water + Q vap + Q steam
   = 2.0 × 10³ J + 17.4 × 10³ J + 21.7 × 10³ J + 118 × 10³ J + 1.6 × 10³ J
   = 160.7 × 10³ (calc) = 161 × 10³ J (corr)

11.61 a) Step 1: Melt solid at 1530°C.
35.2 g × 247 J/g = 8694.4 (calc) = 8.69 × 10³ J (corr)
b) Step 1: Heat solid at 25°C below melting point.
35.2 g × 0.444 J/g °C × 25°C = 390.72 (calc) = 0.39 × 10³ J (corr)
Step 2: Melt solid at 1530°C.
35.2 g × 247 J/g = 8694.4 (calc) = 8.69 × 10³ J (corr)
Q total = 0.39 × 10³ J + 8.69 × 10³ J = 9.08 × 10³ J (calc and corr)
c) Step 1: Heat solid from 1120°C to 1530°C.
35.2 g × 0.444 J/g °C × 410.°C = 6407.808 (calc) = 6.41 × 10³ J (corr)
Step 2: Melt solid at 1530°C.
35.2 g × 247 J/g = 8694.4 (calc) = 8.69 × 10³ J (corr)
Q total = 6.41 × 10³ J + 8.69 × 10³ J = 15.10 × 10³ J (calc and corr)
d) Step 1: Heat solid from 27°C to 1530°C.
35.2 g × 0.444 J/g °C × 1503.°C = 23490.0864 (calc) = 23.4 × 10³ J (corr)
Step 2: Melt solid at 1530°C.
35.2 g × 247 J/g = 8694.4 (calc) = 8.69 × 10³ J (corr)
Q total = 23.4 × 10³ J + 8.69 × 10³ J = 32.09 × 10³ J (calc) = 32.1 × 10³ J (corr)

11.63 Step 1: solid (−135°C) → solid (−117°C)
15.0 g × 18°C × 0.97 J/g °C = 261.9 (calc) = 2.6 × 10² J (corr)
Step 2: solid (−117°C) → liquid (−117°C)
15.0 g × 109 J/g = 1635 J (calc) = 1.64 × 10³ J (corr)
Step 3: liquid (−117°C) → liquid (78°C)
15.0 g × 195°C × 2.3 J/g °C = 6727.5 (calc) = 6.7 × 10³ J (corr)
Step 4: liquid (78°C) → gas (78°C)
15.0 g × 837 J/g = 12,555 (calc) = 1.26 × 10⁴ J (corr)
Step 5: gas (78°C) → gas (95°C)
15.0 g × 17°C × 0.95 J/g °C = 242.25 J (calc) = 2.40 × 10² J (corr)
Total = 2.6 × 10² J + 1.64 × 10³ J + 6.7 × 10³ J + 1.26 × 10⁴ J + 2.40 × 10² J
= 21,440 (calc) = 2.14 × 10⁴ J (corr)

Properties of Liquids (Secs. 11.13–11.15)

11.65 a) boiling point   b) vapor pressure   c) boiling   d) boiling point

11.67 a) Increasing the temperature increases the average kinetic energy of the particles, enabling more molecules to evaporate.
b) The boiling point is lower; reactions occur more slowly at lower temperatures.
c) The boiling point is higher; reactions occur more rapidly at higher temperatures.
d) The particles leaving have higher than average kinetic energy. The particles remaining as liquid have a lower average kinetic energy and a lower temperature.

11.69 a) increase
b) no change—the rate of evaporation depends on the temperature, the cohesive forces present in the liquid and the surface area where evaporation can occur. The concentration of air molecules above the liquid has a slight effect on the rate at which condensation of the vapor can occur. The *net* evaporation rate, the evaporation rate minus the condensation rate, may increase slightly at higher elevations, but the absolute rate of evaporation would not change.
c) increase—the liquid surface area has increased
d) no change

11.71 a) no change b) decrease c) no change d) no change

11.73 a) increase
b) no change—the rate of evaporation depends on the temperature the cohesive forces present in the liquid, and the surface area where evaporation can occur. The concentration of air molecules above the liquid has a slight effect on the rate at which condensation of the vapor can occur. The *net* evaporation rate, the evaporation rate minus the condensation rate, may increase slightly at higher elevations, but the absolute rate of evaporation would not change. This will make the vapor pressure increase very slightly.
c) no change
d) no change

11.75 B must have lower cohesive forces between particles than A to make B evaporate faster.

11.77 In comparing substances, at the same temperature, the substance with the highest vapor pressure is the most volatile. Therefore, CS_2 is more volatile.

Intermolecular Forces in Liquids (Sec. 11.16)

11.79 Since dipole–dipole interactions occur between polar molecules, polar molecules must be present.

11.81 The stronger the intermolecular forces, the higher the boiling point.

11.83 a) London forces   b) London forces
c) dipole–dipole interactions   d) London forces

11.85 a) no hydrogen bonding   b) yes, hydrogen bonding
c) yes, hydrogen bonding   d) no hydrogen bonding

11.87 a) F_2, larger molar mass   b) HF, hydrogen bonding
c) CO, dipole–dipole   d) O_2, larger mass

11.89 a) As larger molecules have greater polarizability, $SiH_4 > CH_4$
b) As larger molecules have greater polarizability, $SiCl_4 > SiH_4$
c) As larger molecules have greater polarizability, $GeBr_4 > SiCl_4$
d) As larger molecules have greater polarizability, $C_2H_4 > N_2$

Hydrogen Bonding and the Properties of Water (Sec. 11.17)

11.91 Each water molecule forms hydrogen bonds to other water molecules. For water to boil, these hydrogen bonds must be broken, allowing the molecules to leave the liquid phase.

11.93 At 4°C, the distance between water molecules maximizes because of "balance" between random motion and hydrogen-bonding effects. Below this temperature, random motion decreases and hydrogen-bonding causes the molecules to move further apart, causing a density decrease.

11.95 The lower density of ice is due to the solid crystal lattice structure, which leaves open spaces between the molecules of H_2O.

11.97 Hydrogen-bonding causes water to have a higher than normal heats of vaporization and condensation. Thus larger amounts of heat are absorbed during evaporation and released during condensation, which produces a temperature-moderating effect.

11.99 A measure of the inward force on the surface of a liquid caused by unbalanced intermolecular forces.

Additional Problems

11.101 a) $SnCl_4$ b) SnI_4 c) SnI_4 d) SnI_4

11.103 $50.0 \text{ g} \times \dfrac{4.18 \text{ J}}{\text{g }°\text{C}} \times 80.0°\text{C} = 16{,}720 \text{ (calc)} = 1.6 \times 10^4 \text{ J (corr)}$

$50.0 \text{ g} \times \dfrac{2260 \text{ J}}{\text{g}} = 113{,}000 \text{ (calc)} = 1.13 \times 10^5 \text{ J (corr)}$

Vaporizing takes longer because it requires more heat energy.

11.105 Let X = grams of ice added
Heat lost by the water is equal in magnitude, but opposite in sign to the heat required to melt the ice.

Heat gained by ice to melt it: $X \text{ g} \times \dfrac{334 \text{ J}}{\text{g}} = 334 \cdot X \text{ J (calc and corr)}$

Heat lost by water: $40.0 \text{ g} \times \dfrac{4.18 \text{ J}}{\text{g }°\text{C}} \times -19.0°\text{C} = 3176.8 \text{ (calc)} = -3180 \text{ J (corr)}$

$-$Heat lost {water} = heat gained {ice}: $3180 \text{ J} = 334 \cdot X \text{ J}$ $X = 9.520958 \text{ (calc)} = 9.52 \text{ g (corr)}$

11.107 Let X = final temperature of water in °C
Heat lost by the water is equal in magnitude but opposite in sign to the heat gained by the ice.
Heat lost by water:
Water (80.0°C) → water (X°C)

$30.0 \text{ g} \times (80.0 - X)°\text{C} \times \dfrac{4.18 \text{ J}}{\text{g }°\text{C}} = (10{,}032 - 125.4 \cdot X) \text{ (calc)} = (10{,}\overline{0}00 - 125 \cdot X) \text{ J (corr)}$

Heat gained by ice:
Step 1: ice (−10°C) → ice (0°C): 10.0 g × 10.0°C × 2.09 J/g °C = 209 J (calc and corr)
Step 2: ice (0°C) → water (0°C): 10.0 g × 334 J/g = 3340 J (calc and corr)
Step 3: water (0°C) → water (X°C): 10.0 g × X°C × 4.18 J/g °C = 41.8 · X J (calc and corr)
Total = (209 J + 3340 J + 41.8 · X J) = ((3349 (calc) or 3550 J (corr)) + 41.8 · X) J (calc and corr)
Heat lost (water) = heat gained (ice): $(10{,}\overline{0}00 - 125 \cdot X) \text{ J} = (3550 + 41.8 \cdot X) \text{ J}$
6450 J = (166.8 · X) J (calc)   6400 J = (167 · X) J (corr)   X = 38.323353 (calc) = 38.3°C (corr)

11.109 Let X = specific heat of metal
Heat lost by the metal is equal in magnitude, but opposite in sign to the heat gained by the water.
Heat lost (metal) = heat gained (water)

$500.0 \text{ g} \times X \dfrac{\text{J}}{\text{g }°\text{C}} \times 26.6°\text{C} = 100.0 \text{ g} \times \dfrac{4.18 \text{ J}}{\text{g }°\text{C}} \times 13.4°\text{C}$

13,300 · X = 5601.2 (calc)   13,300 · X = 56$\overline{0}$0 (corr)   X = 0.42105263 (calc) = 0.421 J/g °C (corr)

11.111 $4.3 \text{ min} \times \dfrac{60 \text{ sec}}{1 \text{ min}} \times \dfrac{136 \text{ J}}{\text{sec}} = 35{,}088 \text{ J (calc)} = 35{,}000 \text{ J (corr)}$

$\dfrac{35{,}000 \text{ J}}{25.3 \text{ g}} = 1383.3992 \text{ (calc)} = 1.4 \times 10^3 \text{ J/g (corr)}$

States of Matter

Cumulative Problems

11.113 $2.50 \text{ moles Ag} \times \dfrac{107.87 \text{ g Ag}}{1 \text{ mole Ag}} \times \dfrac{0.24 \text{ J}}{\text{g °C}} \times 10.0\text{°C} = 647.22 \text{ (calc)} = 6.5 \times 10^2 \text{ J (corr)}$

11.115 $5.7 \text{ L blood} \times \dfrac{1 \text{ mL}}{10^{-3} \text{ L}} \times \dfrac{1.06 \text{ g}}{1 \text{ mL}} = 6042 \text{ (calc)} = 6\bar{0}00 \text{ g blood (corr)}$

heat required $= 4.18 \dfrac{\text{J}}{\text{g °C}} \times 6\bar{0}00 \text{ g} \times 1.0\text{°C} \times \dfrac{1 \text{ kJ}}{10^3 \text{ J}} = 25.08 \text{ (calc)} = 25 \text{ kJ (corr)}$

11.117 specific heat $= \dfrac{59.5 \text{ J}}{35.0 \text{ g} \times 1.0\text{°C}} = 1.7 \dfrac{\text{J}}{\text{g °C}}$ (calc and corr)

The unknown is likely a mixture of A and B since the specific heat is intermediate between that for pure A and that for pure B.

11.119 $\dfrac{10.0 \text{ g}}{1485 \text{ J}} \times \dfrac{1151 \text{ J}}{10.0 \text{ mL}} = 0.77508417 \text{ (calc)} = 0.775 \text{ g/mL (corr)}$

11.121 $\dfrac{1.00 \text{ g}}{602 \text{ J}} \times \dfrac{10^3 \text{ J}}{1 \text{ kJ}} \times \dfrac{18.1 \text{ kJ}}{1 \text{ mole}} = 30.066445 \text{ (calc)} = 30.1 \text{ g/mole (corr)}$ molecular mass $= 30.1$ amu

11.123 $2 \text{ Cu}_2\text{O} \rightarrow 4 \text{ Cu} + \text{O}_2$

$52.0 \text{ g Cu}_2\text{O} \times \dfrac{1 \text{ mole Cu}_2\text{O}}{143.10 \text{ g Cu}_2\text{O}} \times \dfrac{4 \text{ moles Cu}}{2 \text{ moles Cu}_2\text{O}} \times \dfrac{13.0 \text{ kJ}}{1 \text{ mole Cu}} = 9.44794 \text{ (calc)} = 9.45 \text{ kJ (corr)}$

11.125 $6.32 \times 10^{24} \text{ H atoms} \times \dfrac{1 \text{ mole H}}{6.022 \times 10^{23} \text{ H atoms}} \times \dfrac{1 \text{ mole C}_6\text{H}_6}{6 \text{ moles H}} \times \dfrac{78.12 \text{ g C}_6\text{H}_6}{1 \text{ mole C}_6\text{H}_6}$

$\times \dfrac{1.74 \text{ J}}{\text{g(C}_6\text{H}_6)\text{°C}} \times 10\text{°C} = 2377.5878 \text{ (calc)} = 2.38 \times 10^3 \text{ J (corr)}$

Answers to Multiple-Choice Practice Test

11.127 a	**11.128** c	**11.129** a	**11.130** e	**11.131** a	**11.132** e
11.133 c	**11.134** d	**11.135** c	**11.136** d	**11.137** c	**11.138** a
11.139 d	**11.140** d	**11.141** b	**11.142** e	**11.143** d	**11.144** d
11.145 b	**11.146** d				

CHAPTER TWELVE
Gas Laws

Practice Problems

Measurement of Pressure (Sec. 12.2)

12.1 a) $6.20 \text{ atm} \times \dfrac{760 \text{ mm Hg}}{1 \text{ atm}} = 4712 \text{ (calc)} = 4.71 \times 10^3 \text{ mm Hg (corr)}$

b) $6.20 \text{ atm} \times \dfrac{29.92 \text{ in. Hg}}{1 \text{ atm}} = 185.504 \text{ (calc)} = 186 \text{ in. Hg (corr)}$

c) $6.20 \text{ atm} \times \dfrac{14.68 \text{ psi}}{1 \text{ atm}} = 91.016 \text{ (calc)} = 91.0 \text{ psi (corr)}$

d) $6.20 \text{ atm} \times \dfrac{760 \text{ mm Hg}}{1 \text{ atm}} \times \dfrac{1 \times 10^{-3} \text{ m}}{1 \text{ mm}} \times \dfrac{1 \text{ cm}}{1 \times 10^{-2} \text{ m}} = 471.2 \text{ (calc)} = 471 \text{ cm Hg (corr)}$

12.3 a) $578 \text{ mm Hg} \times \dfrac{1 \text{ atm}}{760 \text{ mm Hg}} = 0.7605263158 \text{ (calc)} = 0.761 \text{ atm (corr)}$, smaller than 1.01 atm

b) $14.21 \text{ psi} \times \dfrac{1 \text{ atm}}{14.68 \text{ psi}} \times \dfrac{760 \text{ mm Hg}}{1 \text{ atm}} = 735.6675749 \text{ (calc)} = 735.7 \text{ mm Hg (corr)}$,

smaller than 775 mm Hg

c) $29.92 \text{ in. Hg} \times \dfrac{1 \text{ atm}}{29.92 \text{ in. Hg}} \times \dfrac{760 \text{ mm Hg}}{1 \text{ atm}} = 760. \text{ mm Hg (calc and corr)}$, equal to 760 mm Hg

d) $3.57 \text{ atm} \times \dfrac{14.68 \text{ psi}}{1 \text{ atm}} = 52.4076 \text{ (calc)} = 52.4 \text{ psi (corr)}$, greater than 48.7 psi

12.5 $(762 + 237) \text{ mm Hg} = 999 \text{ mm Hg (calc and corr)}$

Boyle's Law (Sec. 12.3)

12.7 a) increase b) increase c) decrease d) increase

12.9 For this problem, $V_2 = V_1 \times \dfrac{P_1}{P_2}$, where $V_1 = 2.00 \text{ L}$ and $P_1 = 2.00 \text{ atm}$

a) $V_2 = 2.00 \text{ L} \times \dfrac{2.00 \text{ atm}}{2.98 \text{ atm}} = 1.342281879 \text{ (calc)} = 1.34 \text{ L (corr)}$

b) $V_2 = 2.00 \text{ L} \times \dfrac{2.00 \text{ atm}}{10.5 \text{ atm}} = 0.38095238 \text{ (calc)} = 0.381 \text{ L (corr)}$

c) Convert 762 mm Hg to atm: $453 \text{ mm Hg} \times \dfrac{1 \text{ atm}}{760 \text{ mm Hg}} = 0.5960526 \text{ (calc)} = 0.596 \text{ atm (corr)}$

$V_2 = 2.00 \text{ L} \times \dfrac{2.00 \text{ atm}}{0.596 \text{ atm}} = 6.711409396 \text{ (calc)} = 6.71 \text{ L (corr)}$

d) Convert 54.2 mm Hg to atm: $54.2 \text{ mm Hg} \times \dfrac{1 \text{ atm}}{760 \text{ mm Hg}} = 0.07131578944$ (calc)

$= 0.0713$ atm (corr)

$V_2 = 2.00 \text{ L} \times \dfrac{2.00 \text{ atm}}{0.0713 \text{ atm}} = 56.100981767$ (calc) $= 56.1$ L (corr)

12.11 For this problem, V_1 is the unknown original volume, in mL. V_2 is the given final volume.

$P_1 = 4.0$ atm, $P_2 = 2.5$ atm. Solve $P_1V_1 = P_2V_2$ for V_1: $V_1 = V_2 \times \dfrac{P_2}{P_1}$

a) $V_1 = 322 \text{ mL} \times \dfrac{2.5 \text{ atm}}{4.0 \text{ atm}} = 201.25$ (calc) $= 2.0 \times 10^2$ mL (corr)

b) $V_1 = 15.0 \text{ mL} \times \dfrac{2.5 \text{ atm}}{4.0 \text{ atm}} = 9.375$ (calc) $= 9.4$ mL (corr)

c) $V_1 = 2.24 \text{ L} \times \dfrac{2.5 \text{ atm}}{4.0 \text{ atm}} = 1.4$ L (calc and corr) $1.4 \text{ L} \times \dfrac{1 \text{ mL}}{10^{-3} \text{ L}} = 1.4 \times 10^3$ mL

d) $V_1 = 0.88 \text{ L} \times \dfrac{2.5 \text{ atm}}{4.0 \text{ atm}} = 0.55$ L (calc and corr) $0.55 \text{ L} \times \dfrac{1 \text{ mL}}{10^{-3} \text{ L}} = 5.5 \times 10^2$ mL

12.13 Using Boyle's law, $P_1V_1 = P_2V_2$, if we double the pressure, the volume will decrease to one-half its original size. Diagram II represents the decrease in volume to one-half its original size with the same number of gas molecules.

12.15 $P_2 = P_1 \times \dfrac{V_1}{V_2}$, where $V_2 = \dfrac{1}{3}V_1$ or $V_1 = 3V_2$ and $P_2 = 645 \text{ mm Hg} \times \dfrac{3V_2}{V_2} = 1935$ (calc)

$= 1.94 \times 10^3$ mm Hg (corr)

Charles's Law (Sec. 12.4)

12.17 a) decrease   b) increase   c) decrease   d) increase

12.19 Solve $\dfrac{V_1}{T_1} = \dfrac{V_2}{T_2}$ for V_2: $V_2 = V_1 \times \dfrac{T_2}{T_1}$ where $V_1 = 5.00$ L and $T_1 = 35°C + 273 = 308$ K

a) $T_2 = 123°C = 123 + 273 = 396$ K $V_2 = 5.00 \text{ L} \times \dfrac{396 \text{ K}}{308 \text{ K}} = 6.4285714$ (calc) $= 6.43$ L (corr)

b) $T_2 = 223°C + 273 = 496$ K $V_2 = 5.00 \text{ L} \times \dfrac{496 \text{ K}}{308 \text{ K}} = 8.051948$ (calc) $= 8.05$ L (corr)

c) $T_2 = -25°C + 273 = 248$ K $V_2 = 5.00 \text{ L} \times \dfrac{248 \text{ K}}{308 \text{ K}} = 4.025974$ (calc) $= 4.03$ L (corr)

d) $T_2 = 883°C + 273 = 1156$ K $V_2 = 5.00 \text{ L} \times \dfrac{1156 \text{ K}}{308 \text{ K}} = 18.766234$ (calc) $= 18.8$ L (corr)

Gas Laws

12.21 For this problem, V_1 is the unknown original volume, in mL. V_2 is the given final volume, $T_1 = 73°C + 273 = 346$ K, and $T_2 = 273$ K. Solve $\frac{V_1}{T_1} = \frac{V_2}{T_2}$ for V_1: $V_1 = V_2 \times \frac{T_1}{T_2}$

a) $V_1 = 17.5$ mL $\times \frac{346 \text{ K}}{273 \text{ K}} = 22.1794872$ (calc) $= 22.2$ mL (corr)

b) $V_1 = 742$ mL $\times \frac{346 \text{ K}}{273 \text{ K}} = 940.41025641$ (calc) $= 940.$ mL (corr)

c) $V_1 = 3.42$ L $\times \frac{346 \text{ K}}{273 \text{ K}} = 4.334505495$ (calc) $= 4.33$ L (corr)

Convert L to mL: 4.33 L $\times \frac{1 \text{ mL}}{10^{-3} \text{ L}} = 4.33 \times 10^3$ mL

d) $V_1 = 0.90$ L $\times \frac{346 \text{ K}}{273 \text{ K}} = 1.140659341$ (calc) $= 1.1$ L (corr)

Convert L to mL: 1.1 L $\times \frac{1 \text{ mL}}{10^{-3} \text{ L}} = 1.1 \times 10^3$ mL

12.23 Using Charles's law, $V_1/T_1 = V_2/T_2$, decreasing the temperature while the pressure and moles of gas remain constant results in an equal decrease in the volume. Diagram II illustrates the decrease of volume by a factor of 2, while the number of moles of gas remain the same.

12.25 Solve $\frac{V_1}{T_1} = \frac{V_2}{T_2}$ for T_2; $T_2 = T_1 \times \frac{V_2}{V_1}$ where $T_1 = 24°C + 273 = 297$ K, $V_2 = \frac{1}{2}V_1$

$T_2 = 297 \text{ K} \times \frac{\frac{1}{2}V_1}{V_1} = 148.5$ (calc) $= 148$ K (corr): Converting to Celsius: 148 K $- 273 = -125°C$

Gay-Lussac's Law (Sec. 12.5)

12.27 a) decrease   b) increase   c) decrease   d) increase

12.29 Solving $\frac{P_1}{T_1} = \frac{P_2}{T_2}$ for P_2 gives $P_2 = P_1 \times \frac{T_2}{T_1}$ where $P_1 = 1.00$; $T_1 = 22°C + 273 = 295$ K

a) $T_2 = 31°C + 273 = 304$ K;

$P_2 = 1.00$ atm $\times \frac{304 \text{ K}}{295 \text{ K}} = 1.030508471$ (calc) $= 1.03$ atm (corr)

b) $T_2 = 6°C + 273 = 279$ K;

$P_2 = 1.00$ atm $\times \frac{279 \text{ K}}{295 \text{ K}} = 0.9457627$ (calc) $= 0.946$ atm (corr)

c) $T_2 = -37°C + 273 = 236$ K;

$P_2 = 1.00$ atm $\times \frac{236 \text{ K}}{295 \text{ K}} = 0.8$ (calc) $= 0.800$ atm (corr)

d) $T_2 = -137°C + 273 = 136$ K;

$P_2 = 1.00$ atm $\times \frac{136 \text{ K}}{295 \text{ K}} = 0.46101695$ (calc) $= 0.461$ atm (corr)

12.31 For this problem, P_1 is the unknown, original pressure. P_2 is the given final pressure, $T_1 = 97°C + 273 = 370$ K, and $T_2 = 27°C + 273 = 300$ K.

Solve $\dfrac{P_1}{T_1} = \dfrac{P_2}{T_2}$ for P_1; $P_1 = P_2 \times \dfrac{T_1}{T_2}$

a) $P_1 = 762$ mm Hg $\times \dfrac{1 \text{ atm}}{760 \text{ mm Hg}} \times \dfrac{370 \text{ K}}{300 \text{ K}} = 1.23657895$ (calc) $= 1.24$ atm (corr)

b) $P_1 = 662$ mm Hg $\times \dfrac{1 \text{ atm}}{760 \text{ mm Hg}} \times \dfrac{370 \text{ K}}{300 \text{ K}} = 1.074298456$ (calc) $= 1.07$ atm (corr)

c) $P_1 = 1.05$ atm $\times \dfrac{370 \text{ K}}{300 \text{ K}} = 1.295$ (calc) $= 1.30$ atm (corr)

d) $P_1 = 25.0$ atm $\times \dfrac{370 \text{ K}}{300 \text{ K}} = 30.83333$ (calc) $= 30.8$ atm (corr)

12.33 $P_2 = P_1 \times \dfrac{T_2}{T_1}$ where $T_1 = 24°C + 273 = 297$ K and $T_2 = 485°C + 273 = 758$ K

$P_2 = 1.2$ atm $\times \dfrac{758 \text{ K}}{297 \text{ K}} = 3.0626262$ (calc) $= 3.1$ atm (corr)

The Combined Gas Law (Sec. 12.6)

12.35 a) $\dfrac{P_1 V_1}{T_1} = \dfrac{P_2 V_2}{T_2}$ Multiply both sides by T_1 and T_2, and divide by P_1 and V_1.

$\dfrac{T_2 T_1}{P_1 V_1} \times \dfrac{P_1 V_1}{T_1} = \dfrac{P_2 V_2}{T_2} \times \dfrac{T_2 T_1}{P_1 V_1}$ gives $T_2 = T_1 \times \dfrac{P_2}{P_1} \times \dfrac{V_2}{V_1}$

b) $\dfrac{P_1 V_1}{T_1} = \dfrac{P_2 V_2}{T_2}$ Reverse the equation: $\dfrac{P_2 V_2}{T_2} = \dfrac{P_1 V_1}{T_1}$

Multiply both sides by T_2 and divide by P_1 and P_2.

$\dfrac{P_2 V_2}{T_2} \times \dfrac{T_2}{P_1 P_2} = \dfrac{P_1 V_1}{T_1} \times \dfrac{T_2}{P_1 P_2}$ gives $\dfrac{V_2}{P_1} = \dfrac{V_1}{P_2} \times \dfrac{T_2}{T_1}$

12.37 $V_2 = V_1 \times \dfrac{P_1}{P_2} \times \dfrac{T_2}{T_1}$ where $V_1 = 3.00$ mL; $P_1 = 0.980$ atm; $T_1 = 230°C + 273 = 503$ K

a) $P_2 = 2.00$ atm; $T_2 = 25°C + 273 = 298$ K

$V_2 = 3.00$ mL $\times \dfrac{0.980 \text{ atm}}{2.00 \text{ atm}} \times \dfrac{298 \text{ K}}{503 \text{ K}} = 0.87089463$ (calc) $= 0.871$ mL (corr)

b) $P_2 = 3.00$ atm; $T_2 = -25°C + 273 = 248$ K

$V_2 = 3.00$ mL $\times \dfrac{0.980 \text{ atm}}{3.00 \text{ atm}} \times \dfrac{248 \text{ K}}{503 \text{ K}} = 0.483180915$ (calc) $= 0.483$ mL (corr)

c) $P_2 = 5.00$ atm; $T_2 = -75°C + 273 = 198$ K

$V_2 = 3.00$ mL $\times \dfrac{0.980 \text{ atm}}{5.00 \text{ atm}} \times \dfrac{198 \text{ K}}{503 \text{ K}} = 0.23145924$ (calc) $= 0.231$ mL (corr)

d) $P_2 = 7.75$ atm; $T_2 = -125°C + 273 = 148$ K

$V_2 = 3.00$ mL $\times \dfrac{0.980 \text{ atm}}{7.75 \text{ atm}} \times \dfrac{148 \text{ K}}{503 \text{ K}} = 0.111619316$ (calc) $= 0.112$ mL (corr)

Gas Laws

12.39 a) $V_2 = V_1 \times \dfrac{P_1}{P_2} \times \dfrac{T_2}{T_1} = 15.2 \text{ L} \times \dfrac{1.35 \text{ atm}}{3.50 \text{ atm}} \times \dfrac{308 \text{ K}}{306 \text{ K}} = 5.9011764$ (calc) $= 5.90$ L (corr)

b) 15.2 L = 15,200 mL
$V_2 = V_1 \times \dfrac{P_1}{P_2} \times \dfrac{T_2}{T_1} = 15{,}200 \text{ mL} \times \dfrac{1.35 \text{ atm}}{6.70 \text{ atm}} \times \dfrac{370 \text{ K}}{306 \text{ K}} = 3703.2485$ (calc)
$= 3.70 \times 10^3$ mL (corr)

c) $P_2 = P_1 \times \dfrac{V_1}{V_2} \times \dfrac{T_2}{T_1} = 1.35 \text{ atm} \times \dfrac{15.2 \text{ L}}{10.0 \text{ L}} \times \dfrac{315 \text{ K}}{306 \text{ K}} = 2.1123529$ (calc) $= 2.11$ atm (corr)

d) $T_2 = T_1 \times \dfrac{P_2}{P_1} \times \dfrac{V_2}{V_1} = 306 \text{ K} \times \dfrac{7.00 \text{ atm}}{1.35 \text{ atm}} \times \dfrac{0.973 \text{ L}}{15.2 \text{ L}} = 101.56754$ (calc) $= 102$ K (corr)
$102 \text{ K} - 273 = -171°\text{C}$

12.41 a) $375 \text{ mL} \times \dfrac{10^{-3} \text{ L}}{1 \text{ mL}} = 0.375 \text{ L}, \; P_2 = P_1 \times \dfrac{V_1}{V_2} = 1.03 \text{ atm} \times \dfrac{0.375 \text{ L}}{1.25 \text{ L}} = 0.309$ atm (calc and corr)

$0.309 \text{ atm} \times \dfrac{760 \text{ mm Hg}}{1 \text{ atm}} = 234.84$ (calc) $= 235$ mm Hg (corr)

b) $T_1 = 27°\text{C} + 273 = 300.\text{ K}, \; T_2 = T_1 \times \dfrac{V_2}{V_1} = 300.\text{ K} \times \dfrac{1.25 \text{ L}}{0.375 \text{ L}} = 10\overline{0}0$ K
Converting to Celsius, $10\overline{0}0 \text{ K} - 273 = 727°\text{C}$ (calc) $= 730°\text{C}$ (corr)

12.43 $T_2 = T_1 \times \dfrac{P_2}{P_1} \times \dfrac{V_2}{V_1}$ where $T_1 = 33°\text{C} + 273 = 306$ K.

a) $P_2 = 3P_1; V_2 = 3V_1 \quad T_2 = 306 \text{ K} \times \dfrac{3P_1}{P_1} \times \dfrac{3V_1}{V_1} = 2754$ (calc) $= 2750$ K (corr)
Converting to Celsius: $2750 \text{ K} - 273 = 2477$ (calc) $= 2480°\text{C}$ (corr)

b) $P_2 = \dfrac{1}{2}P_1$ or $2P_2 = P_1; V_2 = \dfrac{1}{2}V_1$ or $2V_2 = V_1$
$T_2 = 306 \text{ K} \times \dfrac{\frac{1}{2}P_1}{P_1} \times \dfrac{\frac{1}{2}V_1}{V_1} = 76.5$ K (calc and corr)
Converting to Celsius: $76.5 \text{ K} - 273.2 = -196.7°\text{C}$ (calc and corr)

c) $P_2 = 3P_1, V_2 = \dfrac{1}{2}V_1 \quad T_2 = 306 \text{ K} \times \dfrac{3P_1}{P_1} \times \dfrac{\frac{1}{2}V_1}{V_1} = 459$ K (calc and corr)
Converting to Celsius: $459 \text{ K} - 273 = 186°\text{C}$ (calc and corr)

d) $P_2 = \dfrac{1}{2}P_1, V_2 = 2V_1 \quad T_2 = 306 \text{ K} \times \dfrac{\frac{1}{2}P_1}{P_1} \times \dfrac{2V_1}{V_1} = 306$ K (no change).
Converting to Celsius: $306 \text{ K} - 273 = 33°\text{C}$ (calc and corr)

Avogadro's Law (Sec. 12.7)

12.45 $\dfrac{V_1}{n_1} = \dfrac{V_2}{n_2}$ Solving for $V_2 = n_2 \times \dfrac{V_1}{n_1}$

a) $2.00 \text{ moles} \times \dfrac{24.0 \text{ L}}{1.00 \text{ mole}} = 48$ (calc) $= 48.0$ L (corr)

b) $2.32 \text{ moles} \times \dfrac{24.0 \text{ L}}{1.00 \text{ mole}} = 55.68$ (calc) $= 55.7$ L (corr)

c) $5.40 \text{ moles} \times \dfrac{24.0 \text{ L}}{1.00 \text{ mole}} = 129.6$ (calc) $= 130.$ L (corr)

d) $0.500 \text{ mole} \times \dfrac{24.0 \text{ L}}{1.00 \text{ mole}} = 12$ (calc) $= 12.0$ L (corr)

12.47 $\dfrac{V_1}{n_1} = \dfrac{V_2}{n_2}$ Solving for $n_2 = V_2 \times \dfrac{n_1}{V_1}$

a) $3.00 \text{ L} \times \dfrac{2.00 \text{ moles}}{2.33 \text{ L}} = 2.575107296$ (calc) $= 2.58$ moles (corr)

b) $4.00 \text{ L} \times \dfrac{2.00 \text{ moles}}{2.33 \text{ L}} = 3.4334764$ (calc) $= 3.43$ moles (corr)

c) $5.00 \text{ L} \times \dfrac{2.00 \text{ moles}}{2.33 \text{ L}} = 4.2918455$ (calc) $= 4.29$ moles (corr)

d) $6.00 \text{ L} \times \dfrac{2.00 \text{ moles}}{2.33 \text{ L}} = 5.15021459$ (calc) $= 5.15$ moles (corr)

12.49 $\dfrac{V_1}{n_1} = \dfrac{V_2}{n_2}$ If we double the volume by increasing the moles of gas keeping temperature and pressure constant, the moles of gas will be doubled. Diagram III represents a double volume and double the moles of gas.

12.51 Balloon, 7.83 moles of He, volume of 27.5 L at certain temperature and pressure.

a) $7.83 \text{ moles} + 1.00 \text{ mole} = 8.83 \text{ moles} \times \dfrac{27.5 \text{ L}}{7.83 \text{ moles}} = 31.01213282$ (calc) $= 31.0$ L (corr)

b) $7.83 \text{ moles} + 3.00 \text{ moles} = 10.83 \text{ moles} \times \dfrac{27.5 \text{ L}}{7.83 \text{ moles}} = 38.0363985$ (calc) $= 38.0$ L (corr)

c) $7.83 \text{ moles} - 1.00 \text{ mole} = 6.83 \text{ moles} \times \dfrac{27.5 \text{ L}}{7.83 \text{ moles}} = 23.98786718$ (calc) $= 24.0$ L (corr)

d) $7.83 \text{ moles} - 3.00 \text{ moles} = 4.83 \text{ moles} \times \dfrac{27.5 \text{ L}}{7.83 \text{ moles}} = 16.9636015$ (calc) $= 17.0$ L (corr)

12.53 1.00 mole Cl_2 gas, volume 24.0 L at certain temperature and pressure.

a) $2.00 \text{ g } Cl_2 \times \dfrac{1 \text{ mole } Cl_2}{70.90 \text{ g } Cl_2} \times \dfrac{24.0 \text{ L}}{1.00 \text{ mole } Cl_2} = 0.67700987306$ (calc) $= 0.677$ L (corr)

b) $2.32 \text{ g } Cl_2 \times \dfrac{1 \text{ mole } Cl_2}{70.90 \text{ g } Cl_2} \times \dfrac{24.0 \text{ L}}{1.00 \text{ mole } Cl_2} = 0.785331453$ (calc) $= 0.785$ L (corr)

c) $5.40 \text{ g } Cl_2 \times \dfrac{1 \text{ mole } Cl_2}{70.90 \text{ g } Cl_2} \times \dfrac{24.0 \text{ L}}{1.00 \text{ mole } Cl_2} = 1.827926658$ (calc) $= 1.83$ L (corr)

d) $0.500 \text{ g } Cl_2 \times \dfrac{1 \text{ mole } Cl_2}{70.90 \text{ g } Cl_2} \times \dfrac{24.0 \text{ L}}{1.00 \text{ mole } Cl_2} = 0.16925247$ (calc) $= 0.169$ L (corr)

Gas Laws

12.55 Ideal gas law $PV = nRT$, if the pressure, volume, and temperature remain constant, the moles of gas will also remain constant.
    a) 0.625 mole N_2 gas = 0.625 mole NO gas, therefore the volume will be 10.9 L
    b) 0.625 mole N_2 gas = 0.625 mole N_2O gas, therefore the volume will be 10.9 L
    c) 0.625 mole N_2 gas = 0.625 mole NO_2 gas, therefore the volume will be 10.9 L
    d) 0.625 mole N_2 gas = 0.625 mole N_2O_5 gas, therefore the volume will be 10.9 L

12.57 0.625 mole sample of N_2 gas at 1.50 atm, 45°C, 10.9 liters.

    a) $2.00 \text{ moles } CO_2 \times \dfrac{10.9 \text{ L}}{0.625 \text{ mole}} = 34.88 \text{ (calc)} = 34.9 \text{ L (corr)}$

    b) $3.00 \text{ moles } N_2O \times \dfrac{10.9 \text{ L}}{0.625 \text{ mole}} = 52.32 \text{ (calc)} = 52.3 \text{ L (corr)}$

    c) $2.50 \text{ moles } O_2 \times \dfrac{10.9 \text{ L}}{0.625 \text{ mole}} = 43.6 \text{ L (calc and corr)}$

    d) $6.00 \text{ moles } NH_3 \times \dfrac{10.9 \text{ L}}{0.625 \text{ mole}} = 104.64 \text{ (calc)} = 105 \text{ L (corr)}$

12.59 Balloon = 2.00 moles N_2 + 1.50 moles O_2 = 3.50 total moles; Volume 0.500 L
Remove 0.48 mole N_2 and 0.76 mole O_2 added = 1.52 moles N_2 + 2.26 moles O_2 = 3.78 moles
new volume =
$$V_2 = V_1 \times \dfrac{n_2}{n_1} = 0.500 \text{ L} \times \dfrac{3.78 \text{ moles}}{3.50 \text{ moles}} = 0.54 \text{ (calc)} = 0.540 \text{ L (corr)}$$

12.61 The one with the larger number of moles will have the larger volume.
    a) 0.450 mole O_2 will have the larger volume
    b) 2.32 moles CH_4 will have the larger volume

    c) $100.0 \text{ g } NO_2 \times \dfrac{1 \text{ mole } NO_2}{46.01 \text{ g } NO_2} = 2.1734406 \text{ (calc)} = 2.173 \text{ moles } NO_2 \text{ (corr)}$

        $100.0 \text{ g } N_2O \times \dfrac{1 \text{ mole } N_2O}{44.02 \text{ g } N_2O} = 2.2716947 \text{ (calc)} = 2.272 \text{ moles } N_2O \text{ (corr)}$

        ∴ 100.0 g N_2O will have the larger volume

    d) $100.0 \text{ g CO} \times \dfrac{1 \text{ mole CO}}{28.01 \text{ g CO}} = 3.5701535 \text{ (calc)} = 3.570 \text{ moles CO (corr)}$

        $100.0 \text{ g } CO_2 \times \dfrac{1 \text{ mole } CO_2}{44.01 \text{ g } CO_2} = 2.2722109 \text{ (calc)} = 2.272 \text{ moles } CO_2 \text{ (corr)}$

        ∴ 100 g CO will have the larger volume

12.63 $\dfrac{P_1 \times V_1}{n_1 \times T_1} = \dfrac{P_2 \times V_2}{n_2 \times T_2}$ $P_1 = 1.5$ atm, $V_1 = 2.0$ L, $n_1 = 0.500$ mole, $T_1 = 27°C + 273 \text{ K} = 300.\text{ K}$

                                   $P_2 = 1.5$ atm $-$ 0.50 atm $= 1.00$ atm
                                     $V_2 = ?$
                                     $n_2 = 0.500$ mole $-$ 0.100 mole $= 0.400$ mole
                                     $T_2 = 20°C + 27°C = 47°C$ $47°C + 273 \text{ K} = 320.\text{ K}$

$$\dfrac{1.5 \text{ atm } 2.0 \text{ L}}{0.500 \text{ mole } 300.\text{ K}} = \dfrac{1.00 \text{ atm } ? \text{ L}}{0.400 \text{ mole } 320.\text{ K}} = 2.56 \text{ (calc)} = 2.6 \text{ L (corr)}$$

The Ideal Gas Law (Sec. 12.9)

12.65 Solving $PV = nRT$ for V gives: $V = \dfrac{nRT}{P}$ where $n = 1.20$ moles

a) $P = 1$ atm, $T = 273$ K

$$V = \dfrac{1.20 \text{ moles}}{1.00 \text{ atm}} \times 0.08206 \dfrac{\text{atm L}}{\text{mole K}} \times 273 \text{ K} = 26.882856 \text{ (calc)} = 26.9 \text{ L (corr)}$$

b) $P = 1.54$ atm, $T = 73°C + 273 = 346$ K

$$V = \dfrac{1.20 \text{ moles}}{1.54 \text{ atm}} \times 0.08206 \dfrac{\text{atm L}}{\text{mole K}} \times 346 \text{ K} = 22.124229 \text{ (calc)} = 22.1 \text{ L (corr)}$$

c) $P = 15.0$ atm, $T = 525°C + 273 = 798$ K

$$V = \dfrac{1.20 \text{ moles}}{15.0 \text{ atm}} \times 0.08206 \dfrac{\text{atm L}}{\text{mole K}} \times 798 \text{ K} = 5.2387104 \text{ (calc)} = 5.24 \text{ L (corr)}$$

d) $P = 765$ mm Hg, $T = -23°C + 273 = 250.$ K

$$V = \dfrac{1.20 \text{ moles}}{765 \text{ mm Hg}} \times 62.36 \dfrac{\text{mm Hg} \cdot \text{L}}{\text{mole K}} \times 250. \text{ K} = 24.454902 \text{ (calc)} = 24.5 \text{ L (corr)}$$

12.67 $n = \dfrac{PV}{RT}$, $T = 25°C + 273 = 298$ K

$$n = \dfrac{3.67 \text{ atm} \times 6.00 \text{ L}}{0.08206 \dfrac{\text{atm L}}{\text{mole K}} \cdot 298 \text{ K}} = 0.9004706 \text{ (calc)} = 0.900 \text{ mole (corr)}$$

12.69 $T = \dfrac{PV}{nR}$, Convert volume to liters: $V = 1275 \text{ mL} \times \dfrac{10^{-3} \text{ L}}{1 \text{ mL}} = 1.275$ L

$$T = \dfrac{5.78 \text{ atm} \times 1.275 \text{ L}}{0.332 \text{ mole} \times 0.08206 \dfrac{\text{atm L}}{\text{mole K}}} = 270.5007 \text{ (calc)} = 271 \text{ K (corr)}$$

$\therefore$ Celsius: $271 - 273 = -2°C$

12.71 $P = \dfrac{nRT}{V}$ where $V = 6.00$ L, $T = 25°C + 273 = 298$ K

a) $P = \dfrac{0.30 \text{ mole}}{6.00 \text{ L}} \times 0.08206 \dfrac{\text{atm L}}{\text{mole K}} \times 298 \text{ K} = 1.222694 \text{ (calc)} = 1.2 \text{ atm (corr)}$

b) $P = \dfrac{1.20 \text{ moles}}{6.00 \text{ L}} \times 0.08206 \dfrac{\text{atm L}}{\text{mole K}} \times 298 \text{ K} = 4.890776 \text{ (calc)} = 4.89 \text{ atm (corr)}$

c) $P = \dfrac{0.30 \text{ g } N_2}{6.00 \text{ L}} \times \dfrac{1 \text{ mole } N_2}{28.02 \text{ g } N_2} \times 0.08206 \dfrac{\text{atm L}}{\text{mole K}} \times 298 \text{ K}$

$= 0.043636474 \text{ (calc)} = 0.044 \text{ atm (corr)}$

d) $P = \dfrac{1.20 \text{ g } N_2}{6.00 \text{ L}} \times \dfrac{1 \text{ mole } N_2}{28.02 \text{ g } N_2} \times 0.08206 \dfrac{\text{atm L}}{\text{mole K}} \times 298 \text{ K}$

$= 0.174545989 \text{ (calc)} = 0.175 \text{ atm (corr)}$

12.73 $R = \dfrac{PV}{nT} = \dfrac{13.3 \text{ atm} \times 3.00 \text{ L}}{1.14 \text{ moles} \times 425 \text{ K}} = 0.082352941 \text{ (calc)} = 0.0824 \dfrac{\text{atm L}}{\text{mole K}}$ (corr)

Gas Laws

12.75 $V = \dfrac{nRT}{P} = \dfrac{1.00 \text{ mole} \times 0.08206 \dfrac{\text{atm L}}{\text{mole K}} \times 400.\text{ K}}{0.908 \text{ atm}} = 36.14978 \text{ (calc)} = 36.1 \text{ L (corr)}$

12.77 $m = \dfrac{(\text{MM}) PV}{RT}$

a) for C_2H_6 the molar mass $(\text{MM}) = 30.08$ g/mole. $T = 29°C + 273 = 302$ K.

$m = \dfrac{30.08 \text{ g}}{1 \text{ mole}} \times \dfrac{0.972 \text{ atm} \times 25.0 \text{ L}}{0.08206 \dfrac{\text{atm L}}{\text{mole K}} \times 302 \text{ K}} = 29.494813 \text{ (calc)} = 29.5 \text{ g (corr)}$

b) for HCl, $(\text{MM}) = 36.46$ g/mole, $T = 75°C + 273 = 348$ K.

$m = \dfrac{36.46 \text{ g}}{1 \text{ mole}} \times \dfrac{854 \text{ mm Hg} \times 2.22 \text{ L}}{62.36 \dfrac{\text{mm Hg} \cdot \text{L}}{\text{mole K}} \times 348 \text{ K}} = 3.18524 \text{ (calc)} = 3.19 \text{ g (corr)}$

c) for SO_2, $(\text{MM}) = 64.07$ g/mole, $T = 273$ K.

$m = \dfrac{64.07 \text{ g}}{1 \text{ mole}} \times \dfrac{2 \text{ atm} \times 5.50 \text{ L}}{0.08206 \dfrac{\text{atm L}}{\text{mole K}} \times 273 \text{ K}} = 31.45960385 \text{ (calc)} = 31.5 \text{ g (corr)}$

d) for N_2O, $(\text{MM}) = 44.02$ g/mole, $T = 273°C + 273 = 546$ K

$V = 783 \text{ mL} \times \dfrac{10^{-3} \text{ L}}{\text{mL}} = 0.783 \text{ L} \quad m = \dfrac{44.02 \text{ g}}{1 \text{ mole}} \times \dfrac{359 \text{ mm Hg} \times 0.783 \text{ L}}{62.36 \dfrac{\text{mm Hg} \cdot \text{L}}{\text{mole K}} \times 546 \text{ K}}$

$= 0.3634188 \text{ (calc)} = 0.363 \text{ g (corr)}$

12.79 $m = \dfrac{(\text{MM}) PV}{RT}$, $T = 23°C + 273 = 296$ K. For Cl_2, $(\text{MM}) = 70.90$ g/mole

$m = \dfrac{70.90 \text{ g}}{1 \text{ mole}} \times \dfrac{1.65 \text{ atm} \times 30.0 \text{ L}}{0.08206 \dfrac{\text{atm L}}{\text{mole K}} \times 296 \text{ K}} = 144.48681 \text{ (calc)} = 144 \text{ g } Cl_2 \text{ after adding (corr)}$

If there were 100.0 g Cl_2 initially, then $144 - 100.0 = 44$ g Cl_2 added.

Modified Forms of the Ideal Gas Law Equation (Sec. 12.10)

12.81 $(\text{MM}) = \dfrac{mRT}{PV}$, $P = 741 \text{ mm Hg} \times \dfrac{1 \text{ atm}}{760 \text{ mm Hg}} = 0.975 \text{ atm}$, $T = 33°C + 273 = 306$ K

$\dfrac{1.305 \text{ g} \times 0.08206 \dfrac{\text{atm L}}{\text{mole K}} \times 306 \text{ K}}{1.20 \text{ L} \times 0.975 \text{ atm}} = 28.007709 \text{ (calc)} = 28.0 \text{ g/mole (corr)}$

12.83 $(\text{MM}) = \dfrac{mRT}{PV}$, $V = 125 \text{ mL} \times \dfrac{10^{-3} \text{ L}}{1 \text{ mL}} = 0.125 \text{ L}$, $T = 75°C + 273 = 348$ K.

$\dfrac{0.450 \text{ g} \times 0.08206 \dfrac{\text{atm L}}{\text{mole K}} \times 348 \text{ K}}{0.125 \text{ L} \times 1.00 \text{ atm}} = 102.804768 \text{ (calc)} = 103 \text{ g/mol (corr)}$

12.85 $(MM) = \dfrac{mRT}{PV}$, $V = 125 \text{ mL} \times \dfrac{10^{-3} \text{ L}}{1 \text{ mL}} = 0.125 \text{ L}$, $T = 90°C + 273 = 363 \text{ K}$

$$\dfrac{0.4537 \text{ g} \times 0.08206 \dfrac{\text{atm L}}{\text{mole K}} \times 363 \text{ K}}{0.125 \text{ L} \times 1.20 \text{ atm}} = 90.098105 \text{ (calc)} = 90.1 \text{ g/mol (corr)}$$

12.87 $(MM) = \dfrac{mRT}{PV}$, $T = 20°C + 273 = 293 \text{ K}$

$$(MM) = \dfrac{1.733 \text{ g} \times 0.08206 \dfrac{\text{atm L}}{\text{mole K}} \times 293 \text{ K}}{0.902 \text{ L} \times 1.65 \text{ atm}} = 27.996772387 \text{ (calc)} = 28.0 \text{ g/mole (corr)}$$

CO has a molar mass of 28.01 g/mole; CO_2 has 44.01 g/mole. The gas is CO.

12.89 $d = \dfrac{m}{V} = \dfrac{(MM) P}{RT}$, $(MM \; H_2S) = 34.09 \text{ g/mole}$

a) $T = 24°C + 273 = 297 \text{ K}$.

$$d = \dfrac{34.09 \text{ g}}{1 \text{ mole}} \times \dfrac{675 \text{ mm Hg}}{62.36 \dfrac{\text{mm Hg L}}{\text{mole K}} \times 297 \text{ K}} = 1.24241919383 \text{ (calc)} = 1.24 \text{ g/L (corr)}$$

b) $T = 24°C + 273 = 297 \text{ K}$.

$$d = \dfrac{34.09 \text{ g}}{1 \text{ mole}} \times \dfrac{1.20 \text{ atm}}{0.08206 \dfrac{\text{atm L}}{\text{mole K}} \times 297 \text{ K}} = 1.678495902 \text{ (calc)} = 1.68 \text{ g/L (corr)}$$

c) $T = 370°C + 273 = 643 \text{ K}$.

$$d = \dfrac{34.09 \text{ g}}{1 \text{ mole}} \times \dfrac{1.30 \text{ atm}}{0.08206 \dfrac{\text{atm L}}{\text{mole K}} \times 643 \text{ K}} = 0.83990055 \text{ (calc)} = 0.840 \text{ g/L (corr)}$$

d) $T = -25°C + 273 = 248 \text{ K}$.

$$d = \dfrac{34.09 \text{ g}}{1 \text{ mole}} \times \dfrac{452 \text{ mm Hg}}{62.36 \dfrac{\text{mm Hg} \cdot \text{L}}{\text{mole K}} \times 248 \text{ K}} = 0.99634019 \text{ (calc)} = 0.996 \text{ g/L (corr)}$$

12.91 $P = \dfrac{dRT}{(MM)}$; $T = 47°C + 273 = 320. \text{ K}$; $d = 1.00 \text{ g/L}$

a) $(MM) = \dfrac{28.02 \text{ g}}{1 \text{ mole}}$; $P = \dfrac{1.00 \text{ g}}{\text{L}} \times \dfrac{1 \text{ mole}}{28.02 \text{ g}} \times 0.08206 \dfrac{\text{atm L}}{\text{mole K}} \times 320. \text{ K}$

$= 0.9371591 \text{ (calc)} = 0.937 \text{ atm (corr)}$

b) $(MM) = \dfrac{131.29 \text{ g}}{1 \text{ mole}}$; $P = \dfrac{1.00 \text{ g}}{\text{L}} \times \dfrac{1 \text{ mole}}{131.29 \text{ g}} \times 0.08206 \dfrac{\text{atm L}}{\text{mole K}} \times 320. \text{ K}$

$= 0.2000091 \text{ (calc)} = 0.200 \text{ atm (corr)}$

c) $(MM) = \dfrac{54.45 \text{ g}}{1 \text{ mole}}$; $P = \dfrac{1.00 \text{ g}}{\text{L}} \times \dfrac{1 \text{ mole}}{54.45 \text{ g}} \times 0.08206 \dfrac{\text{atm L}}{\text{mole K}} \times 320. \text{ K}$

$= 0.4822626 \text{ (calc)} = 0.482 \text{ atm (corr)}$

Gas Laws

d) $(MM) = \dfrac{44.02 \text{ g}}{1 \text{ mole}}$; $P = \dfrac{1.00 \text{ g}}{L} \times \dfrac{1 \text{ mole}}{44.02 \text{ g}} \times 0.08206 \dfrac{\text{atm L}}{\text{mole K}} \times 320.\text{ K}$

$= 0.5965288 \text{ (calc)} = 0.597 \text{ atm (corr)}$

12.93 $T = 27°C + 273 = 300.\text{ K}$; $(MM) = \dfrac{dRT}{P}$

$(MM) = \dfrac{2.27 \text{ g}}{L} \times 0.08206 \dfrac{\text{atm L}}{\text{mole K}} \times \dfrac{300.\text{ K}}{2.00 \text{ atm}} = 27.94143 \text{ (calc)} = 27.9 \dfrac{\text{g}}{\text{mole}} \text{ (corr)}$

Since CO has a molar mass of 28.01 g/mole and NO has 30.01 g/mole, the gas is CO.

12.95 $T = 25°C + 273 = 298 \text{ K}$; $(MM) = \dfrac{dRT}{P}$

$(MM) = \dfrac{1.35 \text{ g}}{L} \times 0.08206 \dfrac{\text{atm L}}{\text{mole K}} \times \dfrac{298.\text{ K}}{1.32 \text{ atm}} = 25.00965 \text{ (calc)} = 25.01 \dfrac{\text{g}}{\text{mole}} \text{ (corr)}$

12.97 $T = \dfrac{P(MM)}{dR} = 1.20 \text{ atm} \times \dfrac{30.01 \text{ g}}{1 \text{ mole}} \times \dfrac{1 \text{ L}}{1.25 \text{ g}} \times \dfrac{\text{K mole}}{0.08206 \text{ L atm}} = 351.0796978 \text{ (calc)}$

$= 351 \text{ K (corr)}\quad °C = 351 \text{ K} - 273 = 78°C \text{ (calc and corr)}$

Volumes of Gases in Chemical Reactions (Sec. 12.11)

12.99 The whole-number combining ratio is $\dfrac{2.00 \text{ L O}_2}{1.60 \text{ L NH}_3} = \dfrac{5 \text{ L O}_2}{4 \text{ L NH}_3}$ which matches the coefficients in the second reaction. $4 \text{ NH}_3(g) + 5 \text{ O}_2(g) \rightarrow 4 \text{ NO}(g) + 6 \text{ H}_2\text{O}(g)$

12.101 a) $1.30 \text{ L CO}_2 \times \dfrac{1 \text{ L C}_3\text{H}_4}{3 \text{ L CO}_2} = 0.43333333 \text{ (calc)} = 0.433 \text{ L C}_3\text{H}_8 \text{ (corr)}$

b) $1.30 \text{ L H}_2\text{O} \times \dfrac{1 \text{ L C}_3\text{H}_8}{4 \text{ L H}_2\text{O}} = 0.325 \text{ L C}_3\text{H}_8 \text{ (calc and corr)}$

12.103 0.75 total volume of products
At constant temperature, pressure 1 L of C_3H_8 produces 3 L CO_2 and 4 L of H_2O, 7 L total products

$0.75 \text{ L total products} \times \dfrac{1 \text{ L C}_3\text{H}_8}{7 \text{ L products}} = 0.10714285 \text{ (calc)} = 0.11 \text{ L C}_3\text{H}_8 \text{ (corr)}$

Volumes of Gases and the Limiting Reactant Concept (Sec. 12.12)

12.105 $N_2 + 3 H_2 \rightarrow 2 NH_3$

a) $1.00 \text{ L N}_2 \times \dfrac{2 \text{ L NH}_3}{1 \text{ L N}_2} = 2 \text{ (calc)} = 2.00 \text{ L NH}_3 \text{ (corr)}$

$1.50 \text{ L H}_2 \times \dfrac{2 \text{ L NH}_3}{3 \text{ L H}_2} = 1 \text{ (calc)} = 1.00 \text{ L NH}_3 \text{ (corr)}\qquad \therefore 1.50 \text{ L H}_2 \text{ is the limiting reactant.}$

b) $2.00 \text{ L N}_2 \times \dfrac{2 \text{ L NH}_3}{1 \text{ L N}_2} = 4 \text{ (calc)} = 4.00 \text{ L NH}_3 \text{ (corr)}$

$5.50 \text{ L H}_2 \times \dfrac{2 \text{ L NH}_3}{3 \text{ L H}_2} = 3.66666 \text{ (calc)} = 3.67 \text{ L NH}_3 \text{ (corr)}\quad \therefore 5.50 \text{ L H}_2 \text{ is the limiting reactant.}$

c) $1.00 \text{ L N}_2 \times \dfrac{2 \text{ L NH}_3}{1 \text{ L N}_2} = 2$ (calc) $= 2.00 \text{ L NH}_3$ (corr)

$4.00 \text{ L H}_2 \times \dfrac{2 \text{ L NH}_3}{3 \text{ L H}_2} = 2.66666$ (calc) $= 2.67 \text{ L NH}_3$ (corr) $\therefore 1.00 \text{ L N}_2$ is the limiting reactant.

d) $3.00 \text{ L N}_2 \times \dfrac{2 \text{ L NH}_3}{1 \text{ L N}_2} = 6$ (calc) $= 6.00 \text{ L NH}_3$ (corr)

$1.00 \text{ L H}_2 \times \dfrac{2 \text{ L NH}_3}{3 \text{ L H}_2} = 0.6667$ (calc) $= 0.667 \text{ L NH}_3$ (corr) $\therefore 1.00 \text{ L H}_2$ is the limiting reactant.

12.107 $CH_4 + 2 H_2O \rightarrow CO_2 + 4 H_2$

a) $45.0 \text{ L CH}_4 \times \dfrac{1 \text{ L CO}_2}{1 \text{ L CH}_4} = 45$ (calc) $= 45.0 \text{ L CO}_2$ (corr)

$45.0 \text{ L H}_2\text{O} \times \dfrac{1 \text{ L CO}_2}{2 \text{ L H}_2\text{O}} = 22.5 \text{ L CO}_2$ (calc and corr) $\therefore 45.0 \text{ L H}_2\text{O}$ is the limiting reactant.

$45.0 \text{ L H}_2\text{O} \times \dfrac{4 \text{ L H}_2}{2 \text{ L H}_2\text{O}} = 90$ (calc) $= 90.0 \text{ L H}_2$ (corr)

$\therefore 22.5 \text{ L CO}_2$ and 90.0 L H_2 are produced.

b) $45.0 \text{ L CH}_4 \times \dfrac{1 \text{ L CO}_2}{1 \text{ L CH}_4} = 45$ (calc) $= 45.0 \text{ L CO}_2$ (corr)

$66.0 \text{ L H}_2\text{O} \times \dfrac{1 \text{ L CO}_2}{2 \text{ L H}_2\text{O}} = 33$ (calc) $= 33.0 \text{ L CO}_2$ (corr) $\therefore 66.0 \text{ L H}_2\text{O}$ is the limiting reactant.

$66.0 \text{ L H}_2\text{O} \times \dfrac{4 \text{ L H}_2}{2 \text{ L H}_2\text{O}} = 132 \text{ L H}_2$ (calc and corr) $\therefore 33.0 \text{ L CO}_2$ and 132 L H_2 are produced.

c) $64.0 \text{ L CH}_4 \times \dfrac{1 \text{ L CO}_2}{1 \text{ L CH}_4} = (64)$ calc $= 64.0 \text{ L CO}_2$ (corr)

$134 \text{ L H}_2\text{O} \times \dfrac{1 \text{ L CO}_2}{2 \text{ L H}_2\text{O}} = 67$ (calc) $= 67.0 \text{ L CO}_2$ (corr) $\therefore 64.0 \text{ L CH}_4$ is the limiting reactant.

$64.0 \text{ L CH}_4 \times \dfrac{4 \text{ L H}_2}{1 \text{ L CH}_4} = 256 \text{ L H}_2$ (calc and corr) $\therefore 64.0 \text{ L CO}_2$ and 256 L H_2 are produced.

d) $16.0 \text{ L CH}_4 \times \dfrac{1 \text{ L CO}_2}{1 \text{ L CH}_4} = 16$ (calc) $= 16.0 \text{ L CO}_2$ (corr)

$32.0 \text{ L H}_2\text{O} \times \dfrac{1 \text{ L CO}_2}{2 \text{ L H}_2\text{O}} = 16$ (calc) $= 16.0 \text{ L CO}_2$ (corr)

$\therefore$ Both 16.0 L CH_4 and $32.0 \text{ L H}_2\text{O}$ are limiting reactants.

$16.0 \text{ L CH}_4 \times \dfrac{4 \text{ L H}_2}{1 \text{ L CH}_4} = 64$ (calc) $= 64.0 \text{ L H}_2$ (corr)

or $32.0 \text{ L H}_2\text{O} \times \dfrac{4 \text{ L H}_2}{2 \text{ L H}_2\text{O}} = 64$ (calc) $= 64.0 \text{ L H}_2$ (corr)

$\therefore 16.0 \text{ L CO}_2$ and 64.0 L H_2 are produced.

Gas Laws

Molar Volume of a Gas (Sec. 12.13)

12.109 a) 1.20 atm, $K = 33°C + 273 = 306$ K

$$V = \frac{nRT}{P} = 1 \text{ mole} \times \frac{0.08206 \text{ L atm}}{\text{K mole}} \times \frac{306 \text{ K}}{1.20 \text{ atm}} = 20.9253 \text{ (calc)} = 20.9 \text{ L (corr)}$$

b) 1.20 atm, $K = 45°C + 273 = 318$ K

$$V = \frac{nRT}{P} = 1 \text{ mole} \times \frac{0.08206 \text{ L atm}}{\text{K mole}} \times \frac{318 \text{ K}}{1.20 \text{ atm}} = 21.7459 \text{ (calc)} = 21.7 \text{ L (corr)}$$

c) 1.00 atm, $K = 123°C + 273 = 396$ K

$$V = \frac{nRT}{P} = 1 \text{ mole} \times \frac{0.08206 \text{ L atm}}{\text{K mole}} \times \frac{396 \text{ K}}{1.00 \text{ atm}} = 32.49576 \text{ (calc)} = 32.5 \text{ L (corr)}$$

d) 2.00 atm, $K = 123°C + 273 = 396$ K

$$V = \frac{nRT}{P} = 1 \text{ mole} \times \frac{0.08206 \text{ L atm}}{\text{K mole}} \times \frac{396 \text{ K}}{2.00 \text{ atm}} = 16.2478 \text{ (calc)} = 16.2 \text{ L (corr)}$$

12.111 Molar volume $= V = \frac{nRT}{P}$, $K = 22°C + 273 = 295$ K, 2.00 atm

a) For N_2: $V = \dfrac{1 \text{ mole} \dfrac{0.08206 \text{ L atm}}{\text{K mole}} 295 \text{ K}}{2.00 \text{ atm}} = 12.10385$ (calc) $= 12.1$ L (corr)

b) For Ar: $V = \dfrac{1 \text{ mole} \dfrac{0.08206 \text{ L atm}}{\text{K mole}} 295 \text{ K}}{2.00 \text{ atm}} = 12.10385$ (calc) $= 12.1$ L (corr)

c) For SO_2: $V = \dfrac{1 \text{ mole} \dfrac{0.08206 \text{ L atm}}{\text{K mole}} 295 \text{ K}}{2.00 \text{ atm}} = 12.10385$ (calc) $= 12.1$ L (corr)

d) For SO_3: $V = \dfrac{1 \text{ mole} \dfrac{0.08206 \text{ L atm}}{\text{K mole}} 295 \text{ K}}{2.00 \text{ atm}} = 12.10385$ (calc) $= 12.1$ L (corr)

12.113 Molar volume = 44.01 g/L $P = \dfrac{nRT}{V}$

a) $K = 0°C + 273 = 273$ K,

$$P = \dfrac{1 \text{ mole} \dfrac{0.08206 \text{ L atm}}{\text{K mole}} 273 \text{ K}}{44.01 \text{ L}} = 0.50902931 \text{ (calc)} = 0.509 \text{ atm (corr)}$$

b) $K = 30°C + 273 = 303$ K,

$$P = \dfrac{1 \text{ mole} \dfrac{0.08206 \text{ L atm}}{\text{K mole}} 303 \text{ K}}{44.01 \text{ L}} = 0.564966598 \text{ (calc)} = 0.565 \text{ atm (corr)}$$

c) $K = 275°C + 273 = 548\text{ K}$,

$$P = \frac{1\text{ mole} \cdot \frac{0.08206\text{ L atm}}{\text{K mole}} \cdot 548\text{ K}}{44.01\text{ L}} = 1.021787776\text{ (calc)} = 1.02\text{ atm (corr)}$$

d) $K = 525°C + 273 = 798\text{ K}$,

$$P = \frac{1\text{ mole} \cdot \frac{0.08206\text{ L atm}}{\text{K mole}} \cdot 798\text{ K}}{44.01\text{ L}} = 1.487931834\text{ (calc)} = 1.49\text{ atm (corr)}$$

12.115 Molar volume of any gas at STP = 22.41 L; the molar volume does not depend on the identity of the gas, only the moles of gas.
For 1 mole of N_2, Ar, SO_2 and SO_3 gases:

$$V = \frac{1\text{ mole} \cdot \frac{0.082057\text{ L atm}}{\text{K mole}} \cdot 273.15\text{ K}}{1.000\text{ atm}} = 22.413869\text{ (calc)} = 22.41\text{ L (corr)}$$

12.117 Density $\frac{g}{L} = \frac{\frac{g}{\text{mole}}}{\frac{\text{moles}}{L}}$

a) Density $= \frac{\frac{17.04\text{ g}}{1\text{ mole}}}{\frac{17.21\text{ L}}{1\text{ mole}}} = 0.990122022\text{ (calc)} = 0.9901\,\frac{g}{L}\text{ (corr)}$

b) Density $= \frac{\frac{17.04\text{ g}}{1\text{ mole}}}{\frac{22.41\text{ L}}{1\text{ mole}}} = 0.760374832\text{ (calc)} = 0.7604\,\frac{g}{L}\text{ (corr)}$

c) Density $= \frac{\frac{17.04\text{ g}}{1\text{ mole}}}{\frac{23.34\text{ L}}{1\text{ mole}}} = 0.73007712\text{ (calc)} = 0.7301\,\frac{g}{L}\text{ (corr)}$

d) Density $= \frac{\frac{17.04\text{ g}}{1\text{ mole}}}{\frac{35.00\text{ L}}{1\text{ mole}}} = 0.486857142\text{ (calc)} = 0.4869\,\frac{g}{L}\text{ (corr)}$

12.119 a) $K = 25°C + 273 = 298\text{ K}$, 1.20 atm

$$\text{Molar Volume} = \frac{1\text{ mole} \cdot \frac{0.08206\text{ L atm}}{\text{K mole}} \cdot 298\text{ K}}{1.20\text{ atm}} = 20.3782333\text{ (calc)} = 20.4\,\frac{L}{\text{mole}}\text{ (corr)}$$

$$\text{Density} = \frac{\frac{44.01\text{ g CO}_2}{1\text{ mole}}}{\frac{20.4\text{ L}}{1\text{ mole}}} = 2.1573529\text{ (calc)} = 2.16\,\frac{g}{L}\text{ (corr)}$$

Gas Laws

b) $K = 35°C + 273 = 308 \text{ K}$, 1.33 atm

$$\text{Molar Volume} = \frac{1 \text{ mole} \times \frac{0.08206 \text{ L atm}}{\text{K mole}} \times 308 \text{ K}}{1.33 \text{ atm}} = 19.0033684 \text{ (calc)} = 19.0 \frac{\text{L}}{\text{mole}} \text{ (corr)}$$

$$\text{Density} = \frac{\frac{28.01 \text{ g CO}}{1 \text{ mole}}}{\frac{19.0 \text{ L}}{1 \text{ mole}}} = 1.47421053 \text{ (calc)} = 1.47 \frac{\text{g}}{\text{L}} \text{ (corr)}$$

c) At STP: $\text{Density} = \dfrac{\frac{16.05 \text{ g CH}_4}{1 \text{ mole}}}{\frac{22.41 \text{ L}}{1 \text{ mole}}} = 0.716198125 \text{ (calc)} = 0.7162 \frac{\text{g}}{\text{L}} \text{ (corr)}$

d) At STP $\text{Density} = \dfrac{\frac{30.08 \text{ g C}_2\text{H}_6}{1 \text{ mole}}}{\frac{22.41 \text{ L}}{1 \text{ mole}}} = 1.342257921 \text{ (calc)} = 1.342 \frac{\text{g}}{\text{L}} \text{ (corr)}$

12.121 Since each gas will have the same molar volume at STP, the density will be the greatest for the highest molecular weight.
a) O_3 b) PH_3 c) CO_2 d) F_2

12.123 $\text{Density} = \dfrac{\text{Molar Mass}}{\text{Molar Volume}} = \text{MM} = D \times \text{Molar Volume}$

a) $1.97 \dfrac{\text{g}}{\text{L}} \times 22.41 \dfrac{\text{L}}{\text{mole}} = 44.1477 \text{ (calc)} = 44.1 \dfrac{\text{g}}{\text{mole}} \text{ (corr)}$

b) $1.25 \dfrac{\text{g}}{\text{L}} \times 22.41 \dfrac{\text{L}}{\text{mole}} = 28.0125 \text{ (calc)} = 28.0 \dfrac{\text{g}}{\text{mole}} \text{ (corr)}$

c) $0.714 \dfrac{\text{g}}{\text{L}} \times 22.41 \dfrac{\text{L}}{\text{mole}} = 16.00074 \text{ (calc)} = 16.0 \dfrac{\text{g}}{\text{mole}} \text{ (corr)}$

d) $3.17 \dfrac{\text{g}}{\text{L}} \times 22.41 \dfrac{\text{L}}{\text{mole}} = 71.0397 \text{ (calc)} = 71.0 \dfrac{\text{g}}{\text{mole}} \text{ (corr)}$

Chemical Calculations Using Molar Volume (Sec. 12.14)

12.125 a) $23.7 \text{ L} \times \dfrac{1 \text{ mole Ar}}{22.41 \text{ L}} \times \dfrac{39.95 \text{ g Ar}}{1 \text{ mole Ar}} = 42.2496653 \text{ (calc)} = 42.2 \text{ g Ar (corr)}$

b) $23.7 \text{ L} \times \dfrac{1 \text{ mole N}_2\text{O}}{22.41 \text{ L}} \times \dfrac{44.02 \text{ g N}_2\text{O}}{1 \text{ mole N}_2\text{O}} = 46.55394913 \text{ (calc)} = 46.6 \text{ g N}_2\text{O (corr)}$

c) $23.7 \text{ L} \times \dfrac{1 \text{ mole SO}_3}{22.41 \text{ L}} \times \dfrac{80.07 \text{ g SO}_3}{1 \text{ mole SO}_3} = 84.67911647 \text{ (calc)} = 84.7 \text{ g SO}_3 \text{ (corr)}$

d) $23.7 \text{ L} \times \dfrac{1 \text{ mole PH}_3}{22.41 \text{ L}} \times \dfrac{34.00 \text{ g PH}_3}{1 \text{ mole PH}_3} = 35.95716198 \text{ (calc)} = 36.0 \text{ g SO}_3 \text{ (corr)}$

12.127 $K = 44°C + 273 = 317$ K, 1.15 atm, 23.7 L samples
Moles of gas present in each sample will be the same,

$$n = \frac{1.15 \text{ atm} \cdot 23.7 \text{ L}}{0.08206 \frac{\text{L atm}}{\text{K mole}} \cdot 317 \text{ K}} = 1.04774455 \text{ (calc)} = 1.05 \text{ moles gas (corr)}$$

a) $1.05 \text{ moles} \times \frac{39.95 \text{ g Ar}}{1 \text{ mole Ar}} = 41.9475 \text{ (calc)} = 41.9 \text{ g Ar (corr)}$

b) $1.05 \text{ moles} \times \frac{44.02 \text{ g N}_2\text{O}}{1 \text{ mole N}_2\text{O}} = 46.221 \text{ (calc)} = 46.2 \text{ g N}_2\text{O} \text{ (corr)}$

c) $1.05 \text{ moles} \times \frac{80.07 \text{ g SO}_3}{1 \text{ mole SO}_3} = 84.0735 \text{ (calc)} = 84.1 \text{ g SO}_3 \text{ (corr)}$

d) $1.05 \text{ moles} \times \frac{17.04 \text{ g NH}_3}{1 \text{ mole NH}_3} = 17.892 \text{ (calc)} = 17.9 \text{ g NH}_3 \text{ (corr)}$

12.129 For a, b, c and d: $1.25 \text{ mole gas} \times \frac{22.41 \text{ L gas}}{1 \text{ mole gas}} = 28.0125 \text{ (calc)} = 28.0 \text{ L gas (corr)}$

Given 1.25 mole samples of each gas, at STP, 1 mole of any gas occupies 22.41 L. Therefore each gas will occupy 28.0 liters.

12.131 $46°C + 273 = 319$ K, 1.73 atm

$$V = \frac{1.25 \text{ moles} \cdot 0.08206 \frac{\text{L atm}}{\text{K mole}} \cdot 319 \text{ K}}{1.73 \text{ atm}} = 18.914118497 \text{ (calc)} = 18.9 \text{ L (corr)}$$

The volume will be the same for each sample of gas; it depends only on the temperature and pressure, not the identity of the gas. Each gas will therefore occupy 18.9 liters.

12.133 a) $24.5 \text{ g N}_2 \times \frac{1 \text{ mole N}_2}{28.02 \text{ g N}_2} \times \frac{22.41 \text{ L N}_2}{1 \text{ mole N}_2} = 19.594754 \text{ (calc)} = 19.6 \text{ L N}_2 \text{ (corr)}$

$24.5 \text{ g NH}_3 \times \frac{1 \text{ mole NH}_3}{17.04 \text{ g NH}_3} \times \frac{22.41 \text{ L NH}_3}{1 \text{ mole NH}_3} = 32.220951 \text{ (calc)} = 32.2 \text{ L NH}_3 \text{ (corr)}$

∴ the NH$_3$ has the larger volume.

b) $30.0 \text{ g O}_2 \times \frac{1 \text{ mole O}_2}{32.00 \text{ g O}_2} \times \frac{22.41 \text{ L O}_2}{1 \text{ mole O}_2} = 21.009375 \text{ (calc)} = 21.0 \text{ L O}_2 \text{ (corr)}$

$30.0 \text{ g O}_3 \times \frac{1 \text{ mole O}_3}{48.00 \text{ g O}_3} \times \frac{22.41 \text{ L O}_3}{1 \text{ mole O}_3} = 14.00625 \text{ (calc)} = 14.0 \text{ L O}_3 \text{ (corr)}$

∴ the O$_2$ has the larger volume.

c) $10.0 \text{ g SO}_2 \times \frac{1 \text{ mole SO}_2}{64.07 \text{ g SO}_2} \times \frac{22.41 \text{ L SO}_2}{1 \text{ mole SO}_2} = 3.49773685 \text{ (calc)} = 3.50 \text{ L SO}_2 \text{ (corr)}$

$20.0 \text{ g NO}_2 \times \frac{1 \text{ mole NO}_2}{46.01 \text{ g NO}_2} \times \frac{22.41 \text{ L NO}_2}{1 \text{ mole NO}_2} = 9.7413606 \text{ (calc)} = 9.74 \text{ L NO}_2 \text{ (corr)}$

∴ the NO$_2$ has the larger volume.

Gas Laws

d) $15.0 \text{ g N}_2\text{O} \times \dfrac{1 \text{ mole N}_2\text{O}}{44.02 \text{ g N}_2\text{O}} \times \dfrac{22.41 \text{ L N}_2\text{O}}{1 \text{ mole N}_2\text{O}} = 7.6363017 \text{ (calc)} = 7.64 \text{ L N}_2\text{O (corr)}$

$20.0 \text{ g NO} \times \dfrac{1 \text{ mole NO}}{30.01 \text{ g NO}} \times \dfrac{22.41 \text{ L NO}}{1 \text{ mole NO}} = 14.935022 \text{ (calc)} = 14.9 \text{ L NO (corr)}$

∴ the NO has the larger volume.

Alternate method: The one with the smaller molar mass will have the lower density.
If the masses are equal, the one with the lower density will have the larger volume.

a) Mass equal, the molar mass of NH_3 is less than N_2, ∴ NH_3 has the larger volume.
b) Mass equal, the molar mass of O_2 is less than O_3, ∴ O_2 has the larger volume.
c) Since NO_2 has smaller molar mass than SO_2, 10.0 g NO_2 would have a larger volume than 10.0 g SO_2, ∴ 20.0 g NO_2 would be even larger.
d) Since NO has smaller molar mass than N_2O, 15.0 g NO would have a larger volume than 15.0 g of N_2O, ∴ 20.0 g NO would be even larger.

12.135 $25.0 \text{ g NO} \times \dfrac{1 \text{ mole NO}}{30.01 \text{ g NO}} \times \dfrac{2 \text{ moles NO}_2}{2 \text{ moles NO}} \times \dfrac{22.41 \text{ L NO}_2}{1 \text{ mole NO}_2} = 18.6688 \text{ (calc)} = 18.7 \text{ L NO}_2 \text{ (corr)}$

12.137 Method 1. Find moles of O_2. $T = 27°C + 273 = 300.\text{ K}$

$n = \dfrac{1.00 \text{ atm} \times 25.0 \text{ L}}{0.08206 \dfrac{\text{atm L}}{\text{mole K}} \times 300.\text{ K}} = 1.0155171 \text{ (calc)} = 1.02 \text{ moles O}_2 \text{ (corr)}$

$1.02 \text{ moles O}_2 \times \dfrac{2 \text{ moles NO}}{1 \text{ mole O}_2} \times \dfrac{30.01 \text{ g NO}}{1 \text{ mole NO}} = 61.2204 \text{ (calc)} = 61.2 \text{ g NO (corr)}$

Method 2. Find volume of O_2 at STP (P constant).

$V_2 = \dfrac{V_1 T_2}{T_1}$. $V_2 = 25.0 \text{ L} \times \dfrac{273 \text{ K}}{300.\text{ K}} = 22.75 \text{ (calc)} = 22.8 \text{ L O}_2 \text{ at STP (corr)}$

$22.8 \text{ L O}_2 \times \dfrac{1 \text{ mole O}_2}{22.41 \text{ L O}_2} \times \dfrac{2 \text{ moles NO}}{1 \text{ mole O}_2} \times \dfrac{30.01 \text{ g NO}}{1 \text{ mole NO}} = 61.06453 \text{ (calc)} = 61.1 \text{ g NO (corr)}$

Note: Rounding in intermediate steps causes the difference.

12.139 $12.0 \text{ g Mg} \times \dfrac{1 \text{ mole Mg}}{24.31 \text{ g Mg}} \times \dfrac{1 \text{ mole H}_2}{1 \text{ mole Mg}} = 0.49382716 \text{ (calc)} = 0.494 \text{ mole H}_2 \text{ (corr)}$

$V = \dfrac{0.494 \text{ mole} \times 0.08206 \dfrac{\text{atm L}}{\text{mole K}} \times 296 \text{ K}}{0.980 \text{ atm}} = 12.244022 \text{ (calc)} = 12.2 \text{ L (corr)}$

12.141 $T = 450.°C + 273 = 723 \text{ K}$. Conversion factor: $\dfrac{7 \text{ moles product gases}}{2 \text{ moles NH}_4\text{NO}_3}$

$100.0 \text{ g NH}_4\text{NO}_3 \times \dfrac{1 \text{ mole NH}_4\text{NO}_3}{80.06 \text{ g NH}_4\text{NO}_3} \times \dfrac{7 \text{ moles product}}{2 \text{ moles NH}_4\text{NO}_3}$

$= 4.3717212 \text{ (calc)} = 4.37 \text{ moles product (corr)}$

Method 1: $V = \dfrac{nRT}{P} = \dfrac{4.37 \text{ moles}}{1 \text{ atm}} \times 0.08206 \dfrac{\text{atm L}}{\text{mole K}} \times 723 \text{ K} = 259.26939 \text{ (calc)} = 259 \text{ L (corr)}$

Method 2: $4.37 \text{ moles} \times \dfrac{22.41 \text{ L}}{1 \text{ mole}} \times \dfrac{723 \text{ K}}{273 \text{ K}} = 259.35758 \text{ (calc)} = 259 \text{ L (corr)}$

12.143 $T = 38°C + 273 = 311 \text{ K}$ $\quad n = \dfrac{645 \text{ mm Hg} \times 75.0 \text{ L}}{62.36 \dfrac{\text{mm Hg L}}{\text{mole K}} \times 311 \text{ K}} = 2.494333287$ (calc)

$= 2.49$ moles NO (corr)

2.49 moles NO $\times \dfrac{3 \text{ moles NO}_2}{1 \text{ mole NO}} = 7.47$ moles NO_2 (calc and corr)

$T = 21°C + 273 = 294 \text{ K}$ $\quad V = \dfrac{7.47 \text{ moles} \times 0.08206 \dfrac{\text{atm L}}{\text{mole K}} \times 294 \text{ K}}{2.31 \text{ atm}} = 78.01668$ L (calc)

$= 78.0$ L (corr)

12.145 $T = 125°C + 273 = 398 \text{ K}$ $\quad n = \dfrac{3.00 \text{ atm} \times 10.0 \text{ L}}{0.08206 \dfrac{\text{atm L}}{\text{mole K}} \times 398 \text{ K}} = 0.9185581$ (calc)

$= 0.919$ mole C_2H_2 (corr)

$T = 20.°C + 273 = 293 \text{ K}$ $\quad n = \dfrac{0.750 \text{ atm} \times 2.50 \text{ L}}{0.08206 \dfrac{\text{atm L}}{\text{mole K}} \times 293 \text{ K}} = 0.077983$ (calc)

$= 0.0780$ mole O_2 (corr)

0.0780 mole $O_2 \times \dfrac{4 \text{ moles CO}_2}{5 \text{ moles O}_2} = 0.0624$ mole CO_2 (calc and corr)

0.919 mole $C_2H_2 \times \dfrac{4 \text{ moles CO}_2}{2 \text{ moles C}_2H_2} = 1.838$ (calc) $= 1.84$ moles CO_2 (corr)

∴ 0.0780 mole O_2 is limiting, and 0.0624 mole CO_2 is produced. $\quad T = 75°C + 273 = 348 \text{ K}$

Volume $CO_2 = \dfrac{0.0624 \text{ mole} \times 0.08206 \dfrac{\text{atm L}}{\text{mole K}} \times 348 \text{ K}}{2.00 \text{ atm}} = 0.890974656$ L (calc) $= 0.891$ L (corr)

Mixtures of Gases (Sec. 12.15)

12.147 Moles of gas $= 3.00$ moles $N_2 + 3.00$ moles $O_2 = 6.00$ moles gas mixture
$T = 27°C + 273 = 300.$ K

$V = \dfrac{nRT}{P} = \dfrac{6.00 \text{ mole} \times 0.08206 \dfrac{\text{atm L}}{\text{mole K}} \times 300. \text{ K}}{20.00 \text{ atm}} = 7.3854$ (calc) $= 7.39$ L (corr)

12.149 $T = 27°C + 273 = 300.$ K

a) 3.00 g Ne $\times \dfrac{1 \text{ mole Ne}}{20.18 \text{ g Ne}} = 0.1486620416$ (calc) $= 0.149$ mole Ne (corr)

3.00 g Ar $\times \dfrac{1 \text{ mole Ar}}{39.95 \text{ g Ar}} = 0.07509386733$ (calc) $= 0.0751$ mole Ar (corr)

Total moles of gas $= 0.149 + 0.0751 = 0.2241$ (calc) $= 0.224$ mole (corr)

$V = \dfrac{nRT}{P} = \dfrac{0.224 \text{ mole} \times 0.08206 \dfrac{\text{atm L}}{\text{mole K}} \times 300. \text{ K}}{1.00 \text{ atm}} = 5.514432$ (calc) $= 5.51$ L (corr)

Gas Laws

b) $4.00 \text{ g Ne} \times \dfrac{1 \text{ mole Ne}}{20.18 \text{ g Ne}} = 0.1982160555 \text{ (calc)} = 0.198 \text{ mole Ne (corr)}$

$2.00 \text{ g Ar} \times \dfrac{1 \text{ mole Ar}}{39.95 \text{ g Ar}} = 0.05006257822 \text{ (calc)} = 0.0501 \text{ mole Ar (corr)}$

Total moles of gas = $0.198 + 0.0501 = 0.2481$ (calc) = 0.248 mole (corr)

$V = \dfrac{nRT}{P} = \dfrac{0.248 \text{ mole} \times 0.08206 \dfrac{\text{atm L}}{\text{mole K}} \times 300.\text{ K}}{1.00 \text{ atm}} = 6.105264 \text{ (calc)} = 6.11 \text{ L (corr)}$

c) $5.00 \text{ g Ar} \times \dfrac{1 \text{ mole Ar}}{39.95 \text{ g Ar}} = 0.1251564456 \text{ (calc)} = 0.125 \text{ mole Ar (corr)}$

Total moles of gas = $3.00 + 0.125 = 3.125$ (calc) = 3.12 moles (corr)

$V = \dfrac{nRT}{P} = \dfrac{3.12 \text{ moles} \times 0.08206 \dfrac{\text{atm L}}{\text{mole K}} \times 300.\text{ K}}{1.00 \text{ atm}} = 76.80816 \text{ (calc)} = 76.8 \text{ L (corr)}$

d) Total moles of gas = $4.00 + 4.00 = 8$ (calc) = 8.00 moles (corr)

$V = \dfrac{nRT}{P} = \dfrac{8.00 \text{ moles} \times 0.08206 \dfrac{\text{atm L}}{\text{mole K}} \times 300.\text{ K}}{1.00 \text{ atm}} = 196.944 \text{ (calc)} = 197 \text{ L (corr)}$

12.151 $3.00 \text{ moles Ar} + 3.00 \text{ moles Ne} + 3.00 \text{ moles He} = 9$ (calc) = 9.00 moles (corr)

$T = 20°C + 273 = 293.\text{ K} \quad P = \dfrac{9.00 \text{ moles} \times 0.08206 \dfrac{\text{atm L}}{\text{mole K}} \times 293 \text{ K}}{27.0 \text{ L}} = 8.01452666 \text{ (calc)}$

$= 8.01 \text{ atm (corr)}$

Dalton's Law of Partial Pressures (Sec. 12.16)

12.153 $P_{He} = 9.0$ atm
$P_{Ne} = 14.0 \text{ atm} - 9.0 \text{ atm} = 5$ (corr) = 5.0 atm (corr)
$P_{Ar} = 29.0 \text{ atm} - 14.0 \text{ atm} = 15$ (calc) = 15.0 atm (corr)

12.155 a) $P_{CO_2} = 842 - 675 = 167$ mm Hg (calc and corr)
b) $P_{N_2} = $ same as before $= 354$ mm Hg (calc and corr)
c) $P_{Ar} = $ same as before $= 235$ mm Hg (calc and corr)
d) $P_{H_2} = $ same as before $= 675 - P_{N_2} - P_{Ar} = 675 - 354 - 235 = 86$ mm Hg (calc and corr)

12.157 Total spheres = $5 \text{ Ne} + 3 \text{ Ar} + 2 \text{ Kr} = 10$ spheres

a) $P_{Ne} = \dfrac{5 \text{ spheres}}{10 \text{ spheres}} \times 0.060 \text{ atm} = 0.030$ atm (calc and corr)

b) $P_{Ar} = \dfrac{3 \text{ spheres}}{10 \text{ spheres}} \times 0.060 \text{ atm} = 0.018$ atm (calc and corr)

c) $P_{Kr} = \dfrac{2 \text{ spheres}}{10 \text{ spheres}} \times 0.060 \text{ atm} = 0.012$ atm (calc and corr)

12.159 a) moles CO = 25.0 g CO × $\dfrac{1 \text{ mole CO}}{28.01 \text{ g CO}}$ = 0.8925383792 (calc) = 0.893 mole CO (corr)

moles CO_2 = 25.0 g CO_2 × $\dfrac{1 \text{ mole } CO_2}{44.01 \text{ g } CO_2}$ = 0.5680527153 (calc) = 0.568 mole CO_2 (corr)

moles H_2S = 25.0 g H_2S × $\dfrac{1 \text{ mole } H_2S}{34.09 \text{ g } H_2S}$ = 0.7333528894 (calc) = 0.733 mole H_2S (corr)

Total moles = 0.893 + 0.568 + 0.733 = 2.194 moles total (calc and corr)

Mole fraction CO = $\dfrac{0.893}{2.194}$ = 0.4070191431 (calc) = 0.407 mole fraction CO (corr)

Mole fraction CO_2 = $\dfrac{0.568}{2.194}$ = 0.258887876 (calc) = 0.259 mole fraction CO_2 (corr)

Mole fraction H_2S = $\dfrac{0.733}{2.194}$ = 0.3340929809 (calc) = 0.334 mole fraction H_2S (corr)

b) Partial pressure = mole fraction × total pressure
P_{CO} = 0.407 × 1.72 atm = 0.70004 (calc) = 0.700 atm (corr)
P_{CO_2} = 0.259 × 1.72 atm = 0.44548 (calc) = 0.445 atm (corr)
P_{H_2S} = 0.334 × 1.72 atm = 0.57448 (calc) = 0.574 atm (corr)

12.161 a) $P_{O_2} = \dfrac{0.50 \text{ mole} \times 0.08206 \frac{\text{atm L}}{\text{mole K}} \times 293 \text{ K}}{2.50 \text{ L}}$ = 4.808716 (calc) = 4.8 atm (corr)

b) 4.8 atm. A change in the amount of N_2 does not affect the partial pressure of O_2.

c) 4.8 atm. A change in the amount of N_2 and Ar does not affect the partial pressure of O_2.

d) 0.50 g O_2 × $\dfrac{1 \text{ mole } O_2}{32.00 \text{ g } O_2}$ = 0.015625 (calc) = 0.016 mole O_2 (corr)

$P_{O_2} = \dfrac{0.016 \text{ mole} \times 0.08206 \frac{\text{atm L}}{\text{mole K}} \times 293 \text{ K}}{2.50 \text{ L}}$ = 0.15387891 (calc) = 0.15 atm (corr)

12.163 Partial pressure = mole fraction × total pressure

a) $P_{O_2} = \dfrac{0.40 \text{ mole}}{0.40 + 0.40 \text{ mole}} \times 1.20 \text{ atm}$ = 0.6 (calc) = 0.60 atm (corr)

b) $P_{O_2} = \dfrac{0.40 \text{ mole}}{0.40 + 0.80 \text{ mole}} \times 1.20 \text{ atm}$ = 0.4 (calc) = 0.40 atm (corr)

c) The mole fraction of O_2 is 1/3.

P_{O_2} = 1/3 × 1.20 atm = 0.4 (calc) = 0.400 atm (corr)

d) 0.400 atm. An equal number of molecules means that there is also an equal number of moles present.

12.165 $P_A = P_T \cdot \dfrac{n_A}{n_T}$ or $P_T = P_A \cdot \dfrac{n_T}{n_A}$ P_T = 0.40 atm × $\dfrac{4.0 + 2.0 + 0.50 \text{ moles total}}{0.5 \text{ mole Ne}}$

= 5.2 atm (calc and corr)

Gas Laws

12.167 a) $1.00 - 0.150 = 0.850$ mole fraction of O_2
0.850×6.00 atm $= 5.0$ (calc) $= 5.10$ atm (corr)
b) 2.26 atm $= 0.180$ (X atm) thus $X = 12.55555556$ (calc) $= 12.6$ atm total pressure (corr)
$1.00 - 0.180 = 0.82$ mole fraction O_2
0.82 (12.6 atm) $= 10.332$ (calc) $= 10.3$ atm partial pressure of O_2 gas (corr)

12.169 0.500 atm He $+ 0.250$ atm Ar $+ 0.350$ atm Xe $= 1.100$ atm total pressure

$$X_{He} = \frac{0.500 \text{ atm}}{1.100 \text{ atm}} = 0.454545 \text{ (calc)} = 0.455 \text{ (corr)}$$

$$X_{Ar} = \frac{0.250 \text{ atm}}{1.100 \text{ atm}} = 0.227272 \text{ (calc)} = 0.227 \text{ (corr)}$$

$$X_{xe} = \frac{0.350 \text{ atm}}{1.100 \text{ atm}} = 0.3181818 \text{ (calc)} = 0.318 \text{ (corr)}$$

12.171 Moles $N_2 = 20.0$ g $N_2 \times \dfrac{1 \text{ mole } N_2}{28.02 \text{ g } N_2} = 0.71377588$ (calc) $= 0.714$ mole N_2 (corr)

Moles Ar $= 8.00$ g Ar $\times \dfrac{1 \text{ mole Ar}}{39.95 \text{ g Ar}} = 0.200250$ (calc) $= 0.200$ mole Ar (corr)

Total moles $= 0.714 + 0.200 = 0.914$ mole

$\dfrac{P_1}{n_1} = \dfrac{P_2}{n_2}$ $\dfrac{1.00 \text{ atm}}{0.714 \text{ mole}} = \dfrac{P_2}{0.914 \text{ mole}}$ $P_2 = 1.280112$ (calc) $= 1.28$ atm (corr)

12.173 From the coefficients in the balanced equation, the mole fraction of $N_2 = \dfrac{1}{1+3} = 0.250$ and the mole fraction of $H_2 = \dfrac{3}{1+3} = 0.750$.

$P_{N_2} = 0.250 \times 852$ mm Hg $= 213$ mm Hg (calc and corr)
$P_{H_2} = 0.750 \times 852$ mm Hg $= 639$ mm Hg (calc and corr)

12.175 Water vapor pressure values are obtained from Table 12.6 in the text.
a) $743 - 16.5$ mm Hg $= 726.5$ (calc) $= 726$ mm Hg (corr)
b) $645 - 28.3$ mm Hg $= 616.7$ (calc) $= 617$ mm Hg (corr)
c) $762 - 39.9$ mm Hg $= 722.1$ (calc) $= 722$ mm Hg (corr)
d) 0.933 atm $\times \dfrac{760 \text{ mm Hg}}{1 \text{ atm}} = 709.08$ (calc) $= 709$ mm Hg (corr)
$709 - 18.7$ mm Hg $= 690.3$ (calc) $= 6\overline{90}$ mm Hg (corr)

12.177 a) mole fraction He, $X_{He} = \dfrac{0.100 \text{ mole He}}{0.100 + 0.200 + 0.800 \text{ moles total}} = 0.09090909$ (calc) $= 0.0909$ (corr)

b) mole % Ar $= \dfrac{0.800 \text{ mole Ar}}{1.100 \text{ moles total}} \times 100 = 72.7272727$ (calc) $= 72.7\%$ Ar (corr)

c) $P_{Ne} = X_{Ne} \times P_{total} = \dfrac{0.200 \text{ mole}}{1.100 \text{ moles total}} \times 1.00 \text{ atm} = 0.1818182 \text{ (calc)} = 0.182 \text{ atm (corr)}$

pressure % Ne $= \dfrac{P_{Ne}}{P_{total}} \times 100 = \dfrac{0.182 \text{ atm}}{1.00 \text{ atm}} \times 100 = 18.2\%$ Ne

d) volume He at STP (alone) $= 0.100 \text{ mole He} \times \dfrac{22.41 \text{ L He}}{1 \text{ mole He}} = 2.241 \text{ (calc)} = 2.24 \text{ L He (corr)}$

volume % He $= \dfrac{\text{vol He}}{\text{total volume}} \times 100 = \dfrac{2.24 \text{ L He}}{24.64 \text{ L total}} \times 100$

$= 9.0909091 \text{ (calc)} = 9.09\% \text{ He (corr)}$

12.179 a) at constant temperature and pressure,

vol % $O_2 = \dfrac{V_{O_2}}{V_{total}} \times 100 = \dfrac{2.0 \text{ L } O_2}{8.0 \text{ L total}} \times 100 = 25\% \, O_2$ (calc and corr)

b) mole Ar $= 3.0 \text{ L Ar} \times \dfrac{1 \text{ mole Ar}}{22.41 \text{ L Ar}} = 0.1338688086 \text{ (calc)} = 0.13 \text{ mole Ar (corr)}$

mole $O_2 = 2.0 \text{ L } O_2 \times \dfrac{1 \text{ mole } O_2}{22.41 \text{ L } O_2} = 0.08924587238 \text{ (calc)} = 0.089 \text{ mole } O_2 \text{ (corr)}$

mole Ne $= 3.0 \text{ L Ne} \times \dfrac{1 \text{ mole Ne}}{22.41 \text{ L Ne}} = 0.1338688086 \text{ (calc)} = 0.13 \text{ mole Ne (corr)}$

total moles $= 0.13 + 0.089 + 0.13 = 0.349 \text{ (calc)} = 0.35 \text{ mole (corr)}$

mole % Ar $= \dfrac{\text{moles Ar}}{\text{total moles}} \times 100 = \dfrac{0.13}{0.35} \times 100 = 37.142857 \text{ (calc)} = 37\% \text{ Ar (corr)}$

c) In the final container, $P_{Ne} = X_{Ne} \times P_{total} = \dfrac{0.13 \text{ mole Ne}}{0.35 \text{ mole total}} \times 1.00 \text{ atm}$

$= 0.37142857 \text{ (calc)} = 0.37 \text{ atm (corr)}$

Pressure % Ne $= \dfrac{0.37 \text{ atm Ne}}{1.0 \text{ atm total}} \times 100 = 37\%$ Ne (calc and corr)

d) $P_{O_2} = X_{O_2} \times P_{total}$ in final container $= \dfrac{0.089 \text{ mole } O_2}{0.35 \text{ mole total}} \times 1.0 \text{ atm}$

$= 0.25428571 \text{ (calc)} = 0.25 \text{ atm (corr)}$

Additional Problems

12.181 $1.00 \text{ L } CO_2 \times \dfrac{1 \text{ mole } CO_2}{22.41 \text{ L } CO_2} \times \dfrac{6.022 \times 10^{23} \text{ molecules } CO_2}{1 \text{ mole } CO_2}$

$= 2.6871932 \times 10^{22} \text{ (calc)} = 2.69 \times 10^{22} \text{ molecules } CO_2 \text{ (corr)}$

12.183 $8.00 \text{ g } N_2 \times \dfrac{1 \text{ mole } N_2}{28.02 \text{ g } N_2} = 0.2855103 \text{ (calc)} = 0.286 \text{ mole } N_2 \text{ (corr)}$

$1.00 \times 10^{23} \text{ molecules } NH_3 \times \dfrac{1 \text{ mole } NH_3}{6.022 \times 10^{23} \text{ molecules } NH_3}$

$= 0.1660577 \text{ (calc)} = 0.166 \text{ mole } NH_3 \text{ (corr)}$

$V_{NH_3} = V_{N_2} \times \dfrac{n_{NH_3}}{n_{H_2}} = 6.00 \text{ L} \times \dfrac{0.166 \text{ mole}}{0.286 \text{ mole}} = 3.4825174 \text{ (calc)} = 3.48 \text{ L } NH_3 \text{ (corr)}$

Gas Laws

12.185 $n_1 = \dfrac{1.00 \text{ atm} \times 1.00 \text{ L}}{0.08206 \dfrac{\text{atm L}}{\text{mole K}} \times 298 \text{ K}} = 0.04089330609 \text{ (calc)} = 0.0409 \text{ mole } O_2 \text{ (corr)}$

$n_2 = \dfrac{0.880 \text{ atm} \times 1.00 \text{ L}}{0.08206 \dfrac{\text{atm L}}{\text{mole K}} \times 295 \text{ K}} = 0.0363520698 \text{ (calc)} = 0.0364 \text{ mole } O_2 \text{ (corr)}$

$\Delta n = (0.0409 - 0.0364) \text{ mole } O_2 = 0.0045 \text{ mole } O_2 \text{ (calc and corr)}$

$0.0045 \text{ mole } O_2 \times \dfrac{32.00 \text{ g } O_2}{1 \text{ mole } O_2} = 0.144 \text{ (calc)} = 0.14 \text{ g } O_2 \text{ (corr)}$

12.187 Let V_1 = original volume in mL, $V_2 = V_1 - 25.0$ mL   Applying Boyle's law: $V_2 = V_1 \times \dfrac{P_1}{P_2}$

$V_1 - 25.0 = V_1 \times \dfrac{1.50 \text{ atm}}{3.50 \text{ atm}}$ Rearranging: $V_1 - V_1 \times \dfrac{1.50}{3.50} = 25.0$ or $V_1\left(1 - \dfrac{1.50}{3.50}\right) = 25.0$

or $V_1 \dfrac{3.50 - 1.50}{3.50} = 25.0$ or $V_1 \times \dfrac{2.00}{3.50} = 25.0$

$V_1 = 25.0 \times \dfrac{3.50}{2.00} = 43.75 \text{ (calc)} = 43.8 \text{ mL (corr)}$

12.189 $P_1V_1 = P_2V_2 \qquad V_2 = 0.800\, V_1$

$P_2 = P_1 \times \dfrac{V_1}{V_2} = P_1 \dfrac{V_1}{0.800\, V_1} = 1.25\, P_1 \text{ (calc and corr)}$

$\Delta P = (1.25\, P_1 - P_1) = 0.25\, P_1$

% change $= \dfrac{\Delta P}{P_1} \times 100 = \dfrac{0.25\, P_1}{P_1} \times 100 = 25\% \text{ (calc and corr)}$

12.191 $V_2 = 1.500\, V_1 \qquad T_2 = T_1 \times \dfrac{V_2}{V_1} = T_1 \dfrac{1.500\, V_1}{V_1} = 1.500\, T_1$

$\Delta T = 1.500\, T_1 - T_1 = 0.500\, T_1$

% change $= \dfrac{\Delta T}{T_1} \times 100 = \dfrac{0.500\, T_1}{T_1} \times 100 = 50\% \text{ (calc)} = 50.0\% \text{ (corr)}$

12.193 a) $\dfrac{d_{O_2}}{d_{N_2}} = \dfrac{\dfrac{\text{molar mass}_{O_2}}{\text{molar volume (STP)}}}{\dfrac{\text{molar mass}_{N_2}}{\text{molar volume (STP)}}} = \dfrac{\text{molar mass}_{O_2}}{\text{molar mass}_{N_2}} = \dfrac{32.00 \text{ g/mole}}{28.02 \text{ g/mole}} = 1.1420414 \text{ (calc)} = 1.14\text{:}1 \text{ (corr)}$

b) $\dfrac{d_{O_2}}{d_{N_2}} = \dfrac{\dfrac{(MM)_{O_2} P_{O_2}}{RT}}{\dfrac{(MM)_{N_2} P_{N_2}}{RT}} = \dfrac{(MM)_{O_2} \times 1.25 \text{ atm}}{(MM)_{N_2} \times 1.25 \text{ atm}} = \dfrac{32.00 \text{ g/mole}}{28.02 \text{ g/mole}} = 1.1420414 \text{ (calc)} = 1.14\text{:}1 \text{ (corr)}$

12.195 Add the partial pressures of each gas in the new container.

$$P_{N_2} = 350.0 \text{ mm Hg} \times \frac{1.000 \text{ L}}{2.000 \text{ L}} = 175 \text{ (calc)} = 175.0 \text{ mm Hg (corr)}$$

$$P_{O_2} = 300.0 \text{ mm Hg} \times \frac{6.000 \text{ L}}{2.000 \text{ L}} = 900. \text{ (calc)} = 900.0 \text{ mm Hg (corr)}$$

$$P_{H_2} = 250.0 \text{ mm Hg} \times \frac{1.000 \text{ L}}{2.000 \text{ L}} = 125 \text{ (calc)} = 125.0 \text{ mm Hg (corr)}$$

Total pressure = 175.0 + 900.0 + 125.0 = 1200.0 mm Hg

12.197 Find the pressure of SF$_6$ in the new container. This is its partial pressure.

$$P_{SF_6} = 0.97 \text{ atm} \times \frac{376 \text{ mL}}{275 \text{ mL}} = 1.3262545 \text{ (calc)} = 1.3 \text{ atm (corr)}$$

12.199 $P_{Ar} = \dfrac{g\,RT}{V(MM)} = \dfrac{15.0 \text{ g} \times 0.08206 \frac{\text{atm L}}{\text{mole K}} \times 327 \text{ K}}{4.0 \text{ L} \times 39.95 \text{ g/mole}} = 2.5188004 \text{ (calc)} = 2.5 \text{ atm (corr)}$

12.201 N$_2$ pressure in He container at 37°C (310. K):

$$P_2 = 645 \text{ mm Hg} \times \frac{30.0 \text{ mL}}{40.0 \text{ mL}} \times \frac{310 \text{ K}}{300 \text{ K}} = 499.875 \text{ (calc)} = 5\overline{00} \text{ mm Hg (corr)}$$

Total pressure in 40.0 mL container at 37°C:
$P_{\text{total}} = 765 + 5\overline{00}$ mm Hg = 1265 mm Hg (calc and corr)
Total pressure at 32°C:

$$P_2 = 1265 \text{ mm Hg} \times \frac{305 \text{ K}}{310. \text{ K}} = 1244.5968 \text{ (calc)} = 1240 \text{ mm Hg (corr)}$$

Cumulative Problems

12.203 $2.24 \text{ L} \times \dfrac{1 \text{ mole gas}}{22.41 \text{ L}} = 0.09995537706 \text{ (calc)} = 0.100 \text{ mole (corr)}$

Molar mass = $\dfrac{7.5 \text{ g}}{0.100 \text{ mole}} = 75 \text{ g/mole (calc and corr)}$ Molecular mass = 75 amu

12.205 molar mass = density (STP) × molar volume (STP)

$$= \frac{1.517 \text{ g}}{1 \text{ L}} \times \frac{22.41 \text{ L}}{1 \text{ mole}} = 33.99597 \text{ (calc)} = 34.00 \text{ g/mole (corr)}$$

molar mass P + X(molar mass H) = 34.00 g/mole
30.97 + X(1.01) = 34.00
1.01X = 3.03
X = 3

Gas Laws

12.207 Take 100 g mixture, giving 75.0 g HCl, 5.00 g H$_2$, and 20.0 g He.

a) moles HCl = 75.0 g HCl × $\dfrac{1 \text{ mole}}{36.46 \text{ g}}$ = 2.0570488 (calc) = 2.06 moles HCl (corr)

moles H$_2$ = 5.00 g H$_2$ × $\dfrac{1 \text{ mole H}_2}{2.02 \text{ g H}_2}$ = 2.4752475 (calc) = 2.48 moles H$_2$ (corr)

moles He = 20.0 g He × $\dfrac{1 \text{ mole He}}{4.00 \text{ g He}}$ = 5 (calc) = 5.00 moles He (corr)

total moles = 2.06 + 2.48 + 5.00 = 9.54 moles (calc and corr)

$X_{HCl} = \dfrac{2.06}{9.54}$ = 0.2159329 (calc) = 0.216 (corr)

$X_{H_2} = \dfrac{2.48}{9.54}$ = 0.2599580713 (calc) = 0.260 (corr)

$X_{He} = \dfrac{5.00}{9.54}$ = 0.5241090147 (calc) = 0.524 (corr)

b) $P_{HCl} = X_{HCl} \cdot P_{total}$ = 0.216 × 1.20 atm = 0.2592 (calc) = 0.259 atm (corr)
P_{H_2} = 0.260 × 1.20 atm = 0.312 atm (calc and corr)
P_{He} = 0.524 × 1.20 atm = 0.6288 (calc) = 0.629 atm (corr)

c) weighted average molar mass ($\overline{MM}$) = sum of the product of each mole fraction times the molar mass
($\overline{MM}$) = 0.216 × 36.46 g/mole + 0.260 × 2.02 g/mole + 0.524 × 4.00 g/mole
= 10.49656 (calc) = 10.50 g/mole (corr)

d) density = $\dfrac{10.50 \text{ g/mole}}{22.41 \text{ L/mole}}$ = 0.4685408 (calc) = 0.4685 g/L (corr)

12.209 0.480 g C × $\dfrac{1 \text{ mole C}}{12.01 \text{ g C}}$ = 0.03996669442 (calc) = 0.0400 mole C (corr)

0.101 g H × $\dfrac{1 \text{ mole H}}{1.01 \text{ g H}}$ = 0.1 (calc) = 0.100 mole H (corr)

C: $\dfrac{0.0400}{0.0400}$ = 1.00   H = $\dfrac{0.100}{0.0400}$ = 2.50   C: 1.00 × 2 = 2.00   H: 2.50 × 2 = 5.00

Empirical formula = C$_2$H$_5$ Empirical formula mass = 29.07 amu

$\dfrac{0.0869 \text{ g}}{33.6 \text{ mL}} \times \dfrac{1 \text{ mL}}{10^{-3} \text{ L}} \times \dfrac{22.41 \text{ L}}{1 \text{ mole}}$ = 57.959196 (calc) = 58.0 g/mole (corr)

$\dfrac{\text{formula mass}}{\text{empirical formula mass}} = \dfrac{58.0 \text{ amu}}{29.07 \text{ amu}}$ = 1.9965517 (calc) = 2.00 (corr)

Molecular formula = (C$_2$H$_5$)$_2$ = C$_4$H$_{10}$

12.211 Calculate the limiting reactant.

60.0 L NH$_3$ × $\dfrac{4 \text{ L NO}}{4 \text{ L NH}_3}$ = 60.0 L NO

60.0 L O$_2$ × $\dfrac{4 \text{ L NO}}{5 \text{ L O}_2}$ = 48 (calc) = 48.0 L NO (corr)

∴ O$_2$ is limiting and 60.0 L O$_2$ × $\dfrac{6 \text{ L H}_2\text{O}}{5 \text{ L O}_2}$ = 72 (calc) = 72.0 L H$_2$O (corr)

72.0 L H$_2$O + 48.0 L NO = 120.0 L total gas products

12.213 $22.0 \text{ g NH}_3 \times \dfrac{1 \text{ mole NH}_3}{17.04 \text{ g NH}_3} \times \dfrac{6 \text{ moles HCl}}{2 \text{ moles NH}_3} \times \dfrac{22.41 \text{ L HCl}}{1 \text{ mole HCl}} \times \dfrac{1 \text{ mL HCl}}{10^{-3} \text{ L HCl}}$

$\times \dfrac{1 \text{ cm}^3 \text{ HCl}}{1 \text{ mL HCl}} \times \left(\dfrac{10^{-2} \text{ m HCl}}{1 \text{ cm HCl}}\right)^3 = 0.08679929577 \text{ (calc)} = 0.0868 \text{ m}^3 \text{ HCl (corr)}$

12.215 Calculate the limiting reactant:

$\text{NH}_3: n = 7.00 \text{ g NH}_3 \times \dfrac{1 \text{ mole NH}_3}{17.04 \text{ g NH}_3} \times \dfrac{1 \text{ mole NH}_4\text{Cl}}{1 \text{ mole NH}_3}$

$= 0.4107981221 \text{ (calc)} = 0.411 \text{ mole NH}_4\text{Cl (corr)}$

$\text{HCl}: n = 12.0 \text{ g HCl} \times \dfrac{1 \text{ mole HCl}}{36.46 \text{ g HCl}} \times \dfrac{1 \text{ mole NH}_4\text{Cl}}{1 \text{ mole HCl}}$

$= 0.3291278113 \text{ (calc)} = 0.329 \text{ mole NH}_4\text{Cl (corr)}$

HCl is the limiting reactant.   0.329 mole NH_3 react (one-to-one ratio).

NH_3 originally present: $7.00 \text{ g NH}_3 \times \dfrac{1 \text{ mole NH}_3}{17.04 \text{ g NH}_3} = 0.41080 \text{ (calc)} = 0.411 \text{ mole NH}_3 \text{ (corr)}$

NH_3 is the only gas that remains after the reaction.

$\Delta n = (0.411 - 0.329) \text{ mole NH}_3 = 0.082 \text{ mole NH}_3$ (calc and corr)

$P = \dfrac{0.082 \text{ mole} \times 0.08206 \dfrac{\text{atm L}}{\text{mole K}} \times 298 \text{ K}}{1.00 \text{ L}} = 2.0052182 \text{ (calc)} = 2.0 \text{ atm (corr)}$

Answers to Multiple-Choice Practice Test

12.217 e	**12.218** b	**12.219** d	**12.220** b	**12.221** b	**12.222** b
12.223 d	**12.224** a	**12.225** e	**12.226** b	**12.227** d	**12.228** c
12.229 d	**12.230** d	**12.231** c	**12.232** a	**12.233** b	**12.234** a
12.235 d	**12.236** b				

CHAPTER THIRTEEN
Solutions

Practice Problems

Characteristics of Solutions (Sec. 13.1)

13.1   a) True, solutions may have multiple solutes.
        b) True, if it is a solution, it is homogeneous by definition.
        c) True, all parts of a homogeneous solution have the same properties.
        d) False, solutes will not settle out of solution.

13.3   a) solute = sodium chloride, solvent = water
        b) solute = sucrose, solvent = water
        c) solute = water, solvent = ethyl alcohol
        d) solute = ethyl alcohol, solvent = methyl alcohol

Solubility (Sec. 13.2)

13.5   a) slightly soluble   b) soluble   c) very soluble   d) very soluble

13.7   a) supersaturated   b) saturated   c) unsaturated   d) saturated

13.9   a) $\dfrac{161.4 \text{ g CsCl}}{100 \text{ g water}}$ at 0°C = saturated

        b) $\dfrac{161.4 \text{ g CsCl}}{100 \text{ g water}}$ at 50°C = unsaturated

        c) $\dfrac{161.4 \text{ g CsCl}}{200 \text{ g water}} = \dfrac{80.70 \text{ g CsCl}}{100 \text{ g water}}$ at 50°C = unsaturated

        d) $\dfrac{161.4 \text{ g CsCl}}{50 \text{ g water}} = \dfrac{322.8 \text{ g CsCl}}{100 \text{ g water}}$ at 100°C = supersaturated

13.11   a) $\dfrac{0.50 \text{ g Ag}_2\text{SO}_4}{100 \text{ g water}}$ at 100°C = dilute

        b) $\dfrac{0.50 \text{ g Ag}_2\text{SO}_4}{100 \text{ g water}}$ at 0°C = concentrated

        c) $\dfrac{0.50 \text{ g Ag}_2\text{SO}_4}{50.0 \text{ g water}} = \dfrac{1.00 \text{ g Ag}_2\text{SO}_4}{100 \text{ g water}}$ at 50°C = concentrated

        d) $\dfrac{0.050 \text{ g Ag}_2\text{SO}_4}{10.0 \text{ g water}} = \dfrac{0.50 \text{ g Ag}_2\text{SO}_4}{100 \text{ g water}}$ at 0°C = concentrated

13.13 a) Second, ammonia gas in water with $P = 1$ atm and $T = 50°C$ is more soluble.
b) First, carbon dioxide gas in water with $P = 2$ atm and $T = 75°C$ is more soluble.
c) Second, table salt in water with $P = 1$ atm and $T = 60°C$ is more soluble.
d) Second, table sugar in water with $P = 1$ atm and $T = 70°C$ is more soluble.

13.15 $200 \text{ mL H}_2\text{O} \times \dfrac{1 \text{ g H}_2\text{O}}{1 \text{ mL H}_2\text{O}} = 200 \text{ g H}_2\text{O}$ (calc and corr)

a) At 70°C, solubility $\dfrac{110 \text{ g Pb(NO}_3)_2}{100 \text{ g water}}$ if $\dfrac{200 \text{ g Pb(NO}_3)_2}{200 \text{ g water}}$ present, none will settle out of water.

b) At 40°C, solubility $\dfrac{78 \text{ g Pb(NO}_3)_2}{100 \text{ g water}}$ if $\dfrac{200 \text{ g Pb(NO}_3)_2}{200 \text{ g water}}$ present,
then $200 \text{ g} - 2(78 \text{ g}) = 44 \text{ g Pb(NO}_3)_2$ will settle out of water.

Solution Formation (Sec. 13.3)

13.17 a) hydrated ion   b) hydrated ion   c) oxygen atom   d) hydrogen atom

13.19 a) decrease   b) increase   c) increase   d) increase

Solubility Rules (Sec. 13.4)

13.21 a) ethanol   b) carbon tetrachloride   c) ethanol   d) ethanol

13.23 a) NO_3^- ions are generally soluble.   b) Cl^- ions are generally soluble.
c) S^{2-} ions are generally insoluble.   d) SO_4^{2-} ions are generally soluble.

13.25 a) Carbonates are insoluble with the exception of Group IA and NH_4^+.
b) Sulfides are insoluble with the exception of Groups IA and IIA and NH_4^+.
c) Ammonium is always soluble.
d) Sulfates are soluble with the exception of Ca^{2+}, Sr^{2+}, Ba^{2+}, Pb^{2+}.

13.27 a) Acetates and nitrates ions are soluble.
b) Sulfates and chlorides are soluble with exceptions.
c) Sodium and ammonium ions are soluble.
d) Bromides and sulfates are soluble with exceptions.

13.29 a) both soluble   b) both soluble   c) insoluble and soluble   d) both insoluble

13.31 a) no, $MgSO_4$ soluble   b) yes   c) no, both soluble   d) no, both soluble

Mass Percent (Sec. 13.6)

13.33 a) $\dfrac{7.37 \text{ g NaCl}}{(95.0 + 7.37) \text{ g solution}} \times 100 = 7.199374817$ (calc) $= 7.20\%$ (m/m) NaCl (corr)

b) $\dfrac{3.73 \text{ g KBr}}{(131.00 + 3.73) \text{ g solution}} \times 100 = 2.7684999$ (calc) $= 2.77\%$ (m/m) KBr (corr)

c) $\dfrac{10.3 \text{ g NH}_4\text{NO}_3}{53.0 \text{ g solution}} \times 100 = 19.4339623$ (calc) $= 19.4\%$ (m/m) NH_4NO_3 (corr)

d) $\dfrac{25.0 \text{ g MgSO}_4}{275.0 \text{ g solution}} \times 100 = 9.090909$ (calc) $= 9.09\%$ (m/m) $MgSO_4$ (corr)

Solutions

13.35 a) $\dfrac{35.7 \text{ g NaCl}}{100 \text{ g solution}} \times 100 = 35.7\% \text{ (m/m) NaCl (calc and corr)}$

b) $\dfrac{39.8 \text{ g AgNO}_3}{100 \text{ g solution}} \times 100 = 39.8\% \text{ (m/m) AgNO}_3 \text{ (calc and corr)}$

c) $\dfrac{122 \text{ g AgNO}_3}{100.0 \text{ g solution}} \times 100 = 122\% \text{ (m/m) AgNO}_3 \text{ (calc and corr)}$

d) $\dfrac{952 \text{ g AgNO}_3}{100 \text{ g solution}} \times 100 = 952\% \text{ (m/m) AgNO}_3 \text{ (calc and corr)}$

13.37 a) $125.0 \text{ g solution} \times \dfrac{2.00 \text{ g NaCl}}{100.0 \text{ g solution}} = 2.5 \text{ (calc)} = 2.50 \text{ g NaCl (corr)}$

b) $125.0 \text{ g solution} \times \dfrac{3.50 \text{ g AgNO}_3}{100.0 \text{ g solution}} = 4.375 \text{ (calc)} = 4.38 \text{ g AgNO}_3 \text{ (corr)}$

c) $125.0 \text{ g solution} \times \dfrac{10.0 \text{ g K}_2\text{SO}_4}{100.0 \text{ g solution}} = 12.5 \text{ g K}_2\text{SO}_4 \text{ (calc and corr)}$

d) $125.0 \text{ g solution} \times \dfrac{8.25 \text{ g HCl}}{100.0 \text{ g solution}} = 10.3125 \text{ (calc)} = 10.3 \text{ g HCl (corr)}$

13.39 a) $5.75 \text{ g solution} \times \dfrac{10.00 \text{ g CaCl}_2}{100.0 \text{ g solution}} = 0.575 \text{ g CaCl}_2 \text{ (calc and corr)}$

$5.75 \text{ g solution} - 0.575 \text{ g CaCl}_2 = 5.175 \text{ (calc)} = 5.18 \text{ g H}_2\text{O (corr)}$

Alternate method: If the solution is 10.00% $CaCl_2$, it is 90.00% H_2O.

$5.75 \text{ g solution} \times \dfrac{90.00 \text{ g H}_2\text{O}}{100.0 \text{ g solution}} = 5.175 \text{ (calc)} = 5.18 \text{ g H}_2\text{O (corr)}$

b) $57.5 \text{ g solution} \times \dfrac{10.00 \text{ g CaCl}_2}{100.0 \text{ g solution}} = 5.75 \text{ g CaCl}_2 \text{ (calc and corr)}$

$57.5 \text{ g solution} - 5.75 \text{ g CaCl}_2 = 51.75 \text{ g (calc)} = 51.8 \text{ g H}_2\text{O (corr)}$

c) $57.5 \text{ g solution} \times \dfrac{1.00 \text{ g CaCl}_2}{100 \text{ g solution}} = 0.575 \text{ g CaCl}_2 \text{ (calc and corr)}$

$57.5 \text{ g solution} - 0.575 \text{ g CaCl}_2 = 56.925 \text{ (calc)} = 56.9 \text{ g H}_2\text{O (corr)}$

d) $2.3 \text{ g solution} \times \dfrac{0.80 \text{ g CuCl}_2}{100 \text{ g solution}} = 0.0184 \text{ (calc)} = 0.018 \text{ g CaCl}_2 \text{ (corr)}$

$2.3 \text{ g solution} - 0.018 \text{ g CaCl}_2 = 2.282 \text{ (calc)} = 2.3 \text{ g H}_2\text{O (corr)}$

13.41 a) $50.0 \text{ g NaCl} \times \dfrac{95.00 \text{ g H}_2\text{O}}{5.00 \text{ g NaCl}} = 950 \text{ (calc)} = 9.50 \times 10^2 \text{ g H}_2\text{O (corr)}$

b) $50.0 \text{ g KCl} \times \dfrac{95.00 \text{ g H}_2\text{O}}{5.00 \text{ g KCl}} = 950 \text{ (calc)} = 9.50 \times 10^2 \text{ g H}_2\text{O (corr)}$

c) $50.0 \text{ g Na}_2\text{SO}_4 \times \dfrac{95.00 \text{ g H}_2\text{O}}{5.00 \text{ g Na}_2\text{SO}_4} = 950 \text{ (calc)} = 9.50 \times 10^2 \text{ g H}_2\text{O (corr)}$

d) $50.0 \text{ g LiNO}_3 \times \dfrac{95.00 \text{ g H}_2\text{O}}{5.00 \text{ g LiNO}_3} = 950 \text{ (calc)} = 9.50 \times 10^2 \text{ g H}_2\text{O (corr)}$

Volume Percent (Sec. 13.6)

13.43 a) $\dfrac{257 \text{ mL ethyl alcohol}}{325 \text{ mL solution}} \times 100 = 79.0769231$ (calc) = 79.1% (v/v) ethyl alcohol (corr)

b) $\dfrac{257 \text{ mL ethyl alcohol}}{675 \text{ mL solution}} \times 100 = 38.0740741$ (calc) = 38.1% (v/v) ethyl alcohol (corr)

c) $\dfrac{257 \text{ mL ethyl alcohol}}{1.23 \text{ L} \times \dfrac{1 \text{ mL}}{10^{-3} \text{ L}} \text{ solution}} \times 100 = 20.8943089$ (calc) = 20.9% (v/v) ethyl alcohol (corr)

d) $\dfrac{257 \text{ mL ethyl alcohol}}{5.000 \text{ L} \times \dfrac{1 \text{ mL}}{10^{-3} \text{ L}} \text{ solution}} \times 100 = 5.14\%$ (v/v) ethyl alcohol (calc and corr)

13.45 a) $\dfrac{360.6 \text{ mL methyl alcohol}}{1000.0 \text{ mL solution}} \times 100 = 36.06\%$ (v/v) methyl alcohol (calc and corr)

b) $\dfrac{667.2 \text{ mL H}_2\text{O}}{1000.0 \text{ mL solution}} \times 100 = 66.72\%$ (v/v) H$_2$O (calc and corr)

13.47 $225 \text{ mL solution} \times \dfrac{1.25 \text{ mL C}_3\text{H}_8\text{O}}{100 \text{ mL solution}} = 2.8125$ (calc) = 2.81 mL C$_3$H$_8$O (corr)

13.49 $4.00 \text{ gal solution} \times \dfrac{35.0 \text{ gal H}_2\text{O}}{100.0 \text{ gal solution}} = 1.4$ (calc) = 1.40 gal H$_2$O (corr)

Mass–Volume Percent (Sec. 13.6)

13.51 Mass in grams, volume in mL

a) $\dfrac{0.325 \text{ g NaNO}_3}{375 \text{ mL solution}} \times 100 = 0.086666667$ (calc) = 0.0867% (m/v) NaNO$_3$ (corr)

b) $\dfrac{1.75 \text{ g NaNO}_3}{375 \text{ mL solution}} \times 100 = 0.46666667$ (calc) = 0.467% (m/v) NaNO$_3$ (corr)

c) $\dfrac{8.43 \text{ g NaNO}_3}{375 \text{ mL solution}} \times 100 = 2.248$ (calc) = 2.25% (m/v) NaNO$_3$ (corr)

d) $\dfrac{23.6 \text{ g NaNO}_3}{375 \text{ mL solution}} \times 100 = 6.2933333$ (calc) = 6.29% (m/v) NaNO$_3$ (corr)

13.53 a) $\dfrac{4.00 \text{ g KBr}}{55.0 \text{ mL solution}} \times 100 = 7.272727$ (calc) = 7.27% (m/v) KBr (corr)

b) $\dfrac{15.0 \text{ g KBr}}{1.75 \text{ L solution}} \times \dfrac{10^{-3} \text{ L}}{1 \text{ mL}} \times 100 = 0.857142857$ (calc) = 0.857% (m/v) KBr (corr)

c) $\dfrac{15.0 \text{ mg KBr}}{1.75 \text{ mL solution}} \times \dfrac{10^{-3} \text{ g}}{1 \text{ mg}} \times 100 = 0.857142857$ (calc) = 0.857% (m/v) KBr (corr)

d) $\dfrac{0.0300 \text{ mole KBr}}{52.0 \text{ mL solution}} \times \dfrac{119.00 \text{ g KBr}}{1 \text{ mole KBr}} \times 100 = 6.865384615$ (calc) = 6.87% (m/v) KBr (corr)

Solutions

13.55 a) $5.00 \text{ g NaCl} \times \dfrac{100 \text{ mL solution}}{6.00 \text{ g NaCl}} = 83.33333 \text{ (calc)} = 83.3 \text{ mL solution (corr)}$

b) $7.00 \text{ g NaCl} \times \dfrac{100 \text{ mL solution}}{6.00 \text{ g NaCl}} = 116.666667 \text{ (calc)} = 117 \text{ mL solution (corr)}$

c) $225 \text{ g NaCl} \times \dfrac{100 \text{ mL solution}}{6.00 \text{ g NaCl}} = 3750 \text{ mL solution (calc and corr)}$

d) $225 \text{ mg NaCl} \times \dfrac{10^{-3} \text{ g}}{1 \text{ mg}} \times \dfrac{100 \text{ mL solution}}{6.00 \text{ g NaCl}} = 3.75 \text{ mL solution (calc and corr)}$

13.57 a) $455 \text{ mL solution} \times \dfrac{2.50 \text{ g Na}_3\text{PO}_4}{100 \text{ mL solution}} = 11.375 \text{ (calc)} = 11.4 \text{ g Na}_3\text{PO}_4 \text{ (corr)}$

b) $50.0 \text{ L solution} \times \dfrac{1 \text{ mL}}{10^{-3} \text{ L}} \times \dfrac{7.50 \text{ g Na}_3\text{PO}_4}{100 \text{ mL solution}} = 3750 \text{ (calc)} = 3.75 \times 10^3 \text{ g Na}_3\text{PO}_4 \text{ (corr)}$

13.59 Volume of solution = $5.0 \text{ g CsCl} + 20.0 \text{ g H}_2\text{O} = 25.0 \text{ g solution} \times \dfrac{1 \text{ mL solution}}{1.18 \text{ g solution}}$

$= 21.1864407 \text{ (calc)} = 21.2 \text{ mL solution (corr)}$

$\%\left(\dfrac{m}{v}\right) = \dfrac{5.0 \text{ g CsCl}}{21.2 \text{ mL solution}} \times 100 = 23.5849057 \text{ (calc)} = 24\% \text{ (m/v) CsCl (corr)}$

Parts per Million and Parts per Billion (Sec. 13.7)

13.61 a) $37.5 \text{ mg NaCl} \times \dfrac{10^{-3} \text{ g}}{1 \text{ mg}} = 0.0375 \text{ g NaCl}$

$21.0 \text{ kg H}_2\text{O} \times \dfrac{10^3 \text{ g}}{1 \text{ kg}} = 21{,}\bar{0}00 \text{ g H}_2\text{O}$

solution mass = $0.0375 \text{ g} + 21{,}\bar{0}00 \text{ g} = 21{,}\bar{0}00 \text{ g solution}$

$\dfrac{0.0375 \text{ g NaCl}}{21{,}\bar{0}00 \text{ g solution}} \times 10^6 = 1.7857143 \text{ (calc)} = 1.79 \text{ ppm NaCl (corr)}$

b) $2.12 \text{ cg NaCl} \times \dfrac{10^{-2} \text{ g}}{1 \text{ cg}} = 0.0212 \text{ g NaCl}$

solution mass = $0.0212 \text{ g NaCl} + 125 \text{ g H}_2\text{O} = 125.0212 \text{ (calc)} = 125 \text{ g solution (corr)}$

$\dfrac{0.0212 \text{ g NaCl}}{125 \text{ g solution}} \times 10^6 = 169.6 \text{ (calc)} = 1\bar{7}0 \text{ ppm NaCl}$

c) $1.00 \text{ μg NaCl} = 1.00 \times 10^{-6} \text{ g NaCl}; \quad 32.0 \text{ dg H}_2\text{O} \times \dfrac{10^{-1} \text{ g}}{1 \text{ dg}} = 3.20 \text{ g H}_2\text{O}$

solution mass = $3.20 \text{ g H}_2\text{O} + 1 \times 10^{-6} \text{ g NaCl} = 3.200001 \text{ (calc)} = 3.20 \text{ g solution (corr)}$

$\dfrac{1.00 \times 10^{-6} \text{ g NaCl}}{3.20 \text{ g solution}} \times 10^6 = 0.3125 \text{ (calc)} = 0.312 \text{ ppm NaCl (corr)}$

d) $35.7 \text{ mg NaCl} \times \dfrac{10^{-3} \text{ g}}{1 \text{ mg}} = 0.0357 \text{ g NaCl}$

solution mass = 15.7 g H$_2$O + 0.0325 g NaCl = 15.7325 (calc) = 15.7 g solution (corr)

$\dfrac{0.0357 \text{ g NaCl}}{15.7 \text{ g solution}} \times 10^6 = 2273.88535$ (calc) = 2,270 ppm NaCl (corr)

13.63 The only difference in set-up for ppb compared with ppm is changing the factor 10^6 to 10^9. The ppb value is 1000 times the ppm value.

a) 1790 ppb   b) 17̄0,000 ppb   c) 312 ppb   d) 2,270,000 ppb

13.65 Take the ratio between the mass of O$_2$ (in g) and the volume of water (in mL) times 10^6.

$\dfrac{7 \text{ mg O}_2 \times \dfrac{10^{-3} \text{ g}}{1 \text{ mg}}}{1 \text{ L} \times \dfrac{1 \text{ mL}}{10^{-3} \text{ L}}} \times 10^6 = 7$ ppm O$_2$ (calc and corr)   Yes, 7 ppm is above the minimum.

13.67 $1.00 \text{ mL SO}_2 \text{ air} \times \dfrac{10^6 \text{ mL air}}{0.087 \text{ mL SO}_2} \times \dfrac{10^{-3} \text{ L}}{1 \text{ mL}} = 11494.25287$ (calc) = 11,000 L air (corr)

13.69 a) $523 \text{ mL air} \times \dfrac{3.0 \text{ g CO}_2}{10^9 \text{ mL air}} = 1.569 \times 10^{-6}$ (calc) = 1.6×10^{-6} g CO$_2$ (corr)

b) $523 \text{ mL air} \times \dfrac{6.0 \text{ g CO}_2}{10^6 \text{ mL air}} = 3.138 \times 10^{-3}$ (calc) = 3.1×10^{-3} g CO$_2$ (corr)

c) $523 \text{ mL air} \times \dfrac{2.5 \text{ g CO}_2}{10^2 \text{ mL air}} = 13.075$ (calc) = 13 g CO$_2$ (corr)

d) $523 \text{ mL air} \times \dfrac{5.2 \text{ g CO}_2}{100 \text{ mL air}} = 27.196$ (calc) = 27 g CO$_2$ (corr)

Molarity (Sec. 13.8)

13.71 a) $\dfrac{2.0 \text{ moles NaOH}}{0.50 \text{ L solution}} = 4$ (calc) = 4.0 M NaOH (corr)

b) $\dfrac{13.7 \text{ g NaOH} \times \dfrac{1 \text{ mole NaOH}}{40.00 \text{ g NaOH}}}{90.0 \text{ mL solution} \times \dfrac{10^{-3} \text{ L}}{1 \text{ mL}}} = 3.8055556$ (calc) = 3.81 M NaOH (corr)

c) $\dfrac{53.0 \text{ g NaOH} \times \dfrac{1 \text{ mole NaOH}}{40.00 \text{ g NaOH}}}{1.255 \text{ L solution}} = 1.0557769$ (calc) = 1.06 M NaOH (corr)

d) $\dfrac{0.0020 \text{ mole NaOH}}{5.00 \text{ mL solution} \times \dfrac{10^{-3} \text{ L}}{1 \text{ mL}}} = 0.4$ (calc) = 0.40 M NaOH (corr)

Solutions 143

13.73 a) $35.0 \text{ mL solution} \times \dfrac{10^{-3} \text{ L}}{1 \text{ mL}} \times \dfrac{6.00 \text{ moles Na}_2\text{SO}_4}{1 \text{ L solution}} \times \dfrac{142.05 \text{ g Na}_2\text{SO}_4}{1 \text{ mole Na}_2\text{SO}_4}$
$= 29.8305 \text{ (calc)} = 29.8 \text{ g Na}_2\text{SO}_4 \text{ (corr)}$

b) $10.0 \text{ mL solution} \times \dfrac{10^{-3} \text{ L}}{1 \text{ mL}} \times \dfrac{0.600 \text{ mole Na}_2\text{SO}_4}{1 \text{ L solution}} \times \dfrac{142.05 \text{ g Na}_2\text{SO}_4}{1 \text{ mole Na}_2\text{SO}_4}$
$= 0.8523 \text{ (calc)} = 0.852 \text{ g Na}_2\text{SO}_4 \text{ (corr)}$

c) $375 \text{ L solution} \times \dfrac{1.00 \text{ mole Na}_2\text{SO}_4}{1 \text{ L solution}} \times \dfrac{142.05 \text{ g Na}_2\text{SO}_4}{1 \text{ mole Na}_2\text{SO}_4}$
$= 53268.75 \text{ (calc)} = 5.33 \times 10^4 \text{ g Na}_2\text{SO}_4 \text{ (corr)}$

d) $375 \text{ g solution} \times \dfrac{1 \text{ mL solution}}{1.25 \text{ g solution}} \times \dfrac{10^{-3} \text{ L}}{1 \text{ mL}} \times \dfrac{7.91 \text{ moles Na}_2\text{SO}_4}{1 \text{ L solution}} \times \dfrac{142.05 \text{ g Na}_2\text{SO}_4}{1 \text{ mole Na}_2\text{SO}_4}$
$= 337.08465 \text{ (calc)} = 337 \text{ g Na}_2\text{SO}_4 \text{ (corr)}$

13.75 a) $2.50 \text{ g of HNO}_3 \times \dfrac{1 \text{ mole HNO}_3}{63.02 \text{ g HNO}_3} \times \dfrac{1 \text{ L solution}}{0.468 \text{ mole HNO}_3} \times \dfrac{1 \text{ mL}}{10^{-3} \text{ L}}$
$= 84.76484198 \text{ (calc)} = 84.8 \text{ mL solution (corr)}$

b) $125 \text{ g HNO}_3 \times \dfrac{1 \text{ mole HNO}_3}{63.02 \text{ g HNO}_3} \times \dfrac{1 \text{ L solution}}{3.50 \text{ moles HNO}_3} \times \dfrac{1 \text{ mL}}{10^{-3} \text{ L}}$
$= 566.713515 \text{ (calc)} = 567 \text{ mL solution (corr)}$

c) $4.50 \text{ moles HNO}_3 \times \dfrac{1 \text{ L solution}}{2.50 \text{ moles HNO}_3} \times \dfrac{1 \text{ mL}}{10^{-3} \text{ L}} = 1800 \text{ (calc)} = 1.80 \times 10^3 \text{ mL solution (corr)}$

d) $0.0015 \text{ mole HNO}_3 \times \dfrac{1 \text{ L solution}}{0.990 \text{ mole HNO}_3} \times \dfrac{1 \text{ mL}}{10^{-3} \text{ L}} = 1.5151515 \text{ (calc)} = 1.52 \text{ mL solution (corr)}$

13.77 Assuming the volume of Diagram I is 1.0 L and each dot represents 1 mole,

Diagram I $= \dfrac{14 \text{ moles}}{1 \text{ L}} = 14 \text{ M (calc and corr)}$ Diagram II $= \dfrac{10 \text{ moles}}{1 \text{ L}} = 10. \text{ M (calc and corr)}$

Diagram III $= \dfrac{7 \text{ moles}}{1/2 \text{ L}} = 14 \text{ M (calc and corr)}$ Diagram IV $= \dfrac{7 \text{ moles}}{1/4 \text{ L}} = 28 \text{ M (calc and corr)}$

a) Diagram IV has the highest concentration.
b) The concentration is the same in Diagrams I and III.

13.79 a) $67.0 \text{ g NaNO}_3 \times \dfrac{1 \text{ mole NaNO}_3}{85.00 \text{ g NaNO}_3} \times \dfrac{1 \text{ L solution}}{1.25 \text{ mole NaNO}_3} = 0.630588235 \text{ (calc)}$
$= 0.631 \text{ L solution (corr)}$

b) $67.0 \text{ g KNO}_3 \times \dfrac{1 \text{ mole KNO}_3}{101.11 \text{ g KNO}_3} \times \dfrac{1 \text{ L solution}}{1.25 \text{ mole KNO}_3} = 0.530115715 \text{ (calc)}$
$= 0.530 \text{ L solution (corr)}$

c) $67.0 \text{ g NH}_4\text{NO}_3 \times \dfrac{1 \text{ mole NH}_4\text{NO}_3}{80.06 \text{ g NH}_4\text{NO}_3} \times \dfrac{1 \text{ L solution}}{1.25 \text{ mole NH}_4\text{NO}_3} = 0.669497876 \text{ (calc)}$
$= 0.669 \text{ L solution (corr)}$

d) $67.0 \text{ g Ca(NO}_3)_2 \times \dfrac{1 \text{ mole CsBr}}{164.10 \text{ g Ca(NO}_3)_2} \times \dfrac{1 \text{ L solution}}{1.25 \text{ mole Ca(NO}_3)_2} = 0.326630103 \text{ (calc)}$
$= 0.327 \text{ L solution (corr)}$

13.81 $\dfrac{88.00 \text{ g CH}_4\text{O}}{100.0 \text{ g solution}} \times \dfrac{0.8274 \text{ g solution}}{1 \text{ mL solution}} \times \dfrac{1 \text{ mL}}{10^{-3} \text{ L}} \times \dfrac{1 \text{ mole CH}_4\text{O}}{32.05 \text{ g CH}_4\text{O}} = 22.7180031$ (calc)

$= 22.72$ M CH$_4$O (corr)

13.83 $\dfrac{2.019 \text{ moles NaBr}}{1 \text{ L solution}} \times \dfrac{102.89 \text{ g NaBr}}{1 \text{ mole NaBr}} \times \dfrac{10^{-3} \text{ L solution}}{1 \text{ mL solution}} \times \dfrac{1 \text{ mL solution}}{1.157 \text{ g solution}} \times 100$

$= 17.954616$ (calc) $= 17.96\%$ (m/m) (corr)

13.85 $\dfrac{20.0 \text{ g HCl}}{100 \text{ mL solution}} \times \dfrac{1 \text{ mL solution}}{10^{-3} \text{ L solution}} \times \dfrac{1 \text{ mole HCl}}{36.46 \text{ g HCl}} = 5.485463522$ (calc) $= 5.49$ M HCl (corr)

Molality (Sec. 13.9)

13.87 a) $\dfrac{16.5 \text{ g C}_{12}\text{H}_{22}\text{O}_{11} \times \dfrac{1 \text{ mole C}_{12}\text{H}_{22}\text{O}_{11}}{342.34 \text{ g C}_{12}\text{H}_{22}\text{O}_{11}}}{1.35 \text{ kg H}_2\text{O}} = 0.03570199866$ (calc) $= 0.0357$ m C$_{12}$H$_{22}$O$_{11}$ (corr)

b) $\dfrac{3.15 \text{ moles C}_{12}\text{H}_{22}\text{O}_{11}}{455 \text{ g H}_2\text{O} \times \dfrac{1 \text{ kg}}{10^3 \text{ g}}} = 6.9230769$ (calc) $= 6.92$ m C$_{12}$H$_{22}$O$_{11}$ (corr)

c) $\dfrac{0.0356 \text{ g C}_{12}\text{H}_{22}\text{O}_{11} \times \dfrac{1 \text{ mole C}_{12}\text{H}_{22}\text{O}_{11}}{342.34 \text{ g C}_{12}\text{H}_{22}\text{O}_{11}}}{13.0 \text{ g H}_2\text{O} \times \dfrac{1 \text{ kg}}{10^3 \text{ g}}} = 0.007999245$ (calc) $= 0.00800$ m C$_{12}$H$_{22}$O$_{11}$ (corr)

d) Moles sucrose $= 45.0 \text{ g} \times \dfrac{1 \text{ mole C}_{12}\text{H}_{22}\text{O}_{11}}{342.34 \text{ g C}_{12}\text{H}_{22}\text{O}_{11}} = 0.1314482$ (calc) $= 0.131$ mole C$_{12}$H$_{22}$O$_{11}$ (corr)

Mass solution $= 318 \text{ mL solution} \times \dfrac{1.06 \text{ g solution}}{1 \text{ mL solution}} = 337.08$ g (calc) $= 337$ g solution (corr)

Mass H$_2$O $= 337$ g solution $- 45.0$ g C$_{12}$H$_{22}$O$_{11}$

$= 292$ g H$_2$O (calc and corr) $\times \dfrac{1 \text{ kg H}_2\text{O}}{10^3 \text{ g H}_2\text{O}} = 0.292$ kg H$_2$O

$\dfrac{0.131 \text{ mole C}_{12}\text{H}_{22}\text{O}_{11}}{0.292 \text{ kg H}_2\text{O}} = 0.4486301$ (calc) $= 0.449$ m C$_{12}$H$_{22}$O$_{11}$ (corr)

13.89 a) $234 \text{ g H}_2\text{O} \times \dfrac{1 \text{ kg}}{10^3 \text{ g}} \times \dfrac{0.600 \text{ mole NH}_4\text{NO}_3}{1 \text{ kg H}_2\text{O}} \times \dfrac{80.06 \text{ g NH}_4\text{NO}_3}{1 \text{ mole NH}_4\text{NO}_3}$

$= 11.240424$ (calc) $= 11.2$ g NH$_4$NO$_3$ (corr)

b) $234 \text{ g H}_2\text{O} \times \dfrac{1 \text{ kg}}{10^3 \text{ g}} \times \dfrac{0.600 \text{ mole NaOH}}{1 \text{ kg H}_2\text{O}} \times \dfrac{40.00 \text{ g NaOH}}{1 \text{ mole NaOH}} = 5.616$ (calc) $= 5.62$ g NaOH (corr)

c) $234 \text{ g H}_2\text{O} \times \dfrac{1 \text{ kg}}{10^3 \text{ g}} \times \dfrac{0.600 \text{ mole K}_3\text{PO}_4}{1 \text{ kg H}_2\text{O}} \times \dfrac{212.27 \text{ g K}_3\text{PO}_4}{1 \text{ mole K}_3\text{PO}_4}$

$= 29.802708$ (calc) $= 29.8$ g K$_3$PO$_4$ (corr)

d) $234 \text{ g H}_2\text{O} \times \dfrac{1 \text{ kg}}{10^3 \text{ g}} \times \dfrac{0.600 \text{ mole Al}_2(\text{SO}_4)_3}{1 \text{ kg H}_2\text{O}} \times \dfrac{342.17 \text{ g Al}_2(\text{SO}_4)_3}{1 \text{ mole Al}_2(\text{SO}_4)_3}$

$= 48.040668$ (calc) $= 48.0$ g Al$_2$(SO$_4$)$_3$ (corr)

Solutions

13.91 a) $30.0 \text{ g NaBr} \times \dfrac{1 \text{ mole NaBr}}{102.89 \text{ g NaBr}} \times \dfrac{1 \text{ kg H}_2\text{O}}{0.150 \text{ mole NaBr}} \times \dfrac{10^3 \text{ g}}{1 \text{ kg}}$

$= 1943.823501 \text{ (calc)} = 1940 \text{ g H}_2\text{O (corr)}$

b) $30.0 \text{ g NaBr} \times \dfrac{1 \text{ mole NaBr}}{102.89 \text{ g NaBr}} \times \dfrac{1 \text{ kg H}_2\text{O}}{0.43 \text{ mole NaBr}} \times \dfrac{10^3 \text{ g}}{1 \text{ kg}}$

$= 678.0779654 \text{ (calc)} = 680 \text{ g H}_2\text{O (corr)}$

c) $30.0 \text{ g NaBr} \times \dfrac{1 \text{ mole NaBr}}{102.89 \text{ g NaBr}} \times \dfrac{1 \text{ kg H}_2\text{O}}{2.435 \text{ mole NaBr}} \times \dfrac{10^3 \text{ g}}{1 \text{ kg}}$

$= 119.7427308 \text{ (calc)} = 120. \text{ g H}_2\text{O (corr)}$

d) $30.0 \text{ g NaBr} \times \dfrac{1 \text{ mole NaBr}}{102.89 \text{ g NaBr}} \times \dfrac{1 \text{ kg H}_2\text{O}}{4.0 \text{ mole NaBr}} \times \dfrac{10^3 \text{ g}}{1 \text{ kg}} = 72.19338128 \text{ (calc)} = 73 \text{ g H}_2\text{O (corr)}$

13.93 Mass of C_6H_{12} = $75.0 \text{ mL} \times \dfrac{0.779 \text{ g C}_6\text{H}_{12}}{1 \text{ mL C}_6\text{H}_{12}} = 58.425 \text{ (calc)} = 58.4 \text{ g C}_6\text{H}_{12} \text{ (corr)}$

Mass of C_6H_{14} = $175.0 \text{ mL} \times \dfrac{0.659 \text{ g C}_6\text{H}_{14}}{1 \text{ mL C}_6\text{H}_{14}} = 115.325 \text{ (calc)} = 115 \text{ g C}_6\text{H}_{14} \text{ (corr)}$

Therefore C_6H_{12} is the solute, and C_6H_{14} is the solvent.

Moles of solute = $58.4 \text{ g C}_6\text{H}_{12} \times \dfrac{1 \text{ mole C}_6\text{H}_{12}}{84.18 \text{ g C}_6\text{H}_{12}} = 0.693751484 \text{ (calc)} = 0.694 \text{ mole C}_6\text{H}_{12} \text{ (corr)}$

kg of solvent = $115 \text{ g C}_6\text{H}_{14} \times \dfrac{1 \text{ kg}}{10^3 \text{ g}} = \text{(calc)} = 0.115 \text{ kg C}_6\text{H}_{14} \text{ (corr)}$

Molality = $\dfrac{\text{moles solute}}{\text{kg solvent}} = \dfrac{0.694 \text{ mole C}_6\text{H}_{12}}{0.115 \text{ kg C}_6\text{H}_{14}} = 6.034782609 \text{ (calc)} = 6.03 \text{ m C}_6\text{H}_{12} \text{ (corr)}$

13.95 Basis: 1 L solution
Mass of solute:

$1 \text{ L solution} \times \dfrac{0.568 \text{ mole H}_2\text{C}_2\text{O}_4}{1 \text{ L solution}} \times \dfrac{90.04 \text{ g H}_2\text{C}_2\text{O}_4}{1 \text{ mole H}_2\text{C}_2\text{O}_4} = 51.14272 \text{ (calc)}$

$= 51.1 \text{ g H}_2\text{C}_2\text{O}_4 \text{ (corr)}$

Mass of solution:

$1 \text{ L solution} \times \dfrac{1 \text{ mL solution}}{10^{-3} \text{ L solution}} \times \dfrac{1.022 \text{ g solution}}{1 \text{ mL solution}} = 1022 \text{ g solution (calc and corr)}$

Mass of solvent: $(1022 - 51.1) \text{ g} = 970.9 \text{ (calc)} = 971 \text{ g (corr)}$

Molality: $\dfrac{51.1 \text{ g H}_2\text{C}_2\text{O}_4}{971 \text{ g H}_2\text{O}} \times \dfrac{1 \text{ mole H}_2\text{C}_2\text{O}_4}{90.04 \text{ g H}_2\text{C}_2\text{O}_4} \times \dfrac{10^3 \text{ g H}_2\text{O}}{1 \text{ kg H}_2\text{O}} = 0.5844753 \text{ (calc)}$

$= 0.584 \text{ m H}_2\text{C}_2\text{O}_4 \text{ (corr)}$

13.97 Basis: 1 kg solvent = 1000 g solvent
Mass of solute:

$$1 \text{ kg solvent} \times \frac{0.796 \text{ mole HC}_2\text{H}_3\text{O}_2}{1 \text{ kg solvent}} \times \frac{60.06 \text{ g HC}_2\text{H}_3\text{O}_2}{1 \text{ mole HC}_2\text{H}_3\text{O}_2} = 47.80776 \text{ (calc)} = 47.8 \text{ g HC}_2\text{H}_3\text{O}_2 \text{ (corr)}$$

Mass of solution: $(1000 + 47.8)$ g = 1047.8 g (calc and corr)

Volume of solution: $1047.8 \text{ g solution} \times \frac{1 \text{ mL solution}}{1.004 \text{ g solution}} \times \frac{10^{-3} \text{ L solution}}{1 \text{ mL solution}}$

$$= 1.043625498 \text{ (calc)} = 1.044 \text{ L solution (corr)}$$

Molarity: $\frac{47.8 \text{ g HC}_2\text{H}_3\text{O}_2}{1.044 \text{ L solution}} \times \frac{1 \text{ mole HC}_2\text{H}_3\text{O}_2}{60.06 \text{ g HC}_2\text{H}_3\text{O}_2} = 0.7623283485 \text{ (calc)} = 0.762 \text{ M HC}_2\text{H}_3\text{O}_2 \text{ (corr)}$

13.99 Basis: 1 L of solution = 10^3 mL solution
Mass of solvent:

$$1 \text{ L solution} \times \frac{11.8 \text{ moles HCl}}{1 \text{ L solution}} \times \frac{1 \text{ kg solvent}}{15.5 \text{ moles HCl}} \times \frac{10^3 \text{ g}}{1 \text{ kg}} = 761.2903226 \text{ (calc)}$$

$$= 761 \text{ g solvent (corr)}$$

Mass of solute:

$$1 \text{ L solution} \times \frac{11.8 \text{ moles HCl}}{1 \text{ L solution}} \times \frac{36.46 \text{ g HCl}}{1 \text{ mole HCl}} = 430.228 \text{ (calc)} = 430. \text{ g HCl (corr)}$$

Mass of solution = 761 g solvent + 430. g solute = 1191 g solution (calc and corr)

$$\text{density of solution} = \frac{\text{mass of solution}}{\text{volume of solution}} = \frac{1191 \text{ g solution}}{10^3 \text{ mL solution}} = 1.191 \text{ g/mL (calc and corr)}$$

13.101 Mass of solvent = 100 − 22.0 g solute = 78.0 g solvent

$$\frac{22.0 \text{ g H}_8\text{C}_6\text{O}_7}{78.0 \text{ g solvent}} \times \frac{1 \text{ mole H}_8\text{C}_6\text{O}_7}{192.14 \text{ g H}_8\text{C}_6\text{O}_7} \times \frac{10^3 \text{ g solvent}}{1 \text{ kg solvent}} = 1.467946716 \text{ (calc)} = 1.47 \text{ m H}_8\text{C}_6\text{O}_7 \text{ (corr)}$$

Dilution (Sec. 13.10)

13.103 $M_1V_1 = M_2V_2$ where $M_1 = 0.400$ M, $V_1 = 25.0$ L and

a) $V_2 = 40.0$ mL; $M_2 = 0.400 \text{ M} \times \dfrac{25.0 \text{ mL}}{40.0 \text{ mL}} = 0.25 \text{ (calc)} = 0.250 \text{ M (corr)}$

b) $V_2 = 87.0$ mL; $M_2 = 0.400 \text{ M} \times \dfrac{25.0 \text{ mL}}{87.0 \text{ mL}} = 0.114942528 \text{ (calc)} = 0.115 \text{ M (corr)}$

c) $V_2 = 225$ mL; $M_2 = 0.400 \text{ M} \times \dfrac{25.0 \text{ mL}}{225 \text{ mL}} = 0.04444 \text{ (calc)} = 0.0444 \text{ M (corr)}$

d) $V_2 = 1.45 \text{ L} \times \dfrac{1 \text{ mL}}{10^{-3} \text{ L}} = 1450 \text{ mL}$; $M_2 = 0.400 \text{ M} \times \dfrac{25.0 \text{ mL}}{1450 \text{ mL}} = 0.006896552 \text{ (calc)}$

$$= 0.00690 \text{ M (corr)}$$

Solutions 147

13.105 $M_2 = M_1 \times \dfrac{V_1}{V_2}$ where $M_1 = 0.500$ M, $V_1 = 1353$ mL and

a) $V_2 = 1223$ mL; $M_2 = 0.500$ M $\times \dfrac{1353 \text{ mL}}{1223 \text{ mL}} = 0.553147996$ (calc) $= 0.553$ M (corr)

b) $V_2 = 1.12$ L $\times \dfrac{1 \text{ mL}}{10^{-3} \text{ L}} = 1120$ mL; $M_2 = 0.500$ M $\times \dfrac{1353 \text{ mL}}{1120 \text{ mL}} = 0.604017857$ (calc)
$= 0.604$ M (corr)

c) $V_2 = 853$ mL; $M_2 = 0.500$ M $\times \dfrac{1353 \text{ mL}}{853 \text{ mL}} = 0.793083235$ (calc) $= 0.793$ M (corr)

d) $V_2 = 302.5$ mL; $M_2 = 0.500$ M $\times \dfrac{1353 \text{ mL}}{302.5 \text{ mL}} = 2.2363636$ (calc) $= 2.24$ M (corr)

13.107 $V_1 = V_2 \times \dfrac{M_2}{M_1}$ where $M_1 = 4.05$ M and

a) $V_2 = 45.0$ mL; $M_2 = 3.90$ M; $V_1 = 45.0$ mL $\times \dfrac{3.90 \text{ M}}{4.05 \text{ M}} = 43.3333$ (calc) $= 43.3$ mL (corr)

b) $V_2 = 7.2$ L $\times \dfrac{1 \text{ mL}}{10^{-3} \text{ L}} = 7200$ mL; $M_2 = 2.00$ M;

$V_1 = 7200$ mL $\times \dfrac{2.00 \text{ M}}{4.05 \text{ M}} = 3555.55$ (calc) $= 3600$ mL (corr)

c) $V_2 = 345$ mL; $M_2 = 1.0$ M;

$V_1 = 345$ mL $\times \dfrac{1.0 \text{ M}}{4.05 \text{ M}} = 85.18518519$ (calc) $= 85.2$ mL (corr)

d) $V_2 = 3.0$ mL; $M_2 = 0.20$ M;

$V_1 = 3.0$ mL $\times \dfrac{0.20 \text{ M}}{4.05 \text{ M}} = 0.148148$ (calc) $= 0.15$ mL (corr)

13.109 Diagram III

13.111 a) $V_1 = 20.0$ mL; $M_1 = 2.00$ M; $V_2 = 20.0$ mL $\times \dfrac{2.00 \text{ M}}{0.100 \text{ M}} = 4\overline{0}0$ mL (calc and corr)

$V_{H_2O} = 4\overline{0}0$ mL $- 20.0$ mL $= 38\overline{0}$ mL H_2O

b) $V_1 = 20.0$ mL; $M_1 = 0.250$ M; $V_2 = 20.0$ mL $\times \dfrac{0.250 \text{ M}}{0.100 \text{ M}} = 50$ mL (calc) $= 50.0$ mL (corr)

$V_{H_2O} = 50.0 - 20.0 = 30.0$ mL H_2O

c) $V_1 = 358$ mL; $M_1 = 0.950$ M; $V_2 = 358$ mL $\times \dfrac{0.950 \text{ M}}{0.100 \text{ M}} = 3401$ mL (calc) $= 34\overline{0}0$ mL (corr)

$V_{H_2O} = 34\overline{0}0 - 358 = 3042$ (calc) $= 3040$ mL H_2O (corr)

d) $V_1 = 2.3$ L; $M_1 = 6.00$ M; $V_2 = 2.3$ L $\times \dfrac{6.00 \text{ M}}{0.100 \text{ M}} = 138$ L (calc) $= 140$ L (corr)

$V_{H_2O} = 140$ L $- 2.3$ L $= 137.7$ (calc) $= 140$ (corr) $\times \dfrac{1 \text{ mL}}{10^{-3} \text{ L}} = 1.4 \times 10^5$ mL H_2O

13.113 a) $6.0 \text{ M} \times \dfrac{25.0 \text{ mL}}{45.0 \text{ mL}} = 3.3333333$ (calc) = 3.3 M (corr)

b) $3.0 \text{ M} \times \dfrac{100.0 \text{ mL}}{120.0 \text{ mL}} = 2.5$ M (calc and corr)

c) $10.0 \text{ M} \times \dfrac{155 \text{ mL}}{175 \text{ mL}} = 8.8571429$ (calc) = 8.86 M (corr)

d) $0.100 \text{ M} \times \dfrac{2.00 \text{ mL}}{22.0 \text{ mL}} = 0.009090909$ (calc) = 0.00909 M (corr)

13.115 a) $V_2 = 275 \text{ mL} + 3.254 \text{ L} \times \dfrac{1 \text{ mL}}{10^{-3} \text{ L}} = 3529 \text{ mL}$

$M_2 = 6.00 \text{ M} \times \dfrac{275 \text{ mL}}{3529 \text{ mL}} = 0.4675545$ (calc) = 0.468 M (corr)

b) simple method: 6.00 M. Both solutions are 6.00 M, hence the mixture is 6.00 M. See part c) for a model to work the problem rigorously.

c) moles NaOH (solution 1) = $\dfrac{6.00 \text{ moles NaOH}}{1 \text{ L solution}} \times 275 \text{ mL} \times \dfrac{10^{-3} \text{ L}}{1 \text{ mL}}$

= 1.65 moles NaOH (calc and corr)

moles NaOH (solution 2) = $\dfrac{2.00 \text{ moles NaOH}}{1 \text{ L solution}} \times 125 \text{ mL} \times \dfrac{10^{-3} \text{ L}}{1 \text{ mL}}$

= 0.250 mole NaOH (calc and corr)

total moles = 1.65 + 0.250 = 1.90 moles NaOH

total volume = 275 + 125 = 400 mL $\times \dfrac{1 \text{ L}}{10^{-3} \text{ mL}} = 0.400$ L solution

$M_{\text{final}} = \dfrac{1.90 \text{ moles NaOH}}{0.400 \text{ L solution}} = 4.75$ M NaOH (calc and corr)

d) moles NaOH (solution 1) = $\dfrac{6.00 \text{ moles NaOH}}{1 \text{ L solution}} \times 275 \text{ mL} \times \dfrac{10^{-3} \text{ L}}{1 \text{ mL}}$

= 1.65 moles (calc and corr)

moles NaOH (solution 2) = $\dfrac{5.80 \text{ moles NaOH}}{1 \text{ L solution}} \times 27 \text{ mL} \times \dfrac{10^{-3} \text{ L}}{1 \text{ mL}}$

= 0.1566 (calc) = 0.16 mole NaOH (corr)

total moles = 1.65 + 0.16 = 1.807 (calc) = 1.81 moles NaOH (corr)

total volume (L) = 275 mL + 27 mL = 302 mL $\times \dfrac{10^{-3} \text{ L}}{1 \text{ mL}} = 0.302$ L solution (calc and corr)

$M_{\text{final}} = \dfrac{1.81 \text{ moles NaOH}}{0.302 \text{ L solution}} = 5.9933775$ (calc) = 5.99 M NaOH (corr)

Solutions

13.117 Use a moles balance to solve for amount of 3.00 M NaCl solution

$$25.0 \text{ L} \times \left(\frac{5.00 \text{ moles NaCl}}{1 \text{ L solution}}\right) + X \text{ L} \left(\frac{3.00 \text{ moles NaCl}}{1 \text{ L solution}}\right) = (25.0 \text{ L} + X \text{ L})\left(\frac{4.50 \text{ moles NaCl}}{1 \text{ L solution}}\right)$$

125 moles + 3.00 X moles = 112.5 (calc) = 112 moles (corr) + 4.50 X moles
Collecting like terms: 125 moles − 112 moles = 4.50 X moles − 3.00 X moles
13 moles = 1.50 X moles; X = 8.66667 (calc)
    = 8.7 L (corr) of 3.00 M NaCl solution (calc and corr)

Molarity and Chemical Equations (Sec. 13.11)

13.119 $0.500 \text{ L NaCl} \times \dfrac{4.00 \text{ moles NaCl}}{1 \text{ L NaCl}} \times \dfrac{1 \text{ mole Pb(NO}_3)_2}{2 \text{ moles NaCl}} \times \dfrac{1 \text{ L Pb(NO}_3)_2 \text{ solution}}{1.00 \text{ mole Pb(NO}_3)_2}$

$= 1$ (calc) $= 1.00$ L (corr) Pb(NO$_3$)$_2$ solution

13.121 $30.0 \text{ mL} \times \dfrac{10^{-3} \text{ L}}{1 \text{ mL}} \times \dfrac{12.0 \text{ moles HNO}_3}{1 \text{ L}} \times \dfrac{3 \text{ moles S}}{2 \text{ moles HNO}_3} \times \dfrac{32.07 \text{ g S}}{1 \text{ mole S}}$

$= 17.3178$ (calc) $= 17.3$ g S (corr)

13.123 $18.0 \text{ g Ni} \times \dfrac{1 \text{ mole Ni}}{58.69 \text{ g Ni}} \times \dfrac{1 \text{ mole H}_2\text{SO}_4}{1 \text{ mole Ni}} \times \dfrac{1 \text{ L H}_2\text{SO}_4 \text{ solution}}{0.50 \text{ mole H}_2\text{SO}_4} \times \dfrac{1 \text{ mL}}{10^{-3} \text{ L}}$

$= 613.3924007$ (calc) $= 610$ mL H$_2$SO$_4$ solution (corr)

13.125 moles HNO$_3$ = $23.7 \text{ mL NaOH solution} \times \dfrac{10^{-3} \text{ L}}{1 \text{ mL}} \times \dfrac{0.100 \text{ mole NaOH}}{1 \text{ L solution}} \times \dfrac{1 \text{ mole HNO}_3}{1 \text{ mole NaOH}}$

$= 0.00237$ mole HNO$_3$ (calc and corr)

volume HNO$_3$ = $37.5 \text{ mL} \times \dfrac{10^{-3} \text{ L}}{1 \text{ mL}} = 0.0375$ L HNO$_3$ solution

$M = \dfrac{0.00237 \text{ mole HNO}_3}{0.0375 \text{ L solution}} = 0.0632$ M HNO$_3$ (calc and corr)

13.127 $50.0 \text{ mL solution} \times \dfrac{10^{-3} \text{ L}}{1 \text{ mL}} \times \dfrac{6.0 \text{ moles HNO}_3}{1 \text{ L solution}} \times \dfrac{2 \text{ moles NO}}{8 \text{ moles HNO}_3} \times \dfrac{22.41 \text{ L NO}}{1 \text{ mole NO}}$

$= 1.68075$ (calc) $= 1.7$ L NO (corr)

13.129 $2.00 \text{ L CO}_2 \times \dfrac{1 \text{ mole CO}_2}{22.41 \text{ L CO}_2} \times \dfrac{1 \text{ mole Ca(OH)}_2}{1 \text{ mole CO}_2} = 0.08924587238$ (calc)

$= 0.0892$ mole Ca(OH)$_2$ (corr)

$M = \dfrac{0.0892 \text{ mole Ca(OH)}_2}{1.75 \text{ L solution}} = 0.05097142857$ (calc) $= 0.0510$ M Ca(OH)$_2$ (corr)

Additional Problems

13.131 a) (NH$_4$)$_3$PO$_4$ b) Ca(OH)$_2$ c) AgNO$_3$ d) CaS, Ca(NO$_3$)$_2$, Ca(C$_2$H$_3$O$_2$)$_2$

13.133 Since the solution is 75% saturated, it contains

0.75×33.3 g $CuSO_4$ = 24.975 (calc) = 25 g $CuSO_4$ in 100 g H_2O.

The mass % = $\dfrac{25. \text{ g}}{(100 + 25) \text{ g solution}} \times 100 = 2\bar{0}\%$ (m/m) $CuSO_4$ (calc and corr)

400.0 g solution $\times \dfrac{2\bar{0}. \text{ g } CuSO_4}{100 \text{ g solution}} = 8\bar{0}$ g $CuSO_4$ (calc and corr)

13.135 a) $\%\dfrac{m}{v} = \dfrac{45.2 \text{ g } AgNO_3}{254 \text{ mL solution}} \times 100 = 17.7952756$ (calc) = 17.8% (m/v) (corr)

b) $\dfrac{45.2 \text{ g } AgNO_3 \times \dfrac{1 \text{ mole } AgNO_3}{169.88 \text{ g } AgNO_3}}{254 \text{ mL solution} \times \dfrac{10^{-3} \text{ L}}{1 \text{ mL}}} = 1.0475203$ (calc) = 1.05 M $AgNO_3$ (corr)

13.137 425 mL solution $\times \dfrac{1.02 \text{ g solution}}{1 \text{ mL solution}} \times \dfrac{1.55 \text{ g } Na_2SO_4}{100 \text{ g solution}} = 6.71925$ (calc) = 6.72 g Na_2SO_4 (corr)

13.139 Grams of solvent = grams of solution − grams of solute
Grams of solution:

1375 mL solution $\times \dfrac{1.161 \text{ g solution}}{1 \text{ mL solution}} = 1596.375$ (calc) = 1596 g solution (corr)

Grams of solute:

1.375 L solution $\times \dfrac{3.000 \text{ moles } NaNO_3}{1 \text{ L solution}} \times \dfrac{85.00 \text{ g } NaNO_3}{1 \text{ mole } NaNO_3} = 350.625$ (calc) = 350.6 g $NaNO_3$ (corr)

Grams of solvent: $(1596 − 350.6)$ g = 1245.4 (calc) = 1245 g H_2O (corr)

13.141 3.74 ppm (m/m) = $\dfrac{3.74 \text{ mg solute}}{10^6 \text{ mg solution}}$

$\dfrac{3.74 \text{ mg solute}}{10^6 \text{ mg solution}} \times \dfrac{1 \text{ mg solution}}{10^{-3} \text{ g solution}} \times \dfrac{10^3 \text{ g solution}}{1 \text{ kg solution}} = 3.74 \dfrac{\text{mg solute}}{\text{kg solution}}$ (calc and corr)

13.143 1.00 g NaCl $\times \dfrac{1 \text{ mole NaCl}}{58.44 \text{ g NaCl}} = 0.01711156742$ (calc) = 0.0171 mole NaCl (corr)

$M = \dfrac{0.0171 \text{ mole NaCl}}{0.01000 \text{ L solution}} = 1.71$ M NaCl (calc and corr)

$M_2 = 1.71 \text{ M} \times \dfrac{1.00 \text{ mL}}{10.00 \text{ mL}} = 0.171$ M NaCl (calc and corr)

Solutions

13.145 Molarity of a 38.0% (m/m) HCl solution: Basis: $\overline{1000}$ g of solution

$$\overline{1000} \text{ g solution} \times \frac{38.0 \text{ g HCl}}{100 \text{ g solution}} \times \frac{1 \text{ mole HCl}}{36.46 \text{ g HCl}} = 10.422381 \text{ (calc)} = 10.4 \text{ moles HCl (corr)}$$

$$\overline{1000} \text{ g solution} \times \frac{1 \text{ mL solution}}{1.19 \text{ g solution}} \times \frac{10^{-3} \text{ L solution}}{1 \text{ mL solution}} = 0.84033613 \text{ (calc)} = 0.840 \text{ L solution (corr)}$$

$$M_1 = \frac{10.4 \text{ moles HCl}}{0.840 \text{ L solution}} = 12.380952 \text{ (calc)} = 12.4 \text{ M HCl (corr)}$$

Dilution: $V_1 = \overline{1000} \text{ mL} \times \frac{0.100 \text{ M}}{12.4 \text{ M}} = 8.0645161 \text{ (calc)} = 8.06 \text{ mL solution (corr)}$

13.147 Moles of NaCl in original solution:

$$1.50 \text{ kg H}_2\text{O} \times \frac{1.23 \text{ moles NaCl}}{1 \text{ kg H}_2\text{O}} = 1.845 \text{ (calc)} = 1.84 \text{ moles NaCl (corr)}$$

Mass of H$_2$O in final solution:

$$1.84 \text{ moles NaCl} \times \frac{1 \text{ kg H}_2\text{O}}{1.00 \text{ mole NaCl}} = 1.84 \text{ kg H}_2\text{O (calc and corr)}$$

$$(1.84 - 1.50) \text{ kg H}_2\text{O} = 0.34 \text{ kg H}_2\text{O} \times \frac{10^3 \text{ g H}_2\text{O}}{1 \text{ kg H}_2\text{O}} = 340 \text{ g H}_2\text{O (calc and corr)}$$

13.149 Mass of solution = mass of solute + mass of solvent;
Mass of solute = 52.0 g
Mass of solvent:

$$52.0 \text{ g H}_3\text{PO}_4 \times \frac{1 \text{ mole H}_3\text{PO}_4}{98.00 \text{ g H}_3\text{PO}_4} \times \frac{1 \text{ kg solvent}}{2.16 \text{ moles H}_3\text{PO}_4} \times \frac{10^3 \text{ g solvent}}{1 \text{ kg solvent}}$$

$$= 245.6538171 \text{ (calc)} = 246 \text{ g solvent (corr)}$$

Mass of solution = (52.0 + 246) g = 298 g solution (calc and corr)
Volume of solution:

$$298 \text{ g solution} \times \frac{1 \text{ mL solution}}{1.12 \text{ g solution}} = 266.0714286 \text{ (calc)} = 266 \text{ mL solution (corr)}$$

13.151 a) $11.3 \text{ mL CH}_3\text{OH} \times \frac{0.793 \text{ g CH}_3\text{OH}}{1 \text{ mL CH}_3\text{OH}} = 8.9609 \text{ (calc)} = 8.96 \text{ g CH}_3\text{OH (corr)}$

$$\% \text{ (m/v)} = \frac{8.96 \text{ g CH}_3\text{OH}}{75.0 \text{ mL solution}} \times 100 = 11.946666 \text{ (calc)} = 11.9\% \text{ (m/v) CH}_3\text{OH (corr)}$$

b) $75.0 \text{ mL solution} \times \frac{0.980 \text{ g solution}}{1 \text{ mL solution}} = 73.5 \text{ g solution (calc and corr)}$

$$\% \text{ (m/m)} = \frac{8.96 \text{ g CH}_3\text{OH}}{73.5 \text{ g solution}} \times 100 = 12.19047619 \text{ (calc)} = 12.2\% \text{ (m/m) CH}_3\text{OH (corr)}$$

c) $\% \text{ (v/v)} = \frac{11.3 \text{ mL CH}_3\text{OH}}{75.0 \text{ mL solution}} \times 100 = 15.066666 \text{ (calc)} = 15.1\% \text{ (v/v) CH}_3\text{OH (corr)}$

13.153 Basis: 1 mole KCl

Volume of solution: $1 \text{ mole KCl} \times \dfrac{1 \text{ L solution}}{0.271 \text{ mole KCl}} \times \dfrac{1 \text{ mL solution}}{10^{-3} \text{ L solution}}$

$= 3690.0369 \text{ (calc)} = 3690 \text{ mL solution (corr)}$

Mass of solution = mass of solute + mass of solvent

Mass of solute: $1 \text{ mole KCl} \times \dfrac{74.55 \text{ g KCl}}{1 \text{ mole KCl}} = 74.55 \text{ g KCl (calc and corr)}$

Mass of solvent: $1 \text{ mole KCl} \times \dfrac{1 \text{ kg solvent}}{0.273 \text{ mole KCl}} \times \dfrac{10^3 \text{ g solvent}}{1 \text{ kg solvent}} = 3663.003663 \text{ (calc)}$

$= 3660 \text{ g solvent (corr)}$

Mass of solution: $(3660 + 74.55) \text{ g} = 3734.55 \text{ (calc)} = 3730 \text{ g (corr)}$

$\text{density} = \dfrac{\text{mass solution}}{\text{volume solution}} = \dfrac{3730 \text{ g solution}}{3690 \text{ mL solution}} = 1.0108401 \text{ (calc)} = 1.01 \text{ g/mL (corr)}$

Cumulative Problems

13.155 a) $NaCl + AgNO_3 \rightarrow AgCl + NaNO_3$ AgCl is insoluble.
       b) $3 \text{ Ba}(C_2H_3O_2)_2 + 2 \text{ K}_3PO_4 \rightarrow 6 \text{ KC}_2H_3O_2 + Ba_3(PO_4)_2$ $Ba_3(PO_4)_2$ is insoluble.
       c) $Pb(NO_3)_2 + Ag_2SO_4 \rightarrow 2 \text{ AgNO}_3 + PbSO_4$ $PbSO_4$ is insoluble.
       d) $CuSO_4 + BaS \rightarrow BaSO_4 + CuS$ $BaSO_4$ and CuS are insoluble.

13.157 Calculate moles NH_3 needed:

$2.00 \text{ L NH}_3 \times \dfrac{3.50 \text{ moles NH}_3}{1 \text{ L NH}_3} = 7 \text{ (calc)} = 7.00 \text{ moles NH}_3 \text{ (corr)}$

$V = \dfrac{nRT}{P} = \dfrac{7.00 \text{ moles} \times 0.08206 \dfrac{\text{atm L}}{\text{mole K}} \times 298 \text{ K}}{1.46 \text{ atm}} = 117.24463 \text{ (calc)} = 117 \text{ L NH}_3 \text{ (corr)}$

13.159 Find the limiting reactant:

$6.41 \text{ g ZnCl}_2 \times \dfrac{1 \text{ mole ZnCl}_2}{136.28 \text{ g ZnCl}_2} \times \dfrac{2 \text{ moles AgCl}}{1 \text{ mole ZnCl}_2} = 0.09407103023 \text{ (calc)} = 0.0941 \text{ mole AgCl (corr)}$

$40.0 \text{ mL AgNO}_3 \times \dfrac{10^{-3} \text{ L AgNO}_3}{1 \text{ mL AgNO}_3} \times \dfrac{0.404 \text{ mole AgNO}_3}{1 \text{ L AgNO}_3} \times \dfrac{2 \text{ moles AgCl}}{2 \text{ moles AgNO}_3} = 0.01616 \text{ (calc)}$

$= 0.0162 \text{ mole AgCl (corr)}$

$AgNO_3$ is the limiting reactant.

$0.0162 \text{ mole AgCl} \times \dfrac{143.32 \text{ g AgCl}}{1 \text{ mole AgCl}} = 2.321784 \text{ (calc)} = 2.32 \text{ g AgCl (corr)}$

Solutions 153

13.161 Find the limiting reactant:

$$0.350 \text{ L BaCl}_2 \times \frac{3.25 \text{ moles BaCl}_2}{1 \text{ L BaCl}_2} \times \frac{1 \text{ mole BaCrO}_4}{1 \text{ mole BaCl}_2} = 1.1375 \text{ (calc)} = 1.14 \text{ moles BaCrO}_4 \text{ (corr)}$$

$$0.450 \text{ L K}_2\text{CrO}_4 \times \frac{4.50 \text{ moles K}_2\text{CrO}_4}{1 \text{ L K}_2\text{CrO}_4} \times \frac{1 \text{ mole BaCrO}_4}{1 \text{ mole K}_2\text{CrO}_4} = 2.025 \text{ (calc)} = 2.02 \text{ moles BaCrO}_4 \text{ (corr)}$$

BaCl$_2$ is the limiting reactant.

$$1.14 \text{ moles BaCrO}_4 \times \frac{253.33 \text{ g BaCrO}_4}{1 \text{ mole BaCrO}_4} = 288.7962 \text{ (calc)} = 289 \text{ g BaCrO}_4 \text{ (corr)}$$

13.163 $70.0 \text{ mL HCl} \times \dfrac{10^{-3} \text{ L HCl}}{1 \text{ mL HCl}} \times \dfrac{0.125 \text{ mole HCl}}{1 \text{ L HCl}} \times \dfrac{1 \text{ mole Na}_2\text{CO}_3}{2 \text{ moles HCl}} \times \dfrac{105.99 \text{ g Na}_2\text{CO}_3}{1 \text{ mole Na}_2\text{CO}_3}$

$$= 0.4637062 \text{ (calc)} = 0.464 \text{ g Na}_2\text{CO}_3 \text{ (corr)}$$

$$\% \text{ Na}_2\text{CO}_3 = \frac{0.464 \text{ g Na}_2\text{CO}_3}{1.25 \text{ g sample}} \times 100 = 37.12 \text{ (calc)} = 37.1\% \text{ (m/m) Na}_2\text{CO}_3 \text{ (corr)}$$

13.165 $27.9 \text{ mL HCl} \times \dfrac{10^{-3} \text{ L HCl}}{1 \text{ mL HCl}} \times \dfrac{2.48 \text{ moles HCl}}{1 \text{ L HCl}} \times \dfrac{1 \text{ mole MCl}_2}{2 \text{ moles HCl}} = 0.034596 \text{ (calc)}$

$$= 0.0346 \text{ mole MCl}_2 \text{ (corr)}$$

$$\text{Molar mass of MCl}_2 = \frac{4.72 \text{ g}}{0.0346 \text{ mole}} = 136.41618 \text{ (calc)} = 136 \text{ g/mole (corr)}$$

at. mass M + 2(at. mass Cl) = 136 amu
M + 2(35.45 amu) = 136 amu
M = 65 amu
M = zinc

13.167 $70.0 \text{ mL O}_2 \times \dfrac{10^{-3} \text{ L O}_2}{1 \text{ mL O}_2} \times \dfrac{1 \text{ mole O}_2}{22.41 \text{ L O}_2} \times \dfrac{4 \text{ moles NaOH}}{1 \text{ mole O}_2} = 0.01249442213 \text{ (calc)}$

$$= 0.0125 \text{ mole NaOH (corr)}$$

$$M = \frac{0.0125 \text{ mole NaOH}}{0.150 \text{ L solution}} = 0.083333333 \text{ (calc)} = 0.0833 \text{ M (corr)}$$

Answers to Multiple-Choice Practice Test

13.169	b	**13.170**	b	**13.171**	d	**13.172**	e	**13.173**	b	**13.174**	b
13.175	b	**13.176**	b	**13.177**	d	**13.178**	d	**3.179**	c	**13.180**	c
13.181	d	**13.182**	b	**13.183**	d	**13.184**	d	**13.185**	b	**13.186**	c
13.187	b	**13.188**	d								

CHAPTER FOURTEEN
Acids, Bases, and Salts

Practice Problems

Acid–Base Definitions (Secs. 14.1 and 14.2)

14.1 a) The species responsible for the properties of acidic solutions is the hydrogen ion, $H^+(aq)$.
b) The term used to describe formation of ions, in aqueous solution, from an ionic compound, is *dissociation*.

14.3 a) Arrhenius acid b) Arrhenius acid

14.5 a) ionization, HNO_3, molecular compound   b) dissociation, NaOH, ionic compound
c) dissociation, RbOH, ionic compound   d) ionization, HCN, molecular compound

14.7 a) $HBr \rightarrow H^+ + Br^-$ b) $HClO_2 \rightarrow H^+ + ClO_2^-$
c) $LiOH \rightarrow Li^+ + OH^-$ d) $Ba(OH)_2 \rightarrow Ba^{2+} + 2\,OH^-$

14.9 Brønsted–Lowry = BL
a) $HBr + H_2O \rightarrow H_3O^+ + Br^-$ b) $H_2O + N_3^- \rightarrow HN_3 + OH^-$
   BL acid + BL base   BL acid + BL base
c) $H_2O + H_2S \rightarrow H_3O^+ + HS^-$ d) $HS^- + H_2O \rightarrow H_3O^+ + S^{2-}$
   BL base + BL acid   BL acid + BL base

14.11 a) $HOCl + NH_3 \rightarrow NH_4^+ + ClO^-$ b) $H_2CO_3 + H_2O \rightarrow HCO_3^- + H_3O^+$
c) $H_2O + H_2O \rightarrow H_3O^+ + OH^-$ d) $HC_2O_4^- + H_2O \rightarrow C_2O_4^{2-} + H_3O^+$

Conjugate Acids and Bases (Sec. 14.3)

14.13 a) NO_2^- b) $H_2PO_4^-$ c) PO_4^{3-} d) NH_3

14.15 a) $HClO_2$ b) H_2SO_4 c) HSO_3^- d) PH_4^+

14.17 Conjugate acid–base pairs
a) $H_2C_2O_4$ and $HC_2O_4^-$; HClO and ClO^- b) HSO_4^- and SO_4^{2-}; H_3O^+ and H_2O
c) $H_2PO_4^-$ and HPO_4^{2-}; NH_4^+ and NH_3 d) H_2CO_3 and HCO_3^-; H_2O and OH^-

14.19 a) yes b) no c) no d) yes

14.21 a) (1) $HS^- + H_3O^+ \rightarrow H_2S + H_2O$ (2) $HS^- + OH^- \rightarrow H_2O + S^{2-}$
b) (1) $HPO_4^{2-} + H_3O^+ \rightarrow H_2PO_4^- + H_2O$ (2) $HPO_4^{2-} + OH^- \rightarrow H_2O + PO_4^{3-}$
c) (1) $HCO_3^- + H_3O^+ \rightarrow H_2CO_3 + H_2O$ (2) $HCO_3^- + OH^- \rightarrow H_2O + CO_3^{2-}$
d) (1) $H_2PO_3^- + H_3O^+ \rightarrow H_3PO_3 + H_2O$ (2) $H_2PO_3^- + OH^- \rightarrow H_2O + HPO_3^{2-}$

Mono-, Di-, and Triprotic Acids (Sec. 14.4)

14.23 a) $HClO_3$, monoprotic   b) $HC_3H_5O_4$, monoprotic
c) $HC_2H_3O_2$, monoprotic   d) $H_2C_4H_2O_4$, diprotic

14.25 a) HClO₃, monoprotic, 1 acidic H b) HC₃H₅O₄, monoprotic, 1 acidic H
c) HC₂H₃O₂, monoprotic, 1 acidic H d) H₂C₄H₂O₄, diprotic, 2 acidic H's

14.27 a) HClO₃, 0 nonacidic H's b) HC₃H₅O₄, 5 nonacidic H's
c) HC₂H₃O₂, 3 nonacidic H's d) H₂C₄H₂O₄, 2 nonacidic H's

14.29 a) i) $H_2C_2O_4 + H_2O \rightarrow HC_2O_4^- + H_3O^+$ ii) $HC_2O_4^- + H_2O \rightarrow C_2O_4^{2-} + H_3O^+$
b) i) $H_2C_3H_2O_4 + H_2O \rightarrow HC_3H_2O_4^- + H_3O^+$ ii) $HC_3H_2O_4^- + H_2O \rightarrow C_3H_2O_4^{2-} + H_3O^+$

14.31 The chemical formula for lactic acid, H₆C₃O₃, is written as HC₃H₅O₃, to differentiate between the 5 nonacidic H's and the 1 acidic H in the acid.

14.33 Pyruvic acid contains one acidic H, attached to the oxygen atom, and 3 nonacidic H's, attached to carbon.

Strength of Acids and Bases (Sec. 14.5)

14.35 a) HClO₃, is a strong acid. b) HC₃H₅O₄, is a weak acid.
c) HC₂H₃O₂, is a weak acid. d) H₂C₄H₂O₄, is a weak acid.

14.37 a) H₂SO₄ is stronger than H₂SO₃. b) HClO₄ is stronger than HCN.
c) HF is weaker than HBr. d) HNO₂ is weaker than HClO₃.

14.39 a) strong acid, strong base b) strong acid, strong base
c) weak acid, strong base d) weak acid, strong base

14.41 a) HNO₃ is stronger acid than HNO₂. b) HClO₄ is stronger acid than HC₂H₃O₂.
c) H₃PO₄ is a stronger acid than HCN. d) HF is a stronger acid than H₂CO₃.

Salts (Sec. 14.6)

14.43 The term *weak acid* pertains to the amount of dissociation, while the term *dilute* refers to the concentration of the acid in solution.

14.45 Diagram IV is the strongest acid.

14.47 a) NH₄Cl is a salt.   b) HCl is an acid.   c) KCl is a salt.   d) NaNO₃ is a salt.

14.49 a) CaSO₄, calcium sulfate b) Li₂CO₃, lithium carbonate
c) NaBr, sodium bromide d) Al₂S₃, aluminum sulfide

14.51 a) CaSO₄, insoluble salt b) Li₂CO₃, soluble salt
c) NaBr, soluble salt d) Al₂S₃, insoluble salt

14.53 a) NaNO₃ produces 1 Na⁺ and 1 NO₃⁻, total 2 ions in water.
b) CuCO₃ produces 1 Cu²⁺ and 1 CO₃²⁻, total 2 ions in water.
c) BaCl₂ produces 1 Ba²⁺ and 2 Cl⁻, total 3 ions in water.
d) Al₂(SO₄)₃ produces 2 Al³⁺ and 3 SO₄²⁻, total of 5 ions in water.

14.55 In water the following dissociation takes place.
a) $NaNO_3(s) \rightarrow Na^+(aq) + NO_3^-(aq)$ b) $CuCO_3(s) \rightarrow Cu^{2+}(aq) + CO_3^{2-}(aq)$
c) $BaCl_2(s) \rightarrow Ba^{2+}(aq) + 2\,Cl^-(aq)$ d) $Al_2(SO_4)_3(s) \rightarrow 2\,Al^{3+}(aq) + 3\,SO_4^{2-}(aq)$

Acids, Bases, and Salts

Reactions of Acids and Bases (Secs. 14.7 and 14.8)

14.57   a) Yes, Fe is more reactive than H.   b) Yes, K is more reactive than H.
        c) No, Au is less reactive than H.   d) Yes, Al is more reactive than H.

14.59   a) $Ni + 2 HCl \rightarrow NiCl_2 + H_2$ b) $Ca + 2 H_2O \rightarrow Ca(OH)_2 + H_2$
        c) $Mg + 2 HCl \rightarrow MgCl_2 + H_2$ d) $Zn + 2 H_2O \rightarrow Zn(OH)_2 + H_2$

14.61   a) No, not an acid–base neutralization reaction   b) Yes, an acid–base neutralization reaction
        c) Yes, an acid–base neutralization reaction   d) No, not an acid–base neutralization reaction

14.63   a) 2:1   b) 1:1   c) 1:1   d) 1:3

14.65   a) $2 HBr + Sr(OH)_2 \rightarrow SrBr_2 + 2 H_2O$
        b) $HCN + LiOH \rightarrow LiCN + H_2O$
        c) $H_2SO_4 + Mg(OH)_2 \rightarrow MgSO_4 + 2 H_2O$
        d) $H_3PO_4 + 3 KOH \rightarrow K_3PO_4 + 3 H_2O$

14.67   a) $HC_3H_5O_3 + NaOH \rightarrow NaC_3H_5O_3 + H_2O$
        b) $H_2C_4H_4O_5 + 2 NaOH \rightarrow Na_2C_4H_4O_5 + 2 H_2O$
        c) $H_2C_4H_4O_4 + 2 NaOH \rightarrow Na_2C_4H_4O_4 + 2 H_2O$
        d) $H_3C_6H_5O_7 + 3 NaOH \rightarrow Na_3C_6H_5O_7 + 3 H_2O$

14.69   a) $NaNO_3$ b) KCN c) Li_2SO_4 d) $BaCO_3$

14.71   a) acid = H_2S, base = NaOH   b) acid = HNO_3, base = KOH
        c) acid = $HClO_3$, base = $Al(OH)_3$ d) acid = HBr, base = $Ca(OH)_2$

14.73   a) $Zn(s) + 2 HCl(aq) \rightarrow ZnCl_2(aq) + H_2(g)$
        b) $HCl(aq) + NaOH(aq) \rightarrow NaCl(aq) + H_2O(l)$
        c) $2 HCl(aq) + Na_2CO_3(aq) \rightarrow 2 NaCl(aq) + CO_2(g) + H_2O(l)$
        d) $HCl(aq) + NaHCO_3(aq) \rightarrow NaCl(aq) + CO_2(g) + H_2O(l)$

Reactions of Salts (Sec. 14.9)

14.75   a) No, Cu is less reactive than Ni.   b) Yes, Zn is more reactive than Ni.
        c) No, Au is less reactive than Ni.   d) Yes, Fe is more reactive than Ni.

14.77   a) $Fe(s) + CuSO_4(aq) \rightarrow FeSO_4(aq) + Cu(s)$
        b) $Sn(s) + 2 AgNO_3(aq) \rightarrow Sn(NO_3)_2(aq) + 2 Ag(s)$
        c) $Zn(s) + NiCl_2(aq) \rightarrow ZnCl_2(aq) + Ni(s)$
        d) $Cr(s) + Pb(C_2H_3O_2)_2(aq) \rightarrow Cr(C_2H_3O_2)_2(aq) + Pb(s)$

14.79   a) $BaSO_4$, an insoluble salt is formed.
        b) $Ca_3(PO_4)_2$, an insoluble salt is formed.
        c) AgCl, an insoluble salt is formed, weak acid is formed.
        d) CO_2, a gas, and H_2O liquid are formed.

14.81   a) $2 Al(NO_3)_3 + 3 (NH_4)_2S \rightarrow Al_2S_3 + 6 NH_4NO_3$
        b) $2 HCl + Ba(OH)_2 \rightarrow 2 H_2O + BaCl_2$
        c) no reaction
        d) no reaction

Hydronium Ion and Hydroxide Ion Concentrations (Sec. 14.10)

14.83 a) $[H_3O^+] = \dfrac{1.00 \times 10^{-14}}{3.5 \times 10^{-3}} = 2.857142857 \times 10^{-12}$ (calc) $= 2.9 \times 10^{-12}$ M (corr)

b) $[H_3O^+] = \dfrac{1.00 \times 10^{-14}}{4.7 \times 10^{-6}} = 2.127659574 \times 10^{-9}$ (calc) $= 2.1 \times 10^{-9}$ M (corr)

c) $[H_3O^+] = \dfrac{1.00 \times 10^{-14}}{1.1 \times 10^{-8}} = 9.090909 \times 10^{-7}$ (calc) $= 9.1 \times 10^{-7}$ M (corr)

d) $[H_3O^+] = \dfrac{1.00 \times 10^{-14}}{8.7 \times 10^{-10}} = 1.149425287 \times 10^{-5}$ (calc) $= 1.1 \times 10^{-5}$ M (corr)

14.85 a) $[H_3O^+] = 2.9 \times 10^{-12}$ M, basic b) $[H_3O^+] = 2.1 \times 10^{-9}$ M, basic
c) $[H_3O^+] = 9.1 \times 10^{-7}$ M, acidic d) $[H_3O^+] = 1.1 \times 10^{-5}$ M, acidic

14.87 a) $[OH^-] = \dfrac{1.00 \times 10^{-14}}{5.5 \times 10^{-2}} = 1.81818181 \times 10^{-13}$ (calc) $= 1.8 \times 10^{-13}$ M (corr)

b) $[OH^-] = \dfrac{1.00 \times 10^{-14}}{9.4 \times 10^{-5}} = 1.06382979 \times 10^{-10}$ (calc) $= 1.1 \times 10^{-10}$ M (corr)

c) $[OH^-] = \dfrac{1.00 \times 10^{-14}}{2.3 \times 10^{-7}} = 4.347826087 \times 10^{-8}$ (calc) $= 4.3 \times 10^{-8}$ M (corr)

d) $[OH^-] = \dfrac{1.00 \times 10^{-14}}{6.6 \times 10^{-12}} = 1.5151515151 \times 10^{-3}$ (calc) $= 1.5 \times 10^{-3}$ M (corr)

14.89 a) $[OH^-] = 1.8 \times 10^{-13}$ M, acidic b) $[OH^-] = 1.1 \times 10^{-10}$ M, acidic
c) $[OH^-] = 4.3 \times 10^{-8}$ M, acidic d) $[OH^-] = 1.5 \times 10^{-3}$ M, basic

14.91

	$[H_3O^+]$	$[OH^-]$	Acidic or Basic
	2.2×10^{-2}	4.5×10^{-13}	acidic
a)	3.0×10^{-12}	3.3×10^{-3}	basic
b)	6.8×10^{-8}	1.5×10^{-7}	basic
c)	1.4×10^{-7}	7.2×10^{-8}	acidic
d)	4.5×10^{-6}	2.2×10^{-9}	acidic

The pH Scale (Sec. 14.11)

14.93 a) pH $= -\log(4 \times 10^{-2}) = 4$ (calc) $= 4.0$ (corr)
b) pH $= -\log(1 \times 10^{-11}) = 11$ (calc) $= 11.0$ (corr)
c) pH $= -\log(0.000001) = 5$ (calc) $= 5.0$ (corr)
d) pH $= -\log(0.000000001) = 9$ (calc) $= 9.0$ (corr)

14.95 a) pH $= -\log(6 \times 10^{-3}) = 2.22184875$ (calc) $= 2.2$ (corr)
b) pH $= -\log(6 \times 10^{-4}) = 3.22184815$ (calc) $= 3.2$ (corr)
c) pH $= -\log(3 \times 10^{-8}) = 7.522878745$ (calc) $= 7.5$ (corr)
d) pH $= -\log(7 \times 10^{-10}) = 9.15490196$ (calc) $= 9.2$ (corr)

Acids, Bases, and Salts

14.97 The calculator answer will be the same for all of these: 2.52287875. The difference is the number of significant figures in the mantissa.
  a) 2.5   b) 2.52   c) 2.523   d) 2.5229

14.99 a) both acidic   b) acidic, basic   c) both basic   d) acidic, neutral

14.101 a) $pH = -\log[H_3O^+] = 4.0$, therefore antilog $-4.0 = [H_3O^+] = 1 \times 10^{-4}$ M (calc and corr)
  b) $pH = -\log[H_3O^+] = 4.2$, therefore antilog $-4.2 = [H_3O^+] = 6.309573445 \times 10^{-5}$ (calc)
  $= 6 \times 10^{-5}$ M (corr)
  c) $pH = -\log[H_3O^+] = 4.5$, therefore antilog $-4.5 = [H_3O^+] = 3.16227766 \times 10^{-5}$ (calc)
  $= 3 \times 10^{-5}$ M (corr)
  d) $pH = -\log[H_3O^+] = 4.8$, therefore antilog $-4.8 = [H_3O^+] = 1.584893192 \times 10^{-5}$ (calc)
  $= 2 \times 10^{-5}$ M (corr)

14.103 a) $pH = -\log[H_3O^+] = 2.43$, therefore antilog $-2.43 = [H_3O^+] = 3.715352291 \times 10^{-3}$ (calc)
  $= 3.7 \times 10^{-3}$ M (corr)
  b) $pH = -\log[H_3O^+] = 5.72$, therefore antilog $-5.72 = [H_3O^+] = 1.905460718 \times 10^{-6}$ (calc)
  $= 1.9 \times 10^{-6}$ M (corr)
  c) $pH = -\log[H_3O^+] = 7.73$, therefore antilog $-7.73 = [H_3O^+] = 1.862087137 \times 10^{-8}$ (calc)
  $= 1.9 \times 10^{-8}$ M (corr)
  d) $pH = -\log[H_3O^+] = 8.750$, therefore antilog $-8.750 = [H_3O^+] = 1.77827941 \times 10^{-9}$ (calc)
  $= 1.78 \times 10^{-9}$ M (corr)

14.105 a) $pH = -\log[H_3O^+] = 2.43$, therefore $[H_3O^+] = 3.715352291 \times 10^{-3}$ (calc)
  $= 3.7 \times 10^{-3}$ M (corr)
  $[OH^-] = \dfrac{1.00 \times 10^{-14}}{3.7 \times 10^{-3}} = 2.702702703 \times 10^{-12}$ (calc) $= 2.7 \times 10^{-12}$ M (corr)
  b) $pH = -\log[H_3O^+] = 5.72$, therefore $[H_3O^+] = 1.905460718 \times 10^{-6}$ (calc)
  $= 1.9 \times 10^{-6}$ M (corr)
  $[OH^-] = \dfrac{1.00 \times 10^{-14}}{1.9 \times 10^{-6}} = 5.263157895 \times 10^{-9}$ (calc) $= 5.3 \times 10^{-9}$ M (corr)
  c) $pH = -\log[H_3O^+] = 7.73$, therefore $[H_3O^+] = 1.862087137 \times 10^{-8}$ (calc)
  $= 1.9 \times 10^{-8}$ M (corr)
  $[OH^-] = \dfrac{1.00 \times 10^{-14}}{1.9 \times 10^{-8}} = 5.263157895 \times 10^{-7}$ (calc) $= 5.3 \times 10^{-7}$ M (corr)
  d) $pH = -\log[H_3O^+] = 8.750$, therefore $[H_3O^+] = 1.77827941 \times 10^{-9}$ (calc)
  $= 1.78 \times 10^{-9}$ M (corr)
  $[OH^-] = \dfrac{1.00 \times 10^{-14}}{1.78 \times 10^{-9}} = 5.617977528 \times 10^{-6}$ (calc) $= 5.62 \times 10^{-6}$ M (corr)

14.107

	$[H_3O^+]$	$[OH^-]$	pH	Acidic or Basic
	6.2×10^{-8}	1.6×10^{-7}	7.21	basic
a)	7.2×10^{-10}	1.4×10^{-5}	9.14	basic
b)	5.0×10^{-6}	2.0×10^{-9}	5.30	acidic
c)	1.4×10^{-5}	7.2×10^{-10}	4.86	acidic
d)	5.9×10^{-9}	1.7×10^{-6}	8.23	basic

14.109 $[H_3O^+]_A = \dfrac{1.0 \times 10^{-14}}{4.3 \times 10^{-4}} = 2.325581395 \times 10^{-11}$ (calc) $= 2.3 \times 10^{-11}$ M (corr)

    a) Solution A is more basic. The smaller the $[H_3O^+]$, the more basic the solution.
    b) Solution B has the lower pH. The larger the $[H_3O^+]$, the lower the pH.

14.111 The original solution has a $[H_3O^+] = 3.1622777 \times 10^{-5}$ (calc) $= 3.16 \times 10^{-5}$ M (corr)
    a) New $[H_3O^+] = 2 \times 3.16 \times 10^{-5} = 6.32 \times 10^{-5}$ M (calc and corr)
       pH $= 4.1992829$ (calc) $= 4.199$ (corr)
    b) New $[H_3O^+] = 4 \times 3.16 \times 10^{-5} = 1.264 \times 10^{-4}$ (calc) $= 1.26 \times 10^{-4}$ M (corr)
       pH $= 3.8996295$ (calc) $= 3.900$ (corr)
    c) New $[H_3O^+] = 10 \times 3.16 \times 10^{-5} = 3.16 \times 10^{-4}$ M (calc and corr)
       pH $= 3.5003129$ (calc) $= 3.500$ (corr)
    d) New $[H_3O^+] = 1000 \times 3.16 \times 10^{-5} = 3.16 \times 10^{-2}$ M (calc and corr)
       pH $= 1.5003129$ (calc) $= 1.500$ (corr)

14.113 a) HNO_3 is a strong acid and ionizes to give 1 H_3O^+ per HNO_3.
       $[H_3O^+] = 6.3 \times 10^{-3}$ M; pH $= 2.2006595$ (calc) $= 2.20$ (corr)
    b) HCl is a strong acid and ionizes to give 1 H_3O^+ per HCl.
       $[H_3O^+] = 0.20$ M; pH $= 0.6989700$ (calc) $= 0.70$ (corr)
    c) H_2SO_4 is a strong acid and ionizes to give 2 H_3O^+ per H_2SO_4.
       $[H_3O^+] = 2 \times 0.000021 = 0.000042$ M; pH $= 4.3767507$ (calc) $= 4.38$ (corr)
    d) NaOH is a strong base containing 1 OH^- per NaOH.
       $[OH^-] = 2.3 \times 10^{-4}$ M;
       $[H_3O^+] = \dfrac{1.00 \times 10^{-14}}{2.3 \times 10^{-4}} = 4.3478261 \times 10^{-11}$ (calc) $= 4.3 \times 10^{-11}$ M (corr)
       pH $= 10.3665315$ (calc) $= 10.37$ (corr)

Hydrolysis of Salts (Sec. 14.12)

14.115 a) PO_4^{3-} b) CN^- c) NH_4^+ d) none

14.117 a) neutral   b) basic   c) acidic   d) neutral

14.119 a) $NH_4^+ + H_2O \rightarrow H_3O^+ + NH_3$ b) $C_2H_3O_2^- + H_2O \rightarrow OH^- + HC_2H_3O_2$
       c) $F^- + H_2O \rightarrow OH^- + HF$ d) $CN^- + H_2O \rightarrow OH^- + HCN$

Buffers (Sec. 14.13)

14.121 a) No, have mixture of strong acid and conjugate base
       b) Yes, have mixture of weak acid and conjugate base
       c) No, do not have an acid in mixture
       d) Yes, have mixture of weak acid and conjugate base

14.123 a) HCN and CN^- b) H_3PO_4 and $H_2PO_4^-$ c) H_2CO_3 and HCO_3^- d) HCO_3^- and CO_3^{2-}

14.125 Each diagram contains the weak acid HA and its conjugate base A^- and therefore each diagram is considered a buffer solution.

14.127 a) $HF + OH^- \rightarrow F^- + H_2O$
       b) $PO_4^{3-} + H_3O^+ \rightarrow HPO_4^{2-} + H_2O$
       c) $CO_3^{2-} + H_3O^+ \rightarrow HCO_3^- + H_2O$
       d) $H_3PO_4 + OH^- \rightarrow H_2PO_4^- + H_2O$

Acids, Bases, and Salts

Acid–Base Titrations (Sec. 14.14)

14.129 a) $HCl + NaOH \rightarrow NaCl + H_2O$

$$20.00 \text{ mL} \times \frac{10^{-3} \text{ L}}{1 \text{ mL}} \times \frac{0.100 \text{ mole NaOH}}{1 \text{ L}} \times \frac{1 \text{ mole HCl}}{1 \text{ mole NaOH}} \times \frac{1 \text{ L HCl}}{0.100 \text{ mole HCl}}$$

$$\times \frac{1 \text{ mL}}{10^{-3} \text{ L}} = 20 \text{ (calc)} = 20.0 \text{ mL HCl solution (corr)}$$

b) $HCl + NaOH \rightarrow NaCl + H_2O$

$$20.00 \text{ mL} \times \frac{10^{-3} \text{ L}}{1 \text{ mL}} \times \frac{0.200 \text{ mole NaOH}}{1 \text{ L}} \times \frac{1 \text{ mole HCl}}{1 \text{ mole NaOH}} \times \frac{1 \text{ L HCl}}{0.100 \text{ mole HCl}}$$

$$\times \frac{1 \text{ mL}}{10^{-3} \text{ L}} = 40 \text{ (calc)} = 40.0 \text{ mL HCl solution (corr)}$$

c) $2 HCl + Ba(OH)_2 \rightarrow BaCl_2 + 2 H_2O$

$$20.00 \text{ mL} \times \frac{10^{-3} \text{ L}}{1 \text{ mL}} \times \frac{0.100 \text{ mole Ba(OH)}_2}{1 \text{ L}} \times \frac{2 \text{ moles HCl}}{1 \text{ mole Ba(OH)}_2} \times \frac{1 \text{ L HCl}}{0.100 \text{ mole HCl}}$$

$$\times \frac{1 \text{ mL}}{10^{-3} \text{ L}} = 40 \text{ (calc)} = 40.0 \text{ mL HCl solution (corr)}$$

d) $2 HCl + Ba(OH)_2 \rightarrow BaCl_2 + 2 H_2O$

$$20.00 \text{ mL} \times \frac{10^{-3} \text{ L}}{1 \text{ mL}} \times \frac{0.200 \text{ mole Ba(OH)}_2}{1 \text{ L}} \times \frac{2 \text{ moles HCl}}{1 \text{ mole Ba(OH)}_2} \times \frac{1 \text{ L HCl}}{0.100 \text{ mole HCl}}$$

$$\times \frac{1 \text{ mL}}{10^{-3} \text{ L}} = 80 \text{ (calc)} = 80.0 \text{ mL HCl solution (corr)}$$

14.131 a) $HCl + KOH \rightarrow KCl + H_2O$

$$20.00 \text{ mL} \times \frac{10^{-3} \text{ L}}{1 \text{ mL}} \times \frac{0.200 \text{ moles HCl}}{1 \text{ L}} \times \frac{1 \text{ mole KOH}}{1 \text{ mole HCl}} \times \frac{1 \text{ L KOH}}{0.200 \text{ mole KOH}}$$

$$\times \frac{1 \text{ mL}}{10^{-3} \text{ L}} = 20 \text{ (calc)} = 20.0 \text{ mL KOH solution (corr)}$$

b) $H_2SO_4 + 2 KOH \rightarrow K_2SO_4 + 2 H_2O$

$$20.00 \text{ mL} \times \frac{10^{-3} \text{ L}}{1 \text{ mL}} \times \frac{0.200 \text{ moles H}_2\text{SO}_4}{1 \text{ L}} \times \frac{2 \text{ moles KOH}}{1 \text{ mole H}_2\text{SO}_4} \times \frac{1 \text{ L KOH}}{0.200 \text{ mole KOH}}$$

$$\times \frac{1 \text{ mL}}{10^{-3} \text{ L}} = 40 \text{ (calc)} = 40.0 \text{ mL KOH solution (corr)}$$

c) $HNO_3 + KOH \rightarrow KNO_3 + H_2O$

$$50.00 \text{ mL} \times \frac{10^{-3} \text{ L}}{1 \text{ mL}} \times \frac{0.400 \text{ mole HNO}_3}{1 \text{ L}} \times \frac{1 \text{ mole HNO}_3}{1 \text{ mole KOH}} \times \frac{1 \text{ L KOH}}{0.200 \text{ mole KOH}}$$

$$\times \frac{1 \text{ mL}}{10^{-3} \text{ L}} = 100 \text{ mL HNO}_3 \text{ solution (calc)} = 1.00 \times 10^2 \text{ mL HNO}_3 \text{ (corr)}$$

d) $H_2CO_3 + 2 KOH \rightarrow K_2CO_3 + 2 H_2O$

$$50.00 \text{ mL} \times \frac{10^{-3} \text{ L}}{1 \text{ mL}} \times \frac{0.300 \text{ moles H}_2\text{CO}_3}{1 \text{ L}} \times \frac{2 \text{ moles KOH}}{1 \text{ mole H}_2\text{CO}_3} \times \frac{1 \text{ L KOH}}{0.200 \text{ mole KOH}}$$

$$\times \frac{1 \text{ mL}}{10^{-3} \text{ L}} = 150 \text{ (calc)} = 1.50 \times 10^2 \text{ mL KOH solution (corr)}$$

14.133 a) $H_2SO_4 + 2 NaOH \rightarrow Na_2SO_4 + 2 H_2O$

$$34.5 \text{ mL} \times \frac{10^{-3} \text{ L}}{1 \text{ mL}} \times \frac{0.102 \text{ mole NaOH}}{1 \text{ L}} \times \frac{1 \text{ mole H}_2\text{SO}_4}{2 \text{ moles NaOH}} = \frac{1.7595 \times 10^{-3} \text{ mole H}_2\text{SO}_4}{0.02500 \text{ L solution}}$$

$$= 7.038 \times 10^{-2} \text{ (calc)} = 0.0704 \text{ M H}_2\text{SO}_4 \text{ (corr)}$$

b) $HClO + NaOH \rightarrow NaClO + H_2O$

$$34.5 \text{ mL} \times \frac{10^{-3} \text{ L}}{1 \text{ mL}} \times \frac{0.102 \text{ mole NaOH}}{1 \text{ L}} \times \frac{1 \text{ mole HClO}}{1 \text{ mole NaOH}} = \frac{3.519 \times 10^{-3} \text{ mole HClO}}{0.02000 \text{ L solution}}$$

$$= 0.17595 \text{ (calc)} = 0.176 \text{ M HClO (corr)}$$

c) $H_3PO_4 + 3\ NaOH \rightarrow Na_3PO_4 + 3\ H_2O$

$$34.5\ mL \times \frac{10^{-3}\ L}{1\ mL} \times \frac{0.102\ mole\ NaOH}{1\ L} \times \frac{1\ mole\ H_3PO_4}{3\ moles\ NaOH} = \frac{1.173 \times 10^{-3}\ mole\ H_3PO_4}{0.02000\ L\ solution}$$
$$= 5.865 \times 10^{-2}\ (calc) = 0.0587\ M\ H_3PO_4\ (corr)$$

d) $HNO_3 + NaOH \rightarrow NaNO_3 + H_2O$

$$34.5\ mL \times \frac{10^{-3}\ L}{1\ mL} \times \frac{0.102\ mole\ NaOH}{1\ L} \times \frac{1\ mole\ HNO_3}{1\ mole\ NaOH} = \frac{3.519 \times 10^{-3}\ mole\ HNO_3}{0.01000\ L\ solution}$$
$$= 3.519 \times 10^{-1}\ (calc) = 0.352\ M\ HNO_3\ (corr)$$

Additional Problems

14.135 a) strong b) weak c) weak d) strong

14.137 $[H_3O^+]_1 = $ antilog $(-2.2) = 6.3095734 \times 10^{-3}$ (calc) $= 6 \times 10^{-3}$ (corr)
$[H_3O^+]_2 = $ antilog $(-4.5) = 3.1622776 \times 10^{-5}$ (calc) $= 3 \times 10^{-5}$ (corr)

$$\frac{[H_3O^+]_1}{[H_3O^+]_2} = \frac{6 \times 10^{-3}}{3 \times 10^{-5}} = 200\ \text{(calc and corr)}$$

14.139 If we let the hydroxide concentration, $[OH^-] = x$, then $[H_3O^+] = 3x$
From the ion product constant for water, we know that $[H_3O^+] \times [OH^-] = 1.00 \times 10^{-14}$
so $x \times 3x = 1.00 \times 10^{-14}$
giving $x = 5.773503 \times 10^{-8}$
Therefore $[H_3O^+] = 3x = 1.732051 \times 10^{-7}$ and pH $= -\log[H_3O^+] = 6.76144$ (calc) $= 6.76$ (corr).

14.141 a) The lower the pH value, the more acidic the solution. So in decreasing acidity: A, B, D, C.
b) In increasing $[H_3O^+]$: C, D, B, A.
c) The higher the pH value, the more basic the solution. So in decreasing $[OH^-]$: C, D, B, A.
d) In increasing basicity: A, B, D, C.

14.143 a) $[H_3O^+] \times [OH^-] = 2.6 \times 10^{-14}$ Since the $[H_3O^+] = [OH^-]$, $[H_3O^+] = \sqrt{2.6 \times 10^{-14}}$
$= 1.61245155 \times 10^{-7}$ (calc) $= 1.6 \times 10^{-7}$ (corr)
b) $[H_3O^+] \times [OH^-] = 2.6 \times 10^{-14}$ Since the $[H_3O^+] = [OH^-]$, $[OH^-] = \sqrt{2.6 \times 10^{-14}}$
$= 1.61245155 \times 10^{-7}$ (calc) $= 1.6 \times 10^{-7}$ (corr)
c) pH $= -\log[1.6 \times 10^{-7}] = 6.795880017$ (calc) $= 6.8$ (corr)
d) acidic

14.145 The more concentrated H_3O^+ has a lower pH.
a) no, higher pH b) no, higher pH c) yes d) yes

14.147 NaCl, a soluble salt, does not hydrolyze (no pH effect). HNO_3, a strong acid, produces 0.1 mole H_3O^+ ion. HCl, a strong acid, produces 0.1 mole H_3O^+ ion. NaOH, a strong base, produces 0.1 mole OH^- ion. The 0.1 mole of OH^- ion will neutralize 0.1 mole H_3O^+ ion, with the other 0.1 mole H_3O^+ ion remaining in solution.

$$[H_3O^+] = \frac{0.1\ mole}{3.00\ L\ solution} = 3.3333333 \times 10^{-2}\ (calc) = 3 \times 10^{-2}\ M\ (corr)$$
pH $= -\log(3 \times 10^{-2}) = 1.5228787$ (calc) $= 1.5$ (corr)

14.149 NH_4Br (salt of a weak base) hydrolyzes to produce a slightly acid solution. $Ba(OH)_2$ (strong base) produces a strongly basic solution. $HClO_4$ (strong acid) produces a strongly acidic solution. K_2SO_4 (salt of a strong acid and strong base) does not hydrolyze and produces a neutral solution. LiCN (salt of a weak acid) hydrolyzes to produce a slightly basic solution.

Acids, Bases, and Salts

The order of decreasing pH will put the most basic solution first (highest pH) and the most acidic solution last (lowest pH): $Ba(OH)_2$, LiCN, K_2SO_4, NH_4Br, $HClO_4$.

14.151 a) HCN and NaCN, HCN and KCN b) HF and NaF

14.153 Buffer 1: $H_2PO_4^-$, H_3PO_4 Buffer 2: $H_2PO_4^-$, HPO_4^{2-}

14.155 Balanced equation: $3\ Ca(OH)_2 + 2\ H_3PO_4 \rightarrow Ca_3(PO_4)_2 + 6\ H_2O$ Molar mass H_3PO_4 = 98.00 g/mol

$$0.40 \text{ g } H_3PO_4 \times \frac{1 \text{ mole } H_3PO_4}{98.00 \text{ g } H_3PO_4} \times \frac{3 \text{ moles } Ca(OH)_2}{2 \text{ moles } H_3PO_4} = 0.00612244898 \text{ (calc)}$$
$$= 0.0061 \text{ mole } Ca(OH)_2 \text{ (corr)}$$

14.157 Balanced equation: $HNO_3 + NaOH \rightarrow NaNO_3 + H_2O$
Moles HNO_3 = 0.125 L × 5.00 M = 0.625 mole
Moles NaOH = 0.125 L × 6.00 M = 0.750 mole
Therefore after reaction we have 0.125 mole of NaOH left in the total volume (0.250 L)

Concentration of NaOH left is $\dfrac{0.125 \text{ mole NaOH}}{0.250 \text{ L solution}}$ = 0.500 M

a) As $[H_3O^+] \times [OH^-] = 1.00 \times 10^{-14}$, $[H_3O^+] = 1.00 \times 10^{-14}/0.50 = 2.00 \times 10^{-14}$ M
b) [NaOH] = 0.500 M
c) pH = $-\log[H_3O^+]$ = 13.69897 (calc) = 13.7 (corr)

Cumulative Problems

14.159 a) the iodide, I^-, ion b) the hydrogen phosphate, HPO_4^{2-}, ion
c) the hydroxide, OH^-, ion d) the hydronium, H_3O^+, ion

14.161 a) $M_{HCl} = \dfrac{4.8 \text{ g HCl} \times \dfrac{1 \text{ mole HCl}}{36.46 \text{ g HCl}}}{0.40 \text{ L solution}}$ = 0.3291278 (calc) = 0.33 M (corr)

$[H_3O^+]$ = 0.33 M. pH = $-\log 0.33$ = 0.4814861 (calc) = 0.48 (corr)

b) $M_{LiOH} = \dfrac{12.5 \text{ g LiOH} \times \dfrac{1 \text{ mole LiOH}}{23.95 \text{ g LiOH}}}{255 \text{ mL solution} \times \dfrac{10^{-3} \text{ L}}{1 \text{ mL}}}$ = 2.0467477 (calc) = 2.05 M (corr)

$[OH^-]$ = 2.05 M

$[H_3O^+] = \dfrac{1.00 \times 10^{-14}}{2.05} = 4.8780488 \times 10^{-15}$ (calc) = 4.88×10^{-15} M (corr)

pH = $-\log(4.88 \times 10^{-15})$ = 14.31158 (calc) = 14.312 (corr)

c) $M_2 = 0.10 \text{ M} \times \dfrac{75 \text{ mL}}{125 \text{ mL}}$ = 0.06 (calc) = 0.060 M (corr) = $[H_3O^+]$

pH = $-\log 0.060$ = 1.2218487 (calc) = 1.22 (corr)

d) When you mix two solutions, each is diluted.
Since $V_{HCl} = V_{HNO_3}$, the total $V_{tot} = 2\ V_{HCl} = 2\ V_{HNO_3}$

$M_{HCl} = 0.20 \text{ M} \times \dfrac{V_{HCl}}{2\ V_{HCl}}$ = 0.10 M H_3O^+ from HCl in new solution.

$M_{HNO_3} = 0.50 \text{ M} \times \dfrac{V_B}{2\ V_B}$ = 0.25 M H_3O^+ from HNO_3 in new solution.

Total $[H_3O^+]$ = 0.10 + 0.25 = 0.35 M.
pH = $-\log 0.35$ = 0.4559320 (calc) = 0.46 (corr)

14.163 a) The initial $[H_3O^+] = [HNO_3] = 0.10$ M

b) After dilution, the $[H_3O^+] = 0.10 \times \dfrac{1.00 \text{ mL}}{100.0 \text{ mL}} = 0.0010$ M

c) pH $= -\log [H_3O^+] = -\log (0.10) = 1.00$

d) pH $= -\log [H_3O^+] = -\log (0.0010) = 3.00$

14.165 In pH 5.42, $[H_3O^+] = 3.8018940 \times 10^{-6}$ (calc) $= 3.8 \times 10^{-6}$ M $[H_3O^+]$

$$10.0 \text{ mL solution} \times \dfrac{10^{-3} \text{ L}}{1 \text{ mL}} \times \dfrac{3.8 \times 10^{-6} \text{ mole } H_3O^+}{1 \text{ L solution}} \times \dfrac{6.022 \times 10^{23} \text{ } H_3O^+ \text{ ions}}{1 \text{ mole } H_3O^+}$$
$$= 2.28836 \times 10^{16} \text{ (calc)} = 2.3 \times 10^{16} \text{ } H_3O^+ \text{ ions (corr)}$$

14.167 $[H_3O^+] = $ antilog $(-2.37) = 4.265795188 \times 10^{-3}$ (calc) $= 4.3 \times 10^{-3}$ M H_3O^+ (corr)

$$236 \text{ mL solution} \times \dfrac{10^{-3} \text{ L solution}}{1 \text{ mL solution}} \times \dfrac{4.3 \times 10^{-3} \text{ mole } H_3O^+}{1 \text{ L solution}}$$
$$= 0.0010148 \text{ (calc)} = 0.0010 \text{ mole } H_3O^+ \text{ (corr)}$$

moles NO_3^-: HNO_3 is a strong acid that dissociates 100%. $HNO_3 + H_2O \rightarrow H_3O^+ + NO_3^-$
so moles NO_3^- = moles H_3O^+ = 0.0010 mole NO_3^-

moles Na^+: Na_2SO_4 is a soluble salt that dissociates 100%. $Na_2SO_4 \rightarrow 2 Na^+ + SO_4^{2-}$

0.100 mole $Na_2SO_4 \times \dfrac{2 \text{ moles } Na^+}{1 \text{ mole } Na_2SO_4} = 0.2$ (calc) $= 0.200$ mole Na^+ (corr)

moles SO_4^{2-}: 0.100 mole $Na_2SO_4 \times \dfrac{1 \text{ mole } SO_4^{2-}}{1 \text{ mole } Na_2SO_4} = 0.1$ (calc) $= 0.100$ mole SO_4^{2-} (corr)

total moles ions: $(0.0010 + 0.0010 + 0.200 + 0.100)$ mole $= 0.302$ mole (calc and corr)
total number of ions present:

0.302 mole ions $\times \dfrac{6.022 \times 10^{23} \text{ ions}}{1 \text{ mole ions}} = 1.818644 \times 10^{23}$ (calc) $= 1.82 \times 10^{23}$ ions (corr)

14.169 As HCl is a strong acid, in this solution, $[H_3O^+] = $ antilog $(-2.40) = 3.98107 \times 10^{-3}$ M.
Therefore number of moles of $H_3O^+ = 3.98107 \times 10^{-3}$ M (calc) $= 4.0 \times 10^{-3}$ M (corr)
$= 4.0 \times 10^{-3}$ M $\times 7.50$ L $= 0.03$ mole (calc) $= 0.030$ mole (corr)

Using the ideal gas equation $PV = nRT$ and having STP conditions, we get

$$V = \dfrac{nRT}{P} = \dfrac{0.030 \text{ mole} \times 0.08206 \text{ L atm mol}^{-1} \text{K}^{-1} \times 273 \text{ K}}{1 \text{ atm}} = 0.6720714 \text{ L (calc)}$$
$$= 0.67 \text{ L (corr)}$$

Answers to Multiple-Choice Practice Test

14.171	a	14.172	b	14.173	b	14.174	c	14.175	b	14.176	b
14.177	c	14.178	c	14.179	b	14.180	e	14.181	a	14.182	a
14.183	a	14.184	e	14.185	d	14.186	c	14.187	e	14.188	e
14.189	b	14.190	c								

CHAPTER FIFTEEN
Chemical Equations: Net Ionic and Oxidation–Reduction

Practice Problems

Electrolytes (Sec. 15.2)

15.1 a) H_2CO_3 is a weak electrolyte.   b) KOH is a strong electrolyte.
      c) H_2SO_4 is a strong electrolyte. d) HCN is a weak electrolyte.

15.3 a) Acetic acid, a weak acid, is present in solution in both ionic and molecular forms.
      b) Sucrose, a nonelectrolyte, is a molecular compound present in solution only in a molecular form.
      c) Sodium sulfate, a soluble salt, is present in solution in ionic form.
      d) Hydrofluoric acid, a weak electrolyte, is present in solution in both ionic and molecular forms.

15.5 When dissolved in water:
      a) NaCl produces two ions per formula unit: 1 Na^+ and 1 Cl^-.
      b) $Mg(NO_3)_2$ produces three ions per formula unit: 1 Mg^{2+} and 2 NO_3^-.
      c) NH_4CN produces two ions per formula unit: 1 NH_4^+ and 1 CN^-.
      d) $HClO_4$ produces two ions per formula unit: 1 H^+ and 1 ClO_4^-.

15.7 Each compound dissociates into the following ions in water:
      a) $NaCl \rightarrow Na^+ + Cl^-$
      b) $Mg(NO_3)_2 \rightarrow Mg^{2+} + 2\ NO_3^-$
      c) $NH_4CN \rightarrow NH_4^+ + CN^-$
      d) $HClO_4 \rightarrow H^+ + ClO_4^-$

15.9 Diagram III is the strongest electrolyte.

15.11 In an ionic equation:
      a) Strong electrolytes are written in ionic form.
      b) Weak acids are not written in ionic form.
      c) Soluble salts are written in ionic form.
      d) Nonelectrolytes are not written in ionic form.

15.13 In an ionic equation:
      a) NaBr is a soluble salt and written in an ionic form.
      b) KOH is a soluble salt and written in an ionic form.
      c) $HClO_3$ is a strong acid and written in ionic form.
      d) $HClO_2$ is a weak acid and not written in ionic form.

15.15 a) Molecular equation   b) Net ionic equation
      c) Ionic equation   d) Net ionic equation

Copyright © 2011 Pearson Education, Inc.

15.17 a) $2 NaBr + Pb(NO_3)_2 \rightarrow 2 NaNO_3 + PbBr_2$
$2 Na^+ + 2 Br^- + Pb^{2+} + 2 NO_3^- \rightarrow 2 Na^+ + 2 NO_3^- + PbBr_2$
$Pb^{2+} + 2 Br^- \rightarrow PbBr_2$
b) $FeCl_3 + 3 NaOH \rightarrow Fe(OH)_3 + 3 NaCl$
$Fe^{3+} + 3 Cl^- + 3 Na^+ + 3 OH^- \rightarrow Fe(OH)_3 + 3 Na^+ + 3 Cl^-$
$Fe^{3+} + 3 OH^- \rightarrow Fe(OH)_3$
c) $Zn + 2 HCl \rightarrow ZnCl_2 + H_2$
$Zn + 2 H^+ + 2 Cl^- \rightarrow Zn^{2+} + 2 Cl^- + H_2$
$Zn + 2 H^+ \rightarrow Zn^{2+} + H_2$
d) $H_2S + 2 KOH \rightarrow K_2S + 2 H_2O$
$H_2S + 2 K^+ + 2 OH^- \rightarrow 2 K^+ + S^{2-} + 2 H_2O$
$H_2S + 2 OH^- \rightarrow S^{2-} + 2 H_2O$

15.19 a) $Pb + 2 AgNO_3 \rightarrow 2 Ag + Pb(NO_3)_2$
$Pb + 2 Ag^+ + 2 NO_3^- \rightarrow 2 Ag + Pb^{2+} + 2 NO_3^-$
$Pb + 2 Ag^+ \rightarrow 2 Ag + Pb^{2+}$
b) $Cl_2 + 2 NaBr \rightarrow 2 NaCl + Br_2$
$Cl_2 + 2 Na^+ + 2 Br^- \rightarrow 2 Na^+ + 2 Cl^- + Br_2$
$Cl_2 + 2 Br^- \rightarrow 2 Cl^- + Br_2$
c) $2 Al(NO_3)_3 + 3 Na_2S \rightarrow Al_2S_3 + 6 NaNO_3$
$2 Al^{3+} + 6 NO_3^- + 6 Na^+ + 3 S^{2-} \rightarrow Al_2S_3 + 6 Na^+ + 6 NO_3^-$
$2 Al^{3+} + 3 S^{2-} \rightarrow Al_2S_3$
d) $NaC_2H_3O_2 + NH_4Cl \rightarrow NH_4C_2H_3O_2 + NaCl$
$Na^+ + C_2H_3O_2^- + NH_4^+ + Cl^- \rightarrow NH_4^+ + C_2H_3O_2^- + Na^+ + Cl^-$
Each ionic compound is a soluble salt; therefore no reaction occurs.

Oxidation–Reduction Terminology (Secs. 15.4 and 15.5)

15.21 a) Oxidation occurs when an atom loses electrons.
b) Oxidation occurs when the oxidation number of an atom increases.

15.23 a) An oxidizing agent gains electrons from another substance.
b) An oxidizing agent contains the atom that shows an oxidation number decrease.
c) An oxidizing agent is itself reduced.

15.25 a) oxidized   b) decrease   c) reducing agent   d) loses

Oxidation Numbers (Sec. 15.5)

15.27 a) 0   b) 0   c) 0   d) 0

15.29 a) $PCl_3, P + 3(-1) = 0, P = +3$ b) $SiH_4, Si + 4(-1) = 0, Si = +4$
c) $CO_2, C + 2(-2) = 0, C = +4$ d) $N_2H_4, 2(N) + 4(+1) = 0, N = -2$

15.31 a) Ca in $Ca^{2+} = +2$ b) Br in $Br^- = -1$
c) P in $PO_4^{3-}, P + 4(-2) = -3, P = +5$ d) C in $CO_3^{2-}, C + 3(-2) = -2, C = +4$

15.33 a) $BeCl_2, +2 + 2(Cl) = 0, Cl = -1$ b) $ClF, Cl + -1 = 0, Cl = +1$
c) $AlCl_4^-, +3 + 4(Cl) = -1, Cl = -1$ d) $ClO^-, Cl + -2 = -1, Cl = +1$

15.35 a) $H_3PO_4, H = +1, P = +5, O = -2$ b) $BaCr_2O_7, Ba = +2, Cr = +6, O = -2$
c) $NH_4ClO_4, N = -3, H = +1, Cl = +7, O = -2$ d) $H_4P_2O_7, H = +1, P = +5, O = -2$

Chemical Equations: Net Ionic and Oxidation–Reduction

15.37 In order of increasing oxidation numbers of S:
H_2S (S = −2) → S_8 (S = 0) → H_2SO_3 (S = +4) → SO_3 (S = +6)

15.39
a) $Rh_2(CO_3)_2$, $2(Rh) + 2(+4) + 6(-2) = 0$, Rh = +2
b) $Cr_2(SO_4)_3$, $2(Cr) + 3(+6) + 12(-2) = 0$, Cr = +3
c) $Cu(ClO_2)_2$, $Cu + 2(+3) + 4(-2) = 0$, Cu = +2
d) $Co_3(PO_4)_2$, $3(Co) + 2(+5) + 8(-2) = 0$, Co = +2

15.41 a) Na_2O, O = −2   b) OF_2, O = +2   c) Na_2O_2, O = −1   d) BaO, O = −2

15.43 a) NaH, H = −1   b) CH_4, H = +1   c) HCl, H = +1   d) CaH_2, H = −1

Characteristics of Oxidation–Reduction Reactions (Sec. 15.5)

15.45
a) $N_2 + 3 H_2 \rightarrow 2 NH_3$ Each H in H_2 is oxidized (0 to +1); each N in N_2 is reduced (0 to −3).
b) $Cl_2 + 2 KI \rightarrow 2 KCl + I_2$ I in KI is oxidized (−1 to 0); each Cl in Cl_2 is reduced (0 to −1).
c) $Sb_2O_3 + 3 Fe \rightarrow 2 Sb + 3 FeO$ Fe is oxidized (0 to +2); Sb in Sb_2O_3 is reduced (+3 to 0).
d) $3 H_2SO_3 + 2 HNO_3 \rightarrow 2 NO + H_2O + 3 H_2SO_4$ S in H_2SO_3 is oxidized (+4 to +6); N in HNO_3 is reduced (+5 to +2).

15.47
a) $N_2 + 3 H_2 \rightarrow 2 NH_3$ H_2 is reducing agent; N_2 is oxidizing agent.
b) $Cl_2 + 2 KI \rightarrow 2 KCl + I_2$ I^- is reducing agent; Cl_2 is oxidizing agent.
c) $Sb_2O_3 + 3 Fe \rightarrow 2 Sb + 3 FeO$ Fe is reducing agent; Sb_2O_3 is oxidizing agent.
d) $3 H_2SO_3 + 2 HNO_3 \rightarrow 2 NO + H_2O + 3 H_2SO_4$ H_2SO_3 is reducing agent; HNO_3 is oxidizing agent.

15.49 $2 HNO_3 + SO_2 \rightarrow H_2SO_4 + 2 NO_2$
a) Sulfur SO_2 is oxidized.
b) HNO_3 is oxidizing agent.
c) HNO_3 contains N that decreases from +5 to +2.
d) SO_2 contains S that loses electrons +4 to +6.

15.51
a) $3 Zn + N_2 \rightarrow Zn_3N_2$ Zn metal is oxidized; the nonmetal (N) is reduced.
b) $2 Ca + O_2 \rightarrow 2 CaO$ Ca metal is oxidized; the nonmetal (O) is reduced.
c) $2 Na + S \rightarrow Na_2S$ Na metal is oxidized; the nonmetal (S) is reduced.
d) $Mg + Cl_2 \rightarrow MgCl_2$ Mg metal is oxidized; the nonmetal (Cl) is reduced.

Redox and Nonredox Chemical Reactions (Sec. 15.6)

15.53
a) $2 FeBr_3 \rightarrow 2 FeBr_2 + Br_2$, Redox reaction, Fe and Br are changing their oxidation states.
b) $K_2O + H_2O \rightarrow 2 KOH$, Nonredox reaction; K, H, and O are not changing their oxidation states.
c) $2 KClO_3 \rightarrow 2 KCl + 3 O_2$, Redox reaction, Cl and O are changing their oxidation states.
d) $CH_4 + 2 O_2 \rightarrow CO_2 + 2 H_2O$, Redox reaction, C and O are changing their oxidation states.

15.55
a) $H_2 + Cl_2 \rightarrow 2 HCl$, Synthesis reaction, redox
b) $Mg + 2 HBr \rightarrow MgBr_2 + H_2$, Single replacement, redox
c) $MgCO_3 \rightarrow MgO + CO_2$, Decomposition, nonredox
d) $2 KOH + H_2SO_4 \rightarrow H_2SO_4 + 2 H_2O$, Double replacement (neutralization), nonredox

15.57 a) redox   b) redox   c) cannot classify   d) redox

Balancing Redox Equations: Oxidation Number Method (Sec. 15.8)

15.59 a)
$\overset{0 \quad\;\; 2(+3) \quad +3}{2\,Cr + 6\,HCl \to 2\,CrCl_3 + 3\,H_2}$
$\underset{+1 \quad 3[2(-1)] \quad 0}{}$

b)
$\overset{0 \quad 3(+4) \quad +4}{2\,Cr_2O_3 + 3\,C \to 4\,Cr + 3\,CO_2}$
$\underset{+3 \quad 2[2(-3)] \quad 0}{}$

c)
$\overset{+4 \;\; (+2) \;\; +6}{SO_2 + NO_2 \to SO_3 + NO}$
$\underset{+4 \;\; (-2) \;\; +2}{}$

d)
$\overset{0 \quad 4(+2) \quad +2}{BaSO_4 + 4\,C \to BaS + 4\,CO}$
$\underset{+6 \quad (-8) \quad -2}{}$

15.61 a)
$\overset{+4 \quad (+2) \quad +6}{Br_2 + 2\,H_2O + SO_2 \to 2\,HBr + H_2SO_4}$
$\underset{0 \quad 2(-1) \quad -1}{}$

b)
$\overset{-2 \quad 3(+2) \quad 0}{3\,H_2S + 2\,HNO_3 \to 3\,S + 2\,NO + 4\,H_2O}$
$\underset{+5 \quad 2(-3) \quad +2}{}$

c)
$\overset{+2 \quad 2(+1) \quad +3}{SnSO_4 + 2\,FeSO_4 \to Sn + Fe_2(SO_4)_3}$
$\underset{+2 \quad (-2) \quad 0}{}$

d)
$\overset{-1 \quad\quad 2[2(+1)] \quad\quad 0}{Na_2TeO_3 + 4\,NaI + 6\,HCl \to 6\,NaCl + Te + 3\,H_2O + 2\,I_2}$
$\underset{+4 \quad\quad (-4) \quad\quad 0}{}$

Chemical Equations: Net Ionic and Oxidation–Reduction

15.63 a)   0   1[2(+5)]   +5

$I_2 + 5 Cl_2 + 6 H_2O \rightarrow 2 HIO_3 + 10 Cl^- + 10 H^+$

   0   5[2(1)]   −1

b)   −3   5(+8)   +5

$8 MnO_4^- + 5 AsH_3 + 24 H^+ \rightarrow 5 H_3AsO_4 + 8 Mn^{2+} + 12 H_2O$

   +7   8(−5)   +2

c)   −1   2(+1)   0

$2 Br^- + SO_4^{2-} + 4 H^+ \rightarrow Br_2 + SO_2 + 2 H_2O$

        +6   (−2)   +4

d)   0   (+3)   +3

$Au + 4 Cl^- + 3 NO_3^- + 6 H^+ \rightarrow AuCl_4^- + 3 NO_2 + 3 H_2O$

                  +5   3(−1)   +4

15.65 a)   −2   (+8)   +6

$8 OH^- + S^{2-} + 4 Cl_2 \rightarrow SO_4^{2-} + 8 Cl^- + 4 H_2O$

          0   4[2(−1)]   −1

b)   +4   3(+2)   +6

$5 H_2O + 3 SO_3^{2-} + 2 CrO_4^{2-} \rightarrow 2 Cr(OH)_4^- + 3 SO_4^{2-} + 2 OH^-$

                 +6   2(−3)   +3

c)   +5   3(+2)   +7

$H_2O + 2 MnO_4^- + 3 IO_3^- \rightarrow 2 MnO_2 + 3 IO_4^- + 2 OH^-$

           +7   2(−3)   +4

d)   0   2(+7)   +7

$18 OH^- + I_2 + 7 Cl_2 \rightarrow 2 H_3IO_6^{2-} + 14 Cl^- + 6 H_2O$

           0   7[2(−1)]   −1

Balancing Redox Equations: Half-Reaction Method (Sec. 15.9)

15.67 a) $NO_3^- \rightarrow NO$, N is reduced from +5 to +3, reduction
   b) $Zn \rightarrow Zn^{2+}$, Zn is oxidized from 0 to +2, oxidation
   c) $Ti^{3+} \rightarrow TiO_2$, Ti is oxidized from +3 to +4, oxidation
   d) $Cr_2O_7^{2-} \rightarrow Cr^{3+}$, Cr is reduced from +6 to +3, reduction

15.69 a) $Te + NO_3^- \rightarrow TeO_2 + NO$, oxidation half-reaction: $Te \rightarrow TeO_2$
   b) $H_2O_2 + Fe^{2+} \rightarrow Fe^{3+} + H_2O$, oxidation half-reaction: $Fe^{2+} \rightarrow Fe^{3+}$
   c) $CN^- + ClO_2^- \rightarrow CNO^- + Cl^-$, oxidation half-reaction: $CN^- \rightarrow CNO^-$
   d) $ClO^- + Cl^- \rightarrow Cl_2$; oxidation half-reaction: $Cl^- \rightarrow Cl_2$

15.71 a) $MnO_2 + e^- \rightarrow Mn^{3+}$ (ox. no.: +4 → +3)
   $MnO_2 + 4 H^+ + e^- \rightarrow Mn^{3+}$
   $MnO_2 + 4 H^+ + e^- \rightarrow Mn^{3+} + 2 H_2O$

   b) $H_3MnO_4 + 5 e^- \rightarrow Mn$ (ox. no.: +5 → 0)
   $H_3MnO_4 + 5 H^+ + 5 e^- \rightarrow Mn$
   $H_3MnO_4 + 5 H^+ + 5 e^- \rightarrow Mn + 4 H_2O$

   c) $MnO_4^- + 5 e^- \rightarrow Mn^{2+}$ (ox. no.: +7 → +2)
   $MnO_4^- + 8 H^+ + 5 e^- \rightarrow Mn^{2+}$
   $MnO_4^- + 8 H^+ + 5 e^- \rightarrow Mn^{2+} + 4 H_2O$

   d) $MnO_4^- + 3 e^- \rightarrow MnO_2$ (ox. no.: +7 → +4)
   $MnO_4^- + 4 H^+ + 3 e^- \rightarrow MnO_2$
   $MnO_4^- + 4 H^+ + 3 e^- \rightarrow MnO_2 + 2 H_2O$

15.73 a) $SeO_4^{2-} + 6 e^- \rightarrow Se$ (ox. no.: +6 → 0)
   $SeO_4^{2-} + 6 e^- \rightarrow Se + 8 OH^-$
   $SeO_4^{2-} + 4 H_2O + 6 e^- \rightarrow Se + 8 OH^-$

   b) $Se^{2-} \rightarrow SeO_3^{2-} + 6 e^-$ (ox. no.: −2 → +4)
   $Se^{2-} + 6 OH^- \rightarrow SeO_3^{2-} + 6 e^-$
   $Se^{2-} + 6 OH^- \rightarrow SeO_3^{2-} + 6 e^- + 3 H_2O$

   c) $SeO_4^{2-} + 2 e^- \rightarrow SeO_3^{2-}$ (ox. no.: +6 → +4)
   $SeO_4^{2-} + 2 e^- \rightarrow SeO_3^{2-} + 2 OH^-$
   $SeO_4^{2-} + H_2O + 2 e^- \rightarrow SeO_3^{2-} + 2 OH^-$

   d) $Se \rightarrow SeO_3^{2-} + 4 e^-$ (ox. no.: 0 → +4)
   $Se + 6 OH^- \rightarrow SeO_3^{2-} + 4 e^-$
   $Se + 6 OH^- \rightarrow SeO_3^{2-} + 4 e^- + 3 H_2O$

15.75 a) oxidation:
   $Zn \rightarrow Zn^{2+} + 2 e^-$ (ox. no.: 0 → +2)
   reduction:
   $Cu^{2+} + 2 e^- \rightarrow Cu$ (ox. no.: +2 → 0)
   combining:
   $Zn \rightarrow Zn^{2+} + 2 e^-$ + $Cu^{2+} + 2 e^- \rightarrow Cu$
   $Zn + Cu^{2+} + 2 e^- \rightarrow Zn^{2+} + 2 e^- + Cu$
   $Zn + Cu^{2+} \rightarrow Zn^{2+} + Cu$

Chemical Equations: Net Ionic and Oxidation–Reduction

b) oxidation:
$$2\ I^- \rightarrow I_2 + 2\ e^- \text{ (ox. no.: } -1 \rightarrow 0)$$
reduction:
$$Br_2 + 2\ e^- \rightarrow 2\ Br^- \text{ (ox. no.: } 0 \rightarrow -1)$$
combining:
$$2\ I^- \rightarrow I_2 + 2\ e^- \quad + \quad Br_2 + 2\ e^- \rightarrow 2\ Br^-$$
$$2\ I^- + Br_2 + 2\ e^- \rightarrow I_2 + 2\ e^- + 2\ Br^-$$
$$2\ I^- + Br_2 \rightarrow I_2 + 2\ Br^-$$

c) oxidation:
$$S_2O_3^{2-} \rightarrow 2\ HSO_4^-$$
$$S_2O_3^{2-} \rightarrow 2\ HSO_4^- + 8\ e^- \text{ (ox. no.: } +2 \rightarrow +6 \text{ per S atom)}$$
$$S_2O_3^{2-} \rightarrow 2\ HSO_4^- + 8\ e^- + 8\ H^+$$
$$S_2O_3^{2-} + 5\ H_2O \rightarrow 2\ HSO_4^- + 8\ e^- + 8\ H^+$$
reduction:
$$Cl_2 + 2\ e^- \rightarrow 2\ Cl^- \text{ (ox. no.: } 0 \rightarrow -1)$$
combining:
$$S_2O_3^{2-} + 5\ H_2O \rightarrow 2\ HSO_4^- + 8\ e^- + 8\ H^+ \quad + \quad 4(Cl_2 + 2\ e^- \rightarrow 2\ Cl^-)$$
$$S_2O_3^{2-} + 5\ H_2O + 4\ Cl_2 + 8\ e^- \rightarrow 2\ H_2SO_4 + 8\ e^- + 8\ H^+ + 8\ Cl^-$$
$$S_2O_3^{2-} + 5\ H_2O + 4\ Cl_2 \rightarrow 2\ HSO_4^- + 8\ H^+ + 8\ Cl^-$$

d) oxidation:
$$Zn \rightarrow Zn^{2+} + 2\ e^- \text{ (ox. no.: } 0 \rightarrow +2)$$
reduction:
$$As_2O_3 \rightarrow 2\ AsH_3 \text{ (balance As)}$$
$$As_2O_3 + 12\ e^- \rightarrow 2\ AsH_3 \text{ (ox. no.: } +3 \rightarrow -3 \text{ per As atom)}$$
$$As_2O_3 + 12\ e^- + 12\ H^+ \rightarrow 2\ AsH_3$$
$$As_2O_3 + 12\ e^- + 12\ H^+ \rightarrow 2\ AsH_3 + 3\ H_2O$$
combining:
$$6(Zn \rightarrow Zn^{2+} + 2e) \quad + \quad As_2O_3 + 12\ e^- + 12\ H^+ \rightarrow 2\ AsH_3 + 3\ H_2O$$
$$6\ Zn + As_2O_3 + 12\ e^- + 12\ H^+ \rightarrow 6\ Zn^{2+} + 12\ e^- + 2\ AsH_3 + 3\ H_2O$$
$$6\ Zn + As_2O_3 + 12\ H^+ \rightarrow 6\ Zn^{2+} + 2\ AsH_3 + 3\ H_2O$$

15.77 a) oxidation:
$$I_2 \rightarrow 2\ HIO_3 \text{ (balance I)}$$
$$I_2 \rightarrow 2\ HIO_3 + 10\ e^- \text{ (ox. no.: } 0 \rightarrow +5 \text{ per I atom)}$$
$$I_2 \rightarrow 2\ HIO_3 + 10\ e^- + 10\ H^+$$
$$I_2 + 6\ H_2O \rightarrow 2\ HIO_3 + 10\ e^- + 10\ H^+$$
reduction:
$$Cl_2 + 2\ e^- \rightarrow 2\ Cl^- \text{ (ox. no.: } 0 \rightarrow -1)$$
combining:
$$I_2 + 6\ H_2O \rightarrow 2\ HIO_3 + 10\ e^- + 10\ H^+ \quad + \quad 5(Cl_2 + 2\ e^- \rightarrow 2\ Cl^-)$$
$$I_2 + 6\ H_2O + 5\ Cl_2 + 10\ e^- \rightarrow 2\ HIO_3 + 10\ e^- + 10\ H^+ + 10\ Cl^-$$
$$I_2 + 6\ H_2O + 5\ Cl_2 \rightarrow 2\ HIO_3 + 10\ H^+ + 10\ Cl^-$$

b) oxidation:
$$AsH_3 \rightarrow H_3AsO_4 + 8\ e^- \text{ (ox. no.: } -3 \rightarrow +5)$$
$$AsH_3 \rightarrow H_3AsO_4 + 8\ e^- + 8\ H^+$$
$$AsH_3 + 4\ H_2O \rightarrow H_3AsO_4 + 8\ e^- + 8\ H^+$$
reduction:
$$MnO_4^- + 5\ e^- \rightarrow Mn^{2+} \text{ (ox. no.: } +7 \rightarrow +2)$$
$$MnO_4^- + 5\ e^- + 8\ H^+ \rightarrow Mn^{2+} + 4\ H_2O$$

combining:

$5(AsH_3 + 4 H_2O \rightarrow H_3AsO_4 + 8 e^- + 8 H^+) + 8(MnO_4^- + 5 e^- + 8 H^+ \rightarrow Mn^{2+} + 4 H_2O)$

$5 AsH_3 + 20 H_2O + 8 MnO_4^- + 40 e^- + 64 H^+$
$\rightarrow 5 H_3AsO_4 + 40 e^- + 40 H^+ + 8 Mn^{2+} + 32 H_2O$

$5 AsH_3 + 8 MnO_4^- + 24 H^+ \rightarrow 5 H_3AsO_4 + 8 Mn^{2+} + 12 H_2O$

c) oxidation:

$2 Br^- \rightarrow Br_2 + 2 e^-$ (ox. no.: $-1 \rightarrow 0$)

reduction:

$SO_4^{2-} + 2 e^- \rightarrow SO_2$ (ox. no.: $+6 \rightarrow +4$)
$SO_4^{2-} + 2 e^- + 4 H^+ \rightarrow SO_2$
$SO_4^{2-} + 2 e^- + 4 H^+ \rightarrow SO_2 + 2 H_2O$

combining:

$2 Br^- \rightarrow Br_2 + 2 e^- \;+\; SO_4^{2-} + 2 e^- + 4 H^+ \rightarrow SO_2 + 2 H_2O$
$2 Br^- + SO_4^{2-} + 2 e^- + 4 H^+ \rightarrow Br_2 + 2 e^- + SO_2 + 2 H_2O$
$2 Br^- + SO_4^{2-} + 4 H^+ \rightarrow Br_2 + SO_2 + 2 H_2O$

d) oxidation:

$Au + 4 Cl^- \rightarrow AuCl_4^-$
$Au + 4 Cl^- \rightarrow AuCl_4^- + 3 e^-$ (ox. no.: $0 \rightarrow +3$)

reduction:

$NO_3^- + e^- \rightarrow NO_2$ (ox. no.: $+5 \rightarrow +4$)
$NO_3^- + e^- + 2 H^+ \rightarrow NO_2$
$NO_3^- + e^- + 2 H^+ \rightarrow NO_2 + H_2O$

combining:

$Au + 4 Cl^- \rightarrow AuCl_4^- + 3 e^- \;+\; 3(NO_3^- + e^- + 2 H^+ \rightarrow NO_2 + H_2O)$
$Au + 4 Cl^- + 3 NO_3^- + 3 e^- + 6 H^+ \rightarrow AuCl_4^- + 3 e^- + 3 NO_2 + 3 H_2O$
$Au + 4 Cl^- + 3 NO_3^- + 6 H^+ \rightarrow AuCl_4^- + 3 NO_2 + 3 H_2O$

15.79 a) oxidation:

$2 NH_3 \rightarrow N_2H_4$ (balance N)
$2 NH_3 \rightarrow N_2H_4 + 2 e^-$ (ox. no.: $-3 \rightarrow -2$ per N atom)
$2 NH_3 + 2 OH^- \rightarrow N_2H_4 + 2 e^-$
$2 NH_3 + 2 OH^- \rightarrow N_2H_4 + 2 e^- + 2 H_2O$

reduction:

$ClO^- + 2 e^- \rightarrow Cl^-$ (ox. no.: $+1 \rightarrow -1$)
$ClO^- + 2 e^- \rightarrow Cl^- + 2 OH^-$
$ClO^- + 2 e^- + H_2O \rightarrow Cl^- + 2 OH^-$

combining:

$2 NH_3 + 2 OH^- \rightarrow N_2H_4 + 2 e^- + 2 H_2O \;+\; ClO^- + 2 e^- + H_2O \rightarrow Cl^- + 2 OH^-$
$2 NH_3 + 2 OH^- + ClO^- + 2 e^- + H_2O \rightarrow N_2H_4 + 2 e^- + 2 H_2O + Cl^- + 2 OH^-$
$2 NH_3 + ClO^- \rightarrow N_2H_4 + Cl^- + H_2O$

b) oxidation:

$Cr(OH)_2 \rightarrow CrO_4^{2-} + 4 e^-$ (ox. no.: $+2 \rightarrow +6$)
$Cr(OH)_2 + 6 OH^- \rightarrow Br^- CrO_4^{2-} + 4 e^-$
$Cr(OH)_2 + 6 OH^- \rightarrow CrO_4^{2-} + 4 H_2O + 4 e^-$

reduction:

$BrO^- + 2 e^- \rightarrow Br^-$ (ox. no.: $+1 \rightarrow -1$)
$BrO^- + 2 e^- \rightarrow Br^- + 2 OH^-$
$BrO^- + H_2O + 2 e^- \rightarrow Br^- + 2 OH^-$

Chemical Equations: Net Ionic and Oxidation–Reduction

combining:
$Cr(OH)_2 + 6\ OH^- \rightarrow CrO_4^- + 4\ H_2O + 4\ e^- + 2(BrO^- + H_2O + 2\ e^- \rightarrow Br^- + 2\ OH^-)$
$Cr(OH)_2 + 6\ OH^- + 2\ BrO^- + 2\ H_2O + 4\ e^-$
$\quad\quad\quad\quad\quad\quad\quad\quad\quad\quad \rightarrow CrO_4^{2-} + 4\ H_2O + 4\ e^- + 2\ Br^- + 4\ OH^-$
$Cr(OH)_2 + 2\ OH^- + 2\ BrO^- \quad\quad \rightarrow CrO_4^{2-} + 2\ H_2O + 2\ Br^-$

c) oxidation:
$CrO_2^- \quad\quad\quad\quad\quad\quad\quad\quad \rightarrow CrO_4^{2-} + 3\ e^-$ (ox. no.: $+3 \rightarrow +6$)
$CrO_2^- + 4\ OH^- \quad\quad\quad\quad \rightarrow CrO_4^{2-} + 3\ e^-$
$CrO_2^- + 4\ OH^- \quad\quad\quad\quad \rightarrow CrO_4^{2-} + 3\ e^- + 2\ H_2O$

reduction:
$H_2O_2 + 2\ e^- \quad\quad\quad\quad\quad \rightarrow 2\ OH^-$ (ox. no.: $-1 \rightarrow -2$ per O atom)

combining:
$2(CrO_2^- + 4\ OH^- \rightarrow CrO_4^{2-} + 3\ e^- + 2\ H_2O) + 3(H_2O_2 + 2\ e^- \rightarrow 2\ OH^-)$
$2\ CrO_2^- + 8\ OH^- + 3\ H_2O_2 + 6\ e^- \rightarrow 2\ CrO_4^{2-} + 6\ e^- + 4\ H_2O + 6\ OH^-$
$2\ CrO_2^- + 2\ OH^- + 3\ H_2O_2 \quad\quad \rightarrow 2\ CrO_4^{2-} + 4\ H_2O$

d) oxidation:
$Sn(OH)_3^- \quad\quad\quad\quad\quad\quad\quad \rightarrow Sn(OH)_6^{2-} + 2\ e^-$ (ox. no.: $+2 \rightarrow +4$)
$Sn(OH)_3^- + 3\ OH^- \quad\quad\quad \rightarrow Sn(OH)_6^{2-} + 2\ e^-$

reduction:
$Bi(OH)_3 + 3\ e^- \quad\quad\quad\quad \rightarrow Bi$ (ox. no.: $+3 \rightarrow 0$)
$Bi(OH)_3 + 3\ e^- \quad\quad\quad\quad \rightarrow Bi + 3\ OH^-$

combining:
$3(Sn(OH)_3^- + 3\ OH^- \rightarrow Sn(OH)_6^{2-} + 2\ e^-) + 2(Bi(OH)_3 + 3\ e^- \rightarrow Bi + 3\ OH^-)$
$3\ Sn(OH)_3^- + 9\ OH^- + 2\ Bi(OH)_3 + 6\ e^- \rightarrow 3\ Sn(OH)_6^{2-} + 6\ e^- + 2\ Bi + 6\ OH^-$
$3\ Sn(OH)_3^- + 3\ OH^- + 2\ Bi(OH)_3 \quad\quad \rightarrow 3\ Sn(OH)_6^{2-} + 2\ Bi$

15.81 a) oxidation:
$S^{2-} \quad\quad\quad\quad\quad\quad\quad\quad\quad \rightarrow SO_4^{2-} + 8\ e^-$ (ox. no.: $-2 \rightarrow +6$)
$S^{2-} + 8\ OH^- \quad\quad\quad\quad\quad \rightarrow SO_4^{2-} + 8\ e^-$
$S^{2-} + 8\ OH^- \quad\quad\quad\quad\quad \rightarrow SO_4^{2-} + 8\ e^- + 4\ H_2O$

reduction:
$Cl_2 + 2\ e^- \quad\quad\quad\quad\quad\quad \rightarrow 2\ Cl^-$ (ox. no.: $0 \rightarrow -1$)

combining:
$S^{2-} + 8\ OH^- \rightarrow SO_4^{2-} + 8\ e^- + 4\ H_2O + 4(Cl_2 + 2\ e^- \rightarrow 2\ Cl^-)$
$S^{2-} + 8\ OH^- + 4\ Cl_2 + 8\ e^- \rightarrow SO_4^{2-} + 8\ e^- + 4\ H_2O + 8\ Cl^-$
$S^{2-} + 8\ OH^- + 4\ Cl_2 \quad\quad \rightarrow SO_4^{2-} + 4\ H_2O + 8\ Cl^-$

b) oxidation:
$SO_3^{2-} \quad\quad\quad\quad\quad\quad\quad\quad \rightarrow SO_4^{2-} + 2\ e^-$ (ox. no.: $+4 \rightarrow +6$)
$SO_3^{2-} + 2\ OH^- \quad\quad\quad\quad \rightarrow SO_4^{2-} + 2\ e^-$
$SO_3^{2-} + 2\ OH^- \quad\quad\quad\quad \rightarrow SO_4^{2-} + 2\ e^- + H_2O$

reduction:
$CrO_4^{2-} + 3\ e^- \quad\quad\quad\quad\quad \rightarrow Cr(OH)_4^-$ (ox. no.: $+6 \rightarrow +3$)
$CrO_4^{2-} + 3\ e^- \quad\quad\quad\quad\quad \rightarrow Cr(OH)_4^- + 4\ OH^-$
$CrO_4^{2-} + 3\ e^- + 4\ H_2O \quad\quad \rightarrow Cr(OH)_4^- + 4\ OH^-$

combining:

$$3(SO_3^{2-} + 2\,OH^- \rightarrow SO_4^{2-} + 2\,e^- + H_2O)$$
$$+\ 2(CrO_4^{2-} + 3\,e^- + 4\,H_2O \rightarrow Cr(OH)_4^- + 4\,OH^-)$$
$$3\,SO_3^{2-} + 6\,OH^- + 2\,CrO_4^{2-} + 6\,e^- + 8\,H_2O$$
$$\rightarrow 3\,SO_4^{2-} + 6\,e^- + 3\,H_2O + 2\,Cr(OH)_4^- + 8\,OH^-$$
$$3\,SO_3^{2-} + 2\,CrO_4^{2-} + 5\,H_2O \rightarrow 3\,SO_4^{2-} + 2\,Cr(OH)_4^- + 2\,OH^-$$

c) oxidation:

$IO_3^- \rightarrow IO_4^- + 2\,e^-$ (ox. no.: $+5 \rightarrow +7$)
$IO_3^- + 2\,OH^- \rightarrow IO_4^- + 2\,e^-$
$IO_3^- + 2\,OH^- \rightarrow IO_4^- + 2\,e^- + H_2O$

reduction:

$MnO_4^- + 3\,e^- \rightarrow MnO_2$ (ox. no.: $+7 \rightarrow +4$)
$MnO_4^- + 3\,e^- \rightarrow MnO_2 + 4\,OH^-$
$MnO_4^- + 3\,e^- + 2\,H_2O \rightarrow MnO_2 + 4\,OH^-$

combining:

$$3(IO_3^- + 2\,OH^- \rightarrow IO_4^- + 2\,e^- + H_2O)$$
$$+\ 2(MnO_4^- + 3\,e^- + 2\,H_2O \rightarrow MnO_2 + 4\,OH^-)$$
$$3\,IO_3^- + 6\,OH^- + 2\,MnO_4^- + 6\,e^- + 4\,H_2O$$
$$\rightarrow 3\,IO_4^- + 6\,e^- + 3\,H_2O + 2\,MnO_2 + 8\,OH^-$$
$$3\,IO_3^- + 2\,MnO_4^- + H_2O \rightarrow 3\,IO_4^- + 2\,MnO_2 + 2\,OH^-$$

d) oxidation:

$I_2 \rightarrow 2\,H_3IO_6^{2-}$ (balance I)
$I_2 \rightarrow 2\,H_3IO_6^{2-} + 14\,e^-$ (ox. no.: $0 \rightarrow +7$ per I atom)
$I_2 + 18\,OH^- \rightarrow 2\,H_3IO_6^{2-} + 14\,e^-$
$I_2 + 18\,OH^- \rightarrow 2\,H_3IO_6^{2-} + 14\,e^- + 6\,H_2O$

reduction:

$Cl_2 + 2\,e^- \rightarrow 2\,Cl^-$ (ox. no.: $0 \rightarrow -1$)

combining:

$$I_2 + 18\,OH^- \rightarrow 2\,H_3IO_6^{2-} + 14\,e^- + 6\,H_2O\ +\ 7(Cl_2 + 2\,e^- \rightarrow 2\,Cl^-)$$
$$I_2 + 18\,OH^- + 7\,Cl_2 + 14\,e^- \rightarrow 2\,H_3IO_6^{2-} + 14\,e^- + 6\,H_2O + 14\,Cl^-$$
$$I_2 + 18\,OH^- + 7\,Cl_2 \rightarrow 2\,H_3IO_6^{2-} + 6\,H_2O + 14\,Cl^-$$

Balancing Redox Reactions: Disproportionation Reactions (Sec. 15.10)

15.83 a)

$\qquad\quad +3\quad (+2)\quad +5$
$$2\,HNO_2 + HNO_2 \rightarrow 2\,NO + NO_3^- + H_2O + H^+$$
$\qquad +3\quad 2(-1)\quad +2$

$$3\,HNO_2 \rightarrow 2\,NO + NO_3^- + H_2O + H^+$$

b)

$\quad -1\qquad 2(+1)\qquad 0$
$$2\,Cl^- + 2\,ClO^- + 4\,H^+ \rightarrow Cl_2 + Cl_2 + 2\,H_2O$$
$\qquad +1\qquad 2(-1)\qquad 0$

$$2\,Cl^- + 2\,ClO^- + 4\,H^+ \rightarrow 2\,Cl_2 + 2\,H_2O = Cl^- + ClO^- + 2\,H^+ \rightarrow Cl_2 + H_2O$$

Chemical Equations: Net Ionic and Oxidation–Reduction

c)
$$\quad\quad\quad\quad\quad 0 \;\;(+4) \;\; +4$$
$$6\,OH^- + 2\,S + S \rightarrow 2\,S^{2-} + SO_3^{2-} + 3\,H_2O$$
$$\quad\quad\quad\quad 0 \;\; 2(-2) \;\; -2$$

$$6\,OH^- + 3\,S \rightarrow 2\,S^{2-} + SO_3^{2-} + 3\,H_2O$$

d)
$$\quad\quad\quad\quad\quad 0 \;\; 2(+5) \;\; +5$$
$$12\,OH^- + Br_2 + 5\,Br_2 \rightarrow 2\,BrO_3^- + 10\,Br^- + 6\,H_2O$$
$$\quad\quad\quad\quad 0 \;\; 5[2(-1)] \;\; -1$$

$$12\,OH^- + 6\,Br_2 \rightarrow 2\,BrO_3^- + 10\,Br^- + 6\,H_2O = 6\,OH^- + 3\,Br_2 \rightarrow BrO_3^- + 5\,Br^- + 3\,H_2O$$

15.85 a) oxidation:

HNO_2	$\rightarrow NO_3^- + 2\,e^-$
HNO_2	$\rightarrow NO_3^- + 2\,e^- + 3\,H^+$
$HNO_2 + H_2O$	$\rightarrow NO_3^- + 2\,e^- + 3\,H^+$

reduction:

$HNO_2 + e^-$	$\rightarrow NO$
$HNO_2 + e^- + H^+$	$\rightarrow NO$
$HNO_2 + e^- + H^+$	$\rightarrow NO + H_2O$

combining:
$HNO_2 + H_2O \rightarrow NO_3^- + 3\,H^+ + 2\,e^- + 2(HNO_2 + e^- + H^+ \rightarrow NO + H_2O)$
$HNO_2 + H_2O + 2\,HNO_2 + 2\,H^+ + 2\,e^- \rightarrow NO_3^- + 2\,e^- + 3\,H^+ + 2\,NO + 2\,H_2O$
$3\,HNO_2 \rightarrow NO_3^- + 2\,NO + H^+ + H_2O$

b) oxidation:

$2\,Cl^-$	$\rightarrow Cl_2$
$2\,Cl^-$	$\rightarrow Cl_2 + 2\,e^-$

reduction:

$2\,ClO^-$	$\rightarrow Cl_2$
$2\,ClO^- + 2\,e^-$	$\rightarrow Cl_2$
$2\,ClO^- + 2\,e^- + 4\,H^+$	$\rightarrow Cl_2$
$2\,ClO^- + 2\,e^- + 4\,H^+$	$\rightarrow Cl_2 + 2\,H_2O$

combining:

$2\,Cl^- \rightarrow Cl_2 + 2\,e^- + 2\,ClO^- + 2\,e^- + 4\,H^+$	$\rightarrow Cl_2 + 2\,H_2O$
$2\,Cl^- + 2\,ClO^- + 2\,e^- + 4\,H^+$	$\rightarrow Cl_2 + 2\,e^- + Cl_2 + 2\,H_2O$
$2\,Cl^- + 2\,ClO^- + 4\,H^+$	$\rightarrow 2\,Cl_2 + 2\,H_2O$
$Cl^- + ClO^- + 2\,H^+$	$\rightarrow Cl_2 + H_2O$

c) oxidation:

S	$\rightarrow SO_3^{2-} + 4\,e^-$
$S + 6\,OH^-$	$\rightarrow SO_3^{2-} + 4\,e^-$
$S + 6\,OH^-$	$\rightarrow SO_3^{2-} + 4\,e^- + 3\,H_2O$

reduction:

$S + 2\,e^-$	$\rightarrow S^{2-}$

combining:

$S + 6\,OH^- \rightarrow SO_3^{2-} + 4\,e^- + 3\,H_2O + 2(S + 2\,e^- \rightarrow S^{2-})$	
$S + 6\,OH^- + 2\,S + 4\,e^-$	$\rightarrow SO_3^{2-} + 4\,e^- + 2\,S^{2-} + 3\,H_2O$
$3\,S + 6\,OH^-$	$\rightarrow SO_3^{2-} + 2\,S^{2-} + 3\,H_2O$

d) oxidation:

$$Br_2 \rightarrow 2\ BrO_3^-$$
$$Br_2 \rightarrow 2\ BrO_3^- + 10\ e^-$$
$$Br_2 + 12\ OH^- \rightarrow 2\ BrO_3^- + 10\ e^-$$
$$Br_2 + 12\ OH^- \rightarrow 2\ BrO_3^- + 10\ e^- + 6\ H_2O$$

reduction:

$$Br_2 + 2\ e^- \rightarrow 2\ Br^-$$

combining:

$$Br_2 + 12\ OH^- \rightarrow 2\ BrO_3^- + 10\ e^- + 6\ H_2O + 5(Br_2 + 2\ e^- \rightarrow 2\ Br^-)$$
$$Br_2 + 12\ OH^- + 5\ Br_2 + 10\ e^- \rightarrow 2\ BrO_3^- + 10\ e^- + 6\ H_2O + 10\ Br^-$$
$$6\ Br_2 + 12\ OH^- \rightarrow 2\ BrO_3^- + 10\ Br^- + 6\ H_2O$$
$$3\ Br_2 + 6\ OH^- \rightarrow BrO_3^- + 5\ Br^- + 3\ H_2O$$

Stoichiometric Calculations Involving Ions (Sec. 15.11)

15.87 a) $Cl^- = 35.45\ \dfrac{g}{mole}$ b) $NH_4^+ = 18.05\ \dfrac{g}{mole}$

c) $Cr_2O_7^{2-} = 216.00\ \dfrac{g}{mole}$ d) $MnO_4^- = 118.94\ \dfrac{g}{mole}$

15.89 a) $1.00 \times 10^{24}\ CO_3^{2-}\ \text{ions} \times \dfrac{1\ \text{mole}\ CO_3^{2-}\ \text{ions}}{6.022 \times 10^{23}\ CO_3^{2-}\ \text{ions}} \times \dfrac{1\ \text{mole}\ Mg^{2+}\ \text{ions}}{1\ \text{mole}\ CO_3^{2-}\ \text{ions}} \times \dfrac{24.31\ g\ Mg^{2+}\ \text{ions}}{1\ \text{mole}\ Mg^{2+}\ \text{ions}}$

$= 40.36864829\ \text{(calc)} = 40.4\ g\ Mg^{2+}\ \text{(corr)}$

b) $1.00 \times 10^{24}\ SO_4^{2-}\ \text{ions} \times \dfrac{1\ \text{mole}\ SO_4^{2-}\ \text{ions}}{6.022 \times 10^{23}\ SO_4^{2-}\ \text{ions}} \times \dfrac{1\ \text{mole}\ Mg^{2+}\ \text{ions}}{1\ \text{mole}\ SO_4^{2-}\ \text{ions}} \times \dfrac{24.31\ g\ Mg^{2+}\ \text{ions}}{1\ \text{mole}\ Mg^{2+}\ \text{ions}}$

$= 40.36864829\ \text{(calc)} = 40.4\ g\ Mg^{2+}\ \text{(corr)}$

c) $1.00 \times 10^{24}\ Cl^-\ \text{ions} \times \dfrac{1\ \text{mole}\ Cl^-\ \text{ions}}{6.022 \times 10^{23}\ Cl^-\ \text{ions}} \times \dfrac{1\ \text{mole}\ Mg^{2+}\ \text{ions}}{2\ \text{moles}\ Cl^-\ \text{ions}} \times \dfrac{24.31\ g\ Mg^{2+}\ \text{ions}}{1\ \text{mole}\ Mg^{2+}\ \text{ions}}$

$= 20.18432414\ \text{(calc)} = 20.2\ g\ Mg^{2+}\ \text{(corr)}$

d) $1.00 \times 10^{24}\ \text{total ions} \times \dfrac{1\ \text{mole ions}}{6.022 \times 10^{23}\ \text{ions}} \times \dfrac{1\ \text{mole}\ Mg^{2+}\ \text{ions}}{3\ \text{moles ions}} \times \dfrac{24.31\ g\ Mg^{2+}\ \text{ions}}{1\ \text{mole}\ Mg^{2+}\ \text{ions}}$

$= 13.4562161\ \text{(calc)} = 13.5\ g\ Mg^{2+}\ \text{(corr)}$

15.91 a) $0.634\ \text{mole}\ Ca^{2+}\ \text{ion} \times \dfrac{1\ \text{mole}\ Ca(CN)_2}{1\ \text{mole}\ Ca^{2+}\ \text{ions}} \times \dfrac{1\ L\ \text{solution}}{0.500\ \text{mole}\ Ca(CN)_2} \times \dfrac{1\ mL}{10^{-3}\ L}$

$= 1268\ \text{(calc)} = 1270\ mL\ \text{(corr)}$

b) $0.634\ \text{mole}\ CN^-\ \text{ion} \times \dfrac{1\ \text{mole}\ Ca(CN)_2}{2\ \text{moles}\ CN^-\ \text{ions}} \times \dfrac{1\ L\ \text{solution}}{0.500\ \text{mole}\ Ca(CN)_2} \times \dfrac{1\ mL}{10^{-3}\ L}$

$= 634\ mL\ \text{(calc and corr)}$

c) $0.553\ \text{mole}\ Ca^{2+}\ \text{ion} \times \dfrac{1\ \text{mole}\ Ca(CN)_2}{1\ \text{mole}\ Ca^{2+}\ \text{ions}} \times \dfrac{1\ L\ \text{solution}}{0.500\ \text{mole}\ Ca(CN)_2} \times \dfrac{1\ mL}{10^{-3}\ L}$

$= 1106\ \text{(calc)} = 1110\ mL\ \text{(corr)}$

d) $1.20\ \text{total moles ions} \times \dfrac{1\ \text{mole}\ Ca(CN)_2}{3\ \text{moles ions}} \times \dfrac{1\ L\ \text{solution}}{0.500\ \text{mole}\ Ca(CN)_2} \times \dfrac{1\ mL}{10^{-3}\ L}$

$= 800.\ mL\ \text{(calc and corr)}$

Chemical Equations: Net Ionic and Oxidation–Reduction

15.93 $20.0 \text{ g Fe} \times \dfrac{1 \text{ mole Fe}}{55.84 \text{ g Fe}} \times \dfrac{1 \text{ mole Cu}^{2+}}{1 \text{ mole Fe}} \times \dfrac{1 \text{ L solution}}{1.35 \text{ moles Cu}^{2+}} \times \dfrac{1 \text{ mL}}{10^{-3} \text{ L}}$

$= 265.3082882 \text{ (calc)} = 265 \text{ mL (corr)}$

15.95
a) $0.20 \text{ M NaCl} = 1(0.20 \text{ M}) \text{ Na}^+ = 0.20 \text{ M Na}^+$ (calc and corr)
   $1(0.20 \text{ M}) \text{ Cl}^- = 0.20 \text{ M Cl}^-$ (calc and corr)
b) $0.20 \text{ M K}_2\text{SO}_4 = 2(0.20 \text{ M}) \text{ K}^+ = 0.40 \text{ M K}^+$ (calc and corr)
   $1(0.20 \text{ M}) \text{ SO}_4^{2-} = 0.20 \text{ M SO}_4^{2-}$ (calc and corr)
c) $0.20 \text{ M Al(NO}_3)_3 = 1(0.20 \text{ M}) \text{ Al}^{3+} = 0.20 \text{ M Al}^{3+}$ (calc and corr)
   $3(0.20 \text{ M}) \text{ NO}_3^- = 0.60 \text{ M NO}_3^-$ (calc and corr)
d) $0.20 \text{ M MgCl}_2 = 1(0.20 \text{ M}) \text{ Mg}^{2+} = 0.20 \text{ M Mg}^{2+}$ (calc and corr)
   $2(0.20 \text{ M}) \text{ Cl}^- = 0.40 \text{ M Cl}^-$ (calc and corr)

15.97
a) $8.45 \text{ g NaBr} \times \dfrac{1 \text{ mole NaBr}}{102.89 \text{ g NaBr}} \times \dfrac{2 \text{ moles ions}}{1 \text{ mole NaBr}} \times \dfrac{6.022 \times 10^{23} \text{ ions}}{1 \text{ mole ions}}$

$= 9.891320828 \times 10^{22} \text{ (calc)} = 9.89 \times 10^{22} \text{ ions (corr)}$

b) $3.20 \text{ g K}_2\text{SO}_4 \times \dfrac{1 \text{ mole K}_2\text{SO}_4}{174.27 \text{ g K}_2\text{SO}_4} \times \dfrac{3 \text{ moles ions}}{1 \text{ mole K}_2\text{SO}_4} \times \dfrac{6.022 \times 10^{23} \text{ ions}}{1 \text{ mole ions}}$

$= 3.31733517 \times 10^{22} \text{ (calc)} = 3.32 \times 10^{22} \text{ ions (corr)}$

c) $30.0 \text{ g HCl} \times \dfrac{1 \text{ mole HCl}}{36.46 \text{ g HCl}} \times \dfrac{2 \text{ moles ions}}{1 \text{ mole HCl}} \times \dfrac{6.022 \times 10^{23} \text{ ions}}{1 \text{ mole ions}}$

$= 9.910038398 \times 10^{23} \text{ (calc)} = 9.91 \times 10^{23} \text{ ions (corr)}$

d) $40.0 \text{ g KOH} \times \dfrac{1 \text{ mole KOH}}{56.11 \text{ g KOH}} \times \dfrac{2 \text{ moles ions}}{1 \text{ mole KOH}} \times \dfrac{6.022 \times 10^{23} \text{ ions}}{1 \text{ mole ions}}$

$= 8.585991802 \times 10^{23} \text{ (calc)} = 8.59 \times 10^{23} \text{ ions (corr)}$

15.99
a) $10.0 \text{ g Na}_2\text{SO}_4 \times \dfrac{1 \text{ mole Na}_2\text{SO}_4}{142.05 \text{ g Na}_2\text{SO}_4} \times \dfrac{1 \text{ L solution}}{0.125 \text{ mole Na}_2\text{SO}_4} \times \dfrac{1 \text{ mL}}{10^{-3} \text{ L}}$

$= 563.1819782 \text{ (calc)} = 563 \text{ mLs (corr)}$

b) $2.5 \text{ g Na}^+ \times \dfrac{1 \text{ mole Na}^+}{22.99 \text{ g Na}^+} \times \dfrac{1 \text{ mole Na}_2\text{SO}_4}{2 \text{ moles Na}^+} \times \dfrac{1 \text{ L solution}}{0.125 \text{ mole Na}_2\text{SO}_4} \times \dfrac{1 \text{ mL}}{10^{-3} \text{ L}}$

$= 434.9717268 \text{ (calc)} = 4.3 \times 10^2 \text{ mLs (corr)}$

c) $0.567 \text{ mole Na}_2\text{SO}_4 \times \dfrac{1 \text{ L solution}}{0.125 \text{ mole Na}_2\text{SO}_4} \times \dfrac{1 \text{ mL}}{10^{-3} \text{ L}} = 4536 \text{ (calc)} = 4540 \text{ mLs (corr)}$

d) $0.112 \text{ mole SO}_4^{2-} \times \dfrac{1 \text{ mole Na}_2\text{SO}_4}{1 \text{ mole SO}_4^{2-}} \times \dfrac{1 \text{ L solution}}{0.125 \text{ mole Na}_2\text{SO}_4} \times \dfrac{1 \text{ mL}}{10^{-3} \text{ L}} = 896 \text{ mLs (calc and corr)}$

15.101
a) $27 \text{ mL} \times \dfrac{10^{-3} \text{ L}}{1 \text{ mL}} \times \dfrac{0.200 \text{ mole KCl}}{1 \text{ L solution}} \times \dfrac{1 \text{ mole K}^+}{1 \text{ mole KCl}} = 0.0054 \text{ mole K}^+$ (calc and corr)

$175 \text{ mL} \times \dfrac{10^{-3} \text{ L}}{1 \text{ mL}} \times \dfrac{0.100 \text{ mole K}_3\text{PO}_4}{1 \text{ L solution}} \times \dfrac{3 \text{ mole K}^+}{1 \text{ mole K}_3\text{PO}_4} = 0.0525 \text{ mole K}^+$ (calc and corr)

Total moles of $\text{K}^+ = 0.0525 + 0.0054 = 0.0579 \text{ mole K}^+$

Concentration of $\text{K}^+ = \dfrac{0.0579 \text{ mole K}^+}{(175 + 27) \text{ mL}} \times \dfrac{1 \text{ mL}}{10^{-3} \text{ L}} = 0.286633663 \text{ (calc)} = 0.287 \text{ M K}^+$ (corr)

b) $27 \text{ mL} \times \dfrac{10^{-3} \text{ L}}{1 \text{ mL}} \times \dfrac{0.200 \text{ mole KCl}}{1 \text{ L solution}} \times \dfrac{1 \text{ mole Cl}^-}{1 \text{ mole KCl}} = 0.0054$ mole Cl^- (calc and corr)

$$\text{Concentration of Cl}^- = \dfrac{0.0054 \text{ mole K}^+}{(175 + 27) \text{ mL}} \times \dfrac{1 \text{ mL}}{10^{-3} \text{ L}} = 0.02673267 \text{ (calc)} = 0.027 \text{ M Cl}^- \text{ (corr)}$$

c) $175 \text{ mL} \times \dfrac{10^{-3} \text{ L}}{1 \text{ mL}} \times \dfrac{0.100 \text{ mole K}_3\text{PO}_4}{1 \text{ L solution}} \times \dfrac{1 \text{ mole PO}_4^{3-}}{1 \text{ mole K}_3\text{PO}_4} = 0.0175$ mole PO_4^{3-} (calc and corr)

$$\text{Concentration of PO}_4^{3-} = \dfrac{0.0175 \text{ mole PO}_4^{3-}}{(175 + 27) \text{ mL}} \times \dfrac{1 \text{ mL}}{10^{-3} \text{ L}} = 0.086633663 \text{ (calc)}$$

$$= 0.0866 \text{ M PO}_4^{3-} \text{ (corr)}$$

Additional Problems

15.103 Oxidation numbers of N: N_2O (+1), NO (+2), N_2O_3 (+3), NO_2 (+4), N_2O_5 (+5)

15.105 a) S^{2-} has the maximum amount of electrons; it can only lose electrons not gain them, acting as a reducing agent.
b) SO_4^{2-} ion, S has a +6 charge. S cannot lose any more electrons; it can only gain them, thereby acting as oxidizing agent.
c) SO_2, S has charge of +4. S can either lose or gain electrons, acting as reducing agent or oxidizing agent.
d) SO_3, S has a +6 charge. S cannot lose any more electrons; it can only gain them, thereby acting as oxidizing agent.

15.107 a) +2, +1   b) +3 in both   c) +1 in both   d) +3 in both

15.109 a) two reduction half-reactions   b) one reduction and one oxidation half-reaction
c) two oxidation half-reactions   d) one reduction and one oxidation half-reaction

15.111 1) $5(2 H_2O + PH_3 \rightarrow H_3PO_2 + 4 H^+ + 4 e^-) + 4(MnO_4^- + 8 H^+ + 5 e^- \rightarrow Mn^{2+} + 4 H_2O)$
$10 H_2O + 5 PH_3 + 4 MnO_4^- + 32 H^+ + 20 e^- \rightarrow 5 H_3PO_2 + 20 H^+ + 20 e^- + 4 Mn^{2+} + 16 H_2O$
$5 PH_3 + 4 MnO_4^- + 12 H^+ \rightarrow 5 H_3PO_2 + 4 Mn^{2+} + 6 H_2O$

2) $2 H_2O + PH_3 \rightarrow H_3PO_2 + 4 H^+ + 4 e^- + 2(SO_4^{2-} + 4 H^+ + 2 e^- \rightarrow SO_2 + 2 H_2O)$
$2 H_2O + PH_3 + 2 SO_4^{2-} + 8 H^+ + 4 e^- \rightarrow H_3PO_2 + 4 H^+ + 4 e^- + 2 SO_2 + 4 H_2O$
$PH_3 + 2 SO_4^{2-} + 4 H^+ \rightarrow H_3PO_2 + 2 SO_2 + 2 H_2O$

3) $5(3 H_2O + As \rightarrow H_3AsO_3 + 3 H^+ + 3 e^-) + 3(MnO_4^- + 8 H^+ + 5 e^- \rightarrow Mn^{2+} + 4 H_2O)$
$15 H_2O + 5 As + 3 MnO_4^- + 24 H^+ + 15 e^-$
$\rightarrow 5 H_3AsO_3 + 15 H^+ + 15 e^- + 3 Mn^{2+} + 12 H_2O$
$5 As + 3 MnO_4^- + 3 H_2O + 9 H^+ \rightarrow 5 H_3AsO_3 + 3 Mn^{2+}$

4) $2(3 H_2O + As \rightarrow H_3AsO_3 + 3 H^+ + 3 e^-) + 3(SO_4^{2-} + 4 H^+ + 2 e^- \rightarrow SO_2 + 2 H_2O)$
$6 H_2O + 2 As + 3 SO_4^{2-} + 12 H^+ + 6 e^- \rightarrow 2 H_3AsO_3 + 6 H^+ + 6 e^- + 3 SO_2 + 6 H_2O$
$2 As + 3 SO_4^{2-} + 6 H^+ \rightarrow 2 H_3AsO_3 + 3 SO_2$

Chemical Equations: Net Ionic and Oxidation–Reduction

15.113 $4\,Zn + 10\,H^+ + NO_3^- \rightarrow 4\,Zn^{2+} + NH_4^+ + 3\,H_2O$

oxidation:
$Zn \rightarrow Zn^{2+} + 2\,e^-$

reduction:
$NO_3^- + 8\,e^- \rightarrow NH_4^+$
$NO_3^- + 8\,e^- + 10\,H^+ \rightarrow NH_4^+$
$NO_3^- + 10\,H^+ + 8\,e^- \rightarrow NH_4^+ + 3\,H_2O$

15.115 $MnO_4^- + C_2O_4^{2-} \rightarrow Mn^{2+} + CO_2$

oxidation:
$C_2O_4^{2-} \rightarrow CO_2$
$C_2O_4^{2-} \rightarrow 2\,CO_2$
$C_2O_4^{2-} \rightarrow 2\,CO_2 + 2\,e^-$

reduction:
$MnO_4^- \rightarrow Mn^{2+}$
$MnO_4^- + 5\,e^- \rightarrow Mn^{2+}$
$MnO_4^- + 5\,e^- + 8\,H^+ \rightarrow Mn^{2+}$
$MnO_4^- + 5\,e^- + 8\,H^+ \rightarrow Mn^{2+} + 4\,H_2O$

combining:
$5(C_2O_4^{2-} \rightarrow 2\,CO_2 + 2\,e^-) + 2(MnO_4^- + 5\,e^- + 8\,H^+ \rightarrow Mn^{2+} + 4\,H_2O)$
$5\,C_2O_4^{2-} + 2\,MnO_4^- + 10\,e^- + 16\,H^+ \rightarrow 10\,CO_2 + 10\,e^- + 2\,Mn^{2+} + 8\,H_2O$
$5\,C_2O_4^{2-} + 2\,MnO_4^- + 16\,H^+ \rightarrow 10\,CO_2 + 2\,Mn^{2+} + 8\,H_2O$ net ionic
$5\,K_2C_2O_4 + 2\,KMnO_4 + 16\,HCl \rightarrow 10\,CO_2 + 2\,MnCl_2 + 8\,H_2O + 12\,KCl$ molecular redox

Cumulative Problems

15.117
a) redox, HNO_3 is the oxidizing agent
b) acid–base, H_2S is the acid
c) redox, H_2O_2 is the oxidizing agent
d) acid–base, H_2SO_4 is the acid

15.119
a) $Sn^{2+} + SO_4^{2-} + 2\,Fe^{2+} + 2\,SO_4^{2-} \rightarrow Sn + 2\,Fe^{3+} + 3\,SO_4^{2-}$
   $Sn^{2+} + 2\,Fe^{2+} \rightarrow Sn + 2\,Fe^{3+}$
b) $PH_3 + 2\,NO_2 \rightarrow H_3PO_4 + N_2$
c) $S + 3\,H_2O + 2\,Pb^{2+} + 4\,NO_3^- \rightarrow 2\,Pb + H_2SO_3 + 4\,H^+ + 4\,NO_3^-$
   $S + 3\,H_2O + 2\,Pb^{2+} \rightarrow 2\,Pb + H_2SO_3 + 4\,H^+$
d) $4\,Zn + 10\,H^+ + 10\,NO_3^- \rightarrow 4\,Zn^{2+} + 8\,NO_3^- + NH_4^+ + NO_3^- + 3\,H_2O$
   $4\,Zn + 10\,H^+ + NO_3^- \rightarrow 4\,Zn^{2+} + NH_4^+ + 3\,H_2O$

15.121
a)
$$-2 \quad\quad 3(+2) \quad\quad 0$$
$$8\,H^+ + 3\,H_2S + Cr_2O_7^{2-} \rightarrow 2\,Cr^{3+} + 3\,S + 7\,H_2O$$
$$+6 \quad 2(-3) \quad +3$$

b)
$$0 \quad 3[2(+5)] \quad +5$$
$$3\,H_2O + 5\,ClO_3^- + 3\,I_2 \rightarrow 5\,Cl^- + 6\,IO_3^- + 6\,H^+$$
$$+5 \quad 5(-6) \quad -1$$

c)
$$8\,\overset{-2}{\text{OH}^-} + \overset{(+8)}{\text{S}^{2-}} + 4\,\text{Br}_2 \rightarrow \overset{+6}{\text{SO}_4^{2-}} + 8\,\text{Br}^- + 4\,\text{H}_2\text{O}$$
$$\overset{0}{} \quad \overset{4[2(-1)]}{} \quad \overset{-1}{}$$

d)
$$2\,\text{OH}^- + \overset{+4}{\text{NO}_2} + \overset{(+1)}{\text{NO}_2} \rightarrow \overset{+5}{\text{NO}_3^-} + \text{NO}_2^- + \text{H}_2\text{O}$$
$$\overset{+4}{} \quad \overset{(-1)}{} \quad \overset{+3}{}$$

15.123 a) molarities of Al^{3+} and NO_3^- in the original solution:

$$\frac{0.245\text{ mole Al(NO}_3)_3}{1\text{ L solution}} \times \frac{1\text{ mole Al}^{3+}}{1\text{ mole Al(NO}_3)_3} = 0.245\text{ M Al}^{3+} \text{ (calc and corr)}$$

$$\frac{0.245\text{ mole Al(NO}_3)_3}{1\text{ L solution}} \times \frac{3\text{ mole NO}_3^-}{1\text{ mole Al(NO}_3)_3} = 0.735\text{ M NO}_3^- \text{ (calc and corr)}$$

b) Since we are diluting the solution and the moles of Al^{3+} and NO_3^- will remain the same, only the volume will change and we can use $M_1 \times V_1 = M_2 \times V_2$

$$225\text{ mL solution} \times \frac{10^{-3}\text{ L solution}}{1\text{ mL solution}} \times \frac{0.245\text{ M Al(NO}_3)_3}{0.750\text{ L solution}} = 0.0735\text{ M Al(NO}_3)_3 \text{ (calc and corr)}$$

$$\frac{0.0735\text{ mole Al(NO}_3)_3}{1\text{ L solution}} \times \frac{1\text{ mole Al}^{3+}}{1\text{ mole Al(NO}_3)_3} = 0.0735\text{ M Al}^{3+} \text{ (calc and corr)}$$

$$\frac{0.0735\text{ mole Al(NO}_3)_3}{1\text{ L solution}} \times \frac{3\text{ mole NO}_3^-}{1\text{ mole Al(NO}_3)_3} = 0.2205\text{ (calc)} = 0.220\text{ M NO}_3^- \text{ (corr)}$$

15.125 $Mg(NO_3)_2 + Na_2CO_3 \rightarrow MgCO_3 + 2\,NaNO_3$

Total volume of solution: $(21.03\text{ mL} + 22.51\text{ mL}) = 43.54 \times \dfrac{10^{-3}\text{ L}}{1\text{ mL}} = 0.04354$ L solution

Solving for the amount of $MgCO_3$ produced:

$$2.341\text{ g Mg(NO}_3)_2 \times \frac{1\text{ mole Mg(NO}_3)_2}{148.33\text{ g Mg(NO}_3)_2} \times \frac{1\text{ mole MgCO}_3}{1\text{ mole Mg(NO}_3)_2} = 0.157823377 \text{ (calc)}$$
$$= 0.01578\text{ mole MgCO}_3 \text{ (corr)}$$

$$1.250\text{ g Na}_2\text{CO}_3 \times \frac{1\text{ mole Na}_2\text{CO}_3}{105.99\text{ g Na}_2\text{CO}_3} \times \frac{1\text{ mole MgCO}_3}{1\text{ mole Na}_2\text{CO}_3} = 0.0117935654 \text{ (calc)}$$
$$= 0.01179\text{ mole MgCO}_3 \text{ (corr)}$$

Na_2CO_3 is the limiting reactant and moles of $MgCO_3$ produced $= 0.01179$ mole
Moles of Mg^{2+} remaining in solution:

Initial moles Mg^{2+}: $0.01578\text{ mole Mg(NO}_3)_2 \times \dfrac{1\text{ mole Mg}^{2+}}{1\text{ mole Mg(NO}_3)_2} = 0.01578\text{ mole Mg}^{2+}$

Moles consumed in $MgCO_3$: $0.01179\text{ mole MgCO}_3 \times \dfrac{1\text{ mole Mg}^{2+}}{1\text{ mole MgCO}_3} = 0.01179\text{ mole Mg}^{2+}$

Chemical Equations: Net Ionic and Oxidation–Reduction

Moles of Mg^{2+} remaining $= (0.01578 - 0.01179) = 0.00399$ mole excess Mg^{2+}

Molarity excess $Mg^{2+} = \dfrac{0.00399 \text{ mole } Mg^{2+}}{0.04354 \text{ L solution}} = 0.091639871$ (calc) $= 0.09164$ M Mg^{2+} (corr)

0 mole of CO_3^{2-} remain in solution, since Na_2CO_3 is the limiting reactant.
Moles of NO_3^- in solution $=$ initial moles of NO_3^-

0.01578 mole $Mg(NO_3)_2 \times \dfrac{2 \text{ moles } NO_3^-}{1 \text{ mole } Mg(NO_3)_2} = 0.03156$ mole NO_3^-

Molarity $NO_3^- = \dfrac{0.03156 \text{ mole } NO_3^-}{0.04354 \text{ L solution}} = 0.724850712$ (calc) $= 0.7249$ M NO_3^- (corr)

Moles of Na^+ in solution $=$ initial moles of Na^+

0.01179 mole $Na_2CO_3 \times \dfrac{2 \text{ moles } Na^+}{1 \text{ mole } Na_2CO_3} = 0.02358$ mole Na^+

Molarity $Na^+ = \dfrac{0.02358 \text{ mole } Na^+}{0.04354 \text{ L solution}} = 0.541570969$ (calc) $= 0.5416$ M Na^+ (corr)

Concentration of each ion at end of reaction: $\;0.09164$ M Mg^{2+}
0.7249 M NO_3^-
0.5416 M Na^+

15.127 $\;43.2$ mL solution $\times \dfrac{10^{-3} \text{ L}}{1 \text{ mL}} \times \dfrac{0.300 \text{ mole } S_2O_3^{2-}}{1 \text{ L solution}} \times \dfrac{1 \text{ mole } I_3^-}{2 \text{ moles } S_2O_3^{2-}}$

$= 0.00648$ mole I_3^- (calc and corr)

$M = \dfrac{0.00648 \text{ mole } I_3^-}{20.0 \text{ mL solution}} \times \dfrac{1 \text{ mL}}{10^{-3} \text{ L}} = 0.324$ M I_3^- (calc and corr)

15.129 $\;18.03$ mL $S_2O_3^{2-}$ solution $\times \dfrac{10^{-3} \text{ L}}{1 \text{ mL}} \times \dfrac{0.00200 \text{ mole } S_2O_3^{2-}}{1 \text{ L } S_2O_3^{2-}} \times \dfrac{1 \text{ mole } I_2}{2 \text{ moles } S_2O_3^{2-}}$

$\times \dfrac{1 \text{ mole } O_3}{1 \text{ mole } I_2} \times \dfrac{48.00 \text{ g } O_3}{1 \text{ mole } O_3} = 0.00086544$ (calc) $= 0.000865$ g O_3 (corr)

$\dfrac{0.000865 \text{ g } O_3}{28.09 \text{ g sample}} \times 10^6 = 30.793877$ (calc) $= 30.8$ ppm (m/m) (corr)

Answers to Multiple-Choice Practice Test

15.131 c	**15.132** e	**15.133** a	**15.134** c	**15.135** d	**15.136** e
15.137 d	**15.138** d	**15.139** a	**15.140** b	**15.141** d	**15.142** d
15.143 c	**15.144** c	**15.145** d	**15.146** d	**15.147** b	**15.148** a
15.149 e	**15.150** d				

CHAPTER SIXTEEN
Reaction Rates and Chemical Equilibrium

Practice Problems

Collision Theory (Sec. 16.1)

16.1 The solute molecules have more motion in the solution, allowing more frequent collisions with other reactant molecules throughout the solution. Only those molecules on the surface of a solid can collide with other reactant molecules.

16.3 The reaction with the lower activation energy, 45 kJ/mole, will have the faster rate. At any given temperature, there will be a greater fraction of collisions with a combined energy equal to or exceeding the lower activation energy.

16.5 (1) The combined kinetic energies of the colliding particles must equal or exceed a minimum value, the activation energy, and (2) the orientation of the particles must be favorable.

16.7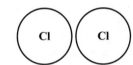

Endothermic and Exothermic Chemical Reactions (Sec. 16.2)

16.9 a) endothermic   b) exothermic   c) exothermic   d) endothermic

16.11 a) (1) The average energy of the reactant molecules is less than the average energy of the products.
b) (3) The average energy of the reactant molecules is greater than the average energy of the products.
c) (3) The average energy of the reactant molecules is greater than the average energy of the products.
d) (1) The average energy of the reactant molecules is less than the average energy of the products.

16.13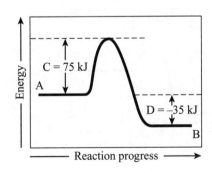

a) average energy of the reactants = A
b) average energy of the products = B
c) the activation energy = C
d) the energy liberated = D

16.15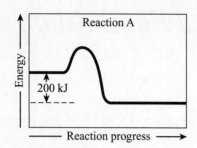

Similarities: Both reactions are exothermic and the energy difference between reactants and products is the same in both reactions. Differences: The activation energy is lower in the reaction that takes place at room temperature, Reaction A.

Factors That Influence Chemical Reaction Rates (Sec. 16.3)

16.17 a) A temperature change causes the collision frequency to change and also causes the molecular energy to change.
b) A catalyst provides an alternate pathway with a lower activation energy.

16.19 a) Increasing the concentration of the reactants increases the rate of reaction.
b) Raising the temperature of the reaction increases the rate of reaction.
c) Removing the catalyst decreases the rate of reaction.
d) Removing some of the reactants decreases the rate of reaction.

16.21 The coal dust has much more surface area exposed to the air than does the same mass of charcoal. The reaction with the O_2 in the air will be much faster for the coal dust.

16.23 a) 1—The lower the activation energy, the faster the reaction.
b) 3—The higher the temperature, the faster the reaction.
c) 4—The higher the concentration of a reactant, the faster the reaction.
d) 3—The higher the temperature and the lower the activation energy, the faster the reaction.

16.25 Diagrams are the same except for the magnitude of the activation energy. A catalyst lowers the activation energy.

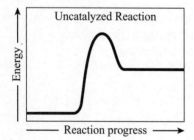 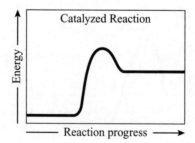

Chemical Equilibrium (Sec. 16.4)

16.27 The rate of the forward reaction must equal the rate of the reverse reaction.

16.29 a) Forward reaction: $CO(g) + H_2O(g) \rightarrow CO_2(g) + H_2(g)$
b) Reverse reaction: $CO_2(g) + H_2(g) \rightarrow CO(g) + H_2O(g)$

Reaction Rates and Chemical Equilibrium

16.31   X ⇌ Y

Diagram I: 10 X   Diagram II: 7 X, 3 Y   Diagram III: 6 X, 4 Y   Diagram IV: 6 X, 4 Y

The system changes in Diagram II; since Diagrams III and IV are the same, the system has possibly reached equilibrium.

16.33   Consider the balanced reaction: $A_2 + 2 B \rightarrow 2 AB$
Diagram I: 6 B, 4 A_2 Diagram II: 2 B, 2 A_2, 4 AB
Diagram III: 4 B, 2 A_2, 4 AB   Diagram IV: 4 B, 3 A_2, 2 AB
Diagrams II and IV represent a possible equilibrium state; they remain balanced.

16.35   As the reactants decrease, the amount of products formed increases.

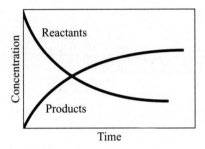

Equilibrium Mixture Stoichiometry (Sec. 16.5)

16.37

	2 SO_3	⇌	2 SO_2	+	O_2
start	0.0200 mole		0 mole		0 mole
change	?		?		?
equilibrium	?		?		0.0029 mole

The change in O_2 = +0.0029 moles to get from 0 at start to 0.0029 at equilibrium.
The change in SO_2 must be twice as great as O_2 or +0.0058 mole (produced).
The change in SO_3 must be twice as great as O_2 or −0.0058 mole (consumed).
At equilibrium, the O_2 is 0.0029 moles (given).
At equilibrium, the SO_2 is 0 + 0.0058 mole (start + change).
At equilibrium, the SO_3 is 0.0200 − 0.0058 = 0.0142 mole (calc and corr).
The full table becomes:

	2 SO_3	⇌	2 SO_2	+	O_2
start	0.0200 mole		0 mole		0 mole
change	−0.0058 mole		+0.0058 mole		0.0029 mole
equilibrium	0.0142 mole		0.0058 mole		0.0029 mole

16.39

	2 $NH_3(g)$	⇌	$N_2(g)$	+	3 $H_2(g)$
start	0.296 mole		0.170 mole		0.095 mole
change	−0.028 mole		+0.014 mole		+0.042 mole
equilibrium	0.268 mole		0.184 mole		0.137 mole

16.41

	2 CO(g)	+	O_2(g)	⇌	2 CO_2(g)
start	0.100 mole		0.200 mole		0 mole
change	−0.006 mole		−0.003 mole		+0.006 mole
equilibrium	0.094 mole		0.197 mole		0.006 mole

Equilibrium Constants (Sec. 16.6)

16.43 a) $K_{eq} = \dfrac{[C_2H_6]}{[H_2]^2[C_2H_2]}$ b) $K_{eq} = \dfrac{[SO_2]^2[O_2]}{[SO_3]^2}$

c) $K_{eq} = \dfrac{[NCl_3][HCl]^3}{[Cl_2]^3[NH_3]}$ d) $K_{eq} = \dfrac{[NO]^2[H_2O]^2}{[N_2H_4][O_2]^2}$

16.45 a) $K_{eq} = \dfrac{[CO_2]}{[SO_2]}$ b) $K_{eq} = [Br_2]$

c) $K_{eq} = [H_2O]$ d) $K_{eq} = \dfrac{[NaNO_3]}{[NaCl][AgNO_3]}$

16.47 a) $K_{eq} = \dfrac{[B]^2[C]}{[A]} = \dfrac{(2.00)^2(5.00)}{3.00} = 6.6666667$ (calc) = 6.67 (corr)

b) $K_{eq} = \dfrac{[C]^2}{[A][B]^3} = \dfrac{(5.00)^2}{(3.00)(2.00)^3} = 1.0416667$ (calc) = 1.04 (corr)

c) $K_{eq} = \dfrac{[A][C]}{[B]^2} = \dfrac{(3.00)(5.00)}{(2.00)^2} = 3.75$ (calc and corr)

d) $K_{eq} = \dfrac{[A]^3}{[C]^4[B]} = \dfrac{(3.00)^3}{(5.00)^4(2.00)} = 0.0216$ (calc and corr)

16.49 Diagram IV has the largest product to reactant ratio: $K_{eq} = \dfrac{[AB]^2}{[A_2][B_2]} = 64$

16.51 Diagram I has the only product to reactant ratio of 64: $K_{eq} = \dfrac{[AB]^2}{[A_2][B_2]} = 64$

64 $([A_2][B_2]) = [AB]^2$ If AB = 8 and $A_2 = 1$ and $B_2 = 1$

16.53 $K_{eq} = \dfrac{[CH_4][H_2S]^2}{[CS_2][H_2]^4}$; or $0.0280 = \dfrac{(0.00100)(1.43)^2}{[CS_2](1.00)^4}$

Solving for $[CS_2]$ gives: $[CS_2] = \dfrac{(0.00100)(1.43)^2}{(0.0280)(1.00)^4} = 0.073032143$ (calc) = 0.0730 M CS_2 (corr)

16.55 Find the molar concentration of each gas by dividing the moles by the volume,

$K_{eq} = \dfrac{[PCl_5]}{[PCl_3][Cl_2]} = \dfrac{\left(\dfrac{0.0189}{6.00}\right)}{\left(\dfrac{0.0222}{6.00}\right)\left(\dfrac{0.1044}{6.00}\right)} = 48.92823858$ (calc) = 48.9 (corr)

Reaction Rates and Chemical Equilibrium

16.57 a) mostly reactants   b) mostly products
c) mostly products   d) significant amounts of both reactants and products

Le Châteliers Principle (Sec. 16.9)

16.59 a) Increasing the concentration of the reactants shifts the equilibrium toward the products.
b) Increasing the concentration of the products shifts the equilibrium toward the reactants.
c) Decreasing the concentration of the products shifts the equilibrium toward the products.
d) Decreasing the concentration of the reactant will shift the equilibrium toward the reactants.

16.61 a) Increasing the concentration of the reactants shifts the equilibrium toward the products.
b) Decreasing the concentration of the reactant will shift the equilibrium toward the reactants.
c) Since this reaction is exothermic, we treat heat as a product. Increasing the temperature will shift the reaction toward the reactants.
d) Increasing the pressure will shift the equilibrium toward the least number of moles of gas, in this case toward the products with 6 moles of gas versus the reactants with 7 moles of gas.

16.63 a) Since this reaction is endothermic, we treat heat as a reactant. Heating the reaction mixture will shift the reaction toward the products.
b) Increasing the concentration of the products shifts the equilibrium toward the reactants.
c) Increasing the pressure by adding an inert gas will have no effect on the equilibrium.
d) Increasing the size of the container will shift the reaction toward the side with the most gas molecules. In this case it shifts toward the products with 6 moles of gas versus the reactants with 4 moles of gas.

16.65 Increased product formation is a shift to the right. Higher temperatures will shift endothermic reactions to the right, so only in c) would high temperature favor product formation.

16.67 If increasing the temperature shifts the equilibrium to the left, heat must be a product and the reaction exothermic. Rewriting the equation: $CO + H_2O \rightleftarrows CO_2 + H_2 + heat$

16.69 Increasing the temperature produces more products; therefore the reaction is endothermic.

Additional Problems

16.71 a) $N_2 + 3 H_2 \rightleftarrows 2 NH_3$ b) $4 NH_3 + 3 O_2 \rightleftarrows 2 N_2 + 6 H_2O$
c) $2 NO \rightleftarrows N_2 + O_2$ d) $N_2 + O_2 \rightleftarrows 2 NO$

16.73 a) $K = \dfrac{[0.350]^2[0.190]}{[0.0300]^2} = 25.861111$ (calc) $= 25.9$ (corr); system is at equilibrium.

b) $K = \dfrac{[0.700]^2[0.380]}{[0.0600]^2} = 51.722222$ (calc) $= 51.7$ (corr)

K value is too high for equilibrium; a shift to the left will reduce the size of K.

c) $K = \dfrac{[0.356]^2[0.160]}{[0.0280]^2} = 25.864489$ (calc) $= 25.9$ (corr); system is at equilibrium.

d) $K = \dfrac{[0.330]^2[0.180]}{[0.0100]^2} = 196.02$ (calc) $= 196$ (corr)

K value is too high for equilibrium; a shift to the left will reduce the size of K.

16.75 Since the numerator and denominator are reversed, the second K is the reciprocal of the first K.

$$K_{\text{reverse reaction}} = \frac{1}{K_{\text{forward reaction}}} = \frac{1}{2 \times 10^3} = 0.0005 \text{ (calc and corr)}$$

16.77
a) no change
b) no change
c) change in equilibrium constant value
d) no change

16.79
a) right b) no change
c) left   d) right

Cumulative Problems

16.81
a) $2 SO_3(g) \rightleftarrows 2 SO_2(g) + O_2(g)$ $K = \dfrac{[SO_2]^2[O_2]}{[SO_3]^2}$

b) $7 H_2(g) + 2 NO_2(g) \rightleftarrows 2 NH_3(g) + 4 H_2O(g)$ $K = \dfrac{[NH_3]^2[H_2O]^4}{[H_2]^7[NO_2]^2}$

c) $FeO(s) + CO(g) \rightleftarrows Fe(s) + CO_2(g)$ $K = \dfrac{[CO_2]}{[CO]}$

d) $MgCO_3(s) \rightleftarrows MgO(s) + CO_2(g)$ $K = [CO_2]$

16.83 $M_{CH_4} = \dfrac{17.6 \text{ g } CH_4 \times \dfrac{1 \text{ mole } CH_4}{16.05 \text{ g } CH_4}}{1.725 \text{ L}} = 0.6356946 \text{ (calc)} = 0.636 \text{ M (corr)}$

$M_{H_2S} = \dfrac{50.8 \text{ g } H_2S \times \dfrac{1 \text{ mole } H_2S}{34.08 \text{ g } H_2S}}{1.725 \text{ L}} = 0.8638684471 \text{ (calc)} = 0.864 \text{ M (corr)}$

$M_{CS_2} = \dfrac{83.8 \text{ g } CS_2 \times \dfrac{1 \text{ mole } CS_2}{76.13 \text{ g } CS_2}}{1.725 \text{ L}} = 0.6379476053 \text{ (calc)} = 0.638 \text{ M (corr)}$

$M_{H_2} = \dfrac{8.10 \text{ g } H_2 \times \dfrac{1 \text{ mole } H_2}{2.02 \text{ g } H_2}}{1.725 \text{ L}} = 2.3245803 \text{ (calc)} = 2.32 \text{ M (corr)}$

$K = \dfrac{[CS_2][H_2]^4}{[CH_4][H_2S]^2} = \dfrac{[0.638][2.32]^4}{[0.636][0.864]^2} = 38.93032396 \text{ (calc)} = 38.9 \text{ (corr)}$

Reaction Rates and Chemical Equilibrium

16.85 $M_{SbCl_5} = \dfrac{2.48 \times 10^{20} \text{ molecules SbCl}_5 \times \dfrac{1 \text{ mole SbCl}_5}{6.022 \times 10^{23} \text{ molecules SbCl}_5}}{1.00 \text{ L}}$

$= 0.00041182331 \text{ (calc)} = 0.000412 \text{ M (corr)}$

$M_{SbCl_3} = \dfrac{0.723 \text{ g SbCl}_3 \times \dfrac{1 \text{ mole SbCl}_3}{228.11 \text{ g SbCl}_3}}{1.00 \text{ L}} = 0.0031695235 \text{ (calc)} = 0.00317 \text{ M (corr)}$

$M_{Cl_2} = \dfrac{0.00317 \text{ mole Cl}_2}{1.00 \text{ L}} = 0.00317 \text{ M (calc and corr)}$

$K = \dfrac{[SbCl_3][Cl_2]}{[SbCl_5]} = \dfrac{[0.00317][0.00317]}{[0.000412]} = 0.024390534 \text{ (calc)} = 0.0244 \text{ (corr)}$

16.87 Pressures may be treated in the same way as moles since the number of moles present determines the pressure.

	$N_2O_4(g)$	$\rightleftarrows$	$2 NO_2(g)$
start	1.50 atm		0 atm
change	-0.70 atm		$+1.40$ atm
equilibrium	0.80 atm		1.40 atm

Total pressure = 0.80 atm + 1.40 atm = 2.2 (calc) = 2.20 atm (corr)

16.89 Basis: 100 g of gaseous mixture (87.43% CO and 12.57% CO_2)

$100 \text{ g mixture} \times \dfrac{87.43 \text{ g CO}}{100 \text{ g mixture}} \times \dfrac{1 \text{ mole CO}}{28.01 \text{ g CO}} = 3.1213852 \text{ (calc)} = 3.121 \text{ moles CO (corr)}$

$100 \text{ g mixture} \times \dfrac{12.57 \text{ g CO}_2}{100 \text{ g mixture}} \times \dfrac{1 \text{ mole CO}_2}{44.01 \text{ g CO}_2} = 0.2856169 \text{ (calc)} = 0.2856 \text{ mole CO}_2 \text{ (corr)}$

Total moles = 3.121 moles CO + 0.2856 mole CO_2 = 3.4066 (calc) = 3.407 moles (corr)

$V_T = \dfrac{nRT}{P} = \dfrac{3.407 \text{ moles} \times 0.08206 \dfrac{\text{atm L}}{\text{mole K}} \times 1023 \text{ K}}{1.000 \text{ atm}} = 286.0087237 \text{ (calc)} = 286.0 \text{ L (corr)}$

$M_{CO} = \dfrac{3.121 \text{ moles}}{286.0 \text{ L}} = 0.010912587 \text{ (calc)} = 0.01091 \text{ M (corr)}$

$M_{CO_2} = \dfrac{0.2856 \text{ mole}}{286.0 \text{ L}} = 0.0009986013986 \text{ (calc)} = 0.0009986 \text{ M (corr)}$

$K = \dfrac{[CO]^2}{[CO_2]} = \dfrac{(0.01091)^2}{0.0009986} = 0.119194973 \text{ (calc)} = 0.1192 \text{ (corr)}$

16.91 $2 H_2O \rightleftarrows H_3O^+ + OH^-$

a) Increasing the concentration of H_3O^+ shifts the equilibrium to the left toward the reactants.
b) Decreasing the concentration of OH^- shifts the equilibrium to the right toward the products.
c) pH increase by 2 units: At equilibrium the pH for this reaction is 7. Increasing the pH to 9 increases the concentration of OH^-, which will shift the equilibrium to the left toward the reactants.
d) Decreasing the pH by 2 units, from pH 7 to pH 5, increases the concentration of H_3O^+; thus the reaction equilibrium will shift to the left toward the reactants.

Answers to Multiple-Choice Practice Test

16.93	c	**16.94**	c	**16.95**	c	**16.96**	e	**16.97**	e	**16.98**	b
16.99	b	**16.100**	d	**16.101**	b	**16.102**	b	**16.103**	a	**16.104**	a
16.105	b	**16.106**	b	**16.107**	b	**16.108**	e	**16.109**	b	**16.110**	c
16.111	c	**16.112**	c								